Odors
and
Deodorization
in
the Environment

Odors and Deodorization in the Environment

EDITED BY

Guy Martin and Paul Laffort

Translated by

Kathe M. Bersillon

Guy Martin
Professor
Laboratoire Chimie des
 Nuisances et Génie de
 l'Environnement
ENSCR-Université de Rennes
35700 Rennes Beaulieu
FRANCE

Paul Laffort
Directeur de Recherches
Laboratoire de Physiologie
 de la Chimioréception
CNRS
91190 Gif-sur-Yvette
FRANCE

This book is printed on acid-free paper. ⊖™

Library of Congress Cataloging-in-Publication-Data

Odeurs et desodorisation dans l'environement. English.
    Odors and deodorization in the environment / edited by
    Guy Martin
and Paul Laffort ; translated by Kathe M. Bersillon.
    p. cm.
Includes bibliographical references and index.
    ISBN 1-56081-666-X (acid-free paper)
    1. Deodorization. 2. Odor control. 3. Smell. 4. Air—Pollution.
I. Martin, Guy, 1934–   . II. Laffort, Paul. III. Title.
RA576.03313 1994
628.5—dc20
                                                94-5422
                                                    CIP

Originally published as: "Odeurs et désodorisation dans l'environnement" by TEC&DOC
Lavoisier, 11, rue Lavoisier, 75384 Paris, CEDEX 08, FRANCE; and translated by Kathe
M. Bersillon.
© 1994 VCH Publishers, Inc.

Printed in the United States of America

ISBN 1-5-6081-666-X VCH Publishers, Inc.

Printing History:
10 9 8 7 6 5 4 3 2 1

Published jointly by

VCH Publishers, Inc.
220 East 23rd Street
New York, New York 10010

VCH Verlagsgesellschaft mbH
P.O. Box 10 11 61
69451 Weinheim
Germany

VCH Publishers (UK) Ltd.
8 Wellington Court
Cambridge CB1 1HZ
United Kingdom

Preface

Sensory receptors and their cerebral projections, the pathways by which living organisms inform themselves about their environments, effectively guide the responses that organisms make to those environments. These "windows on the world" are essential to individual and group survival through physiological regulation, and form the basis of more elaborate behaviors, including social behavior, and in human beings, esthetic behavior.

Throughout the animal world, natural selection has differentiated sensory receptors for the physical stimuli of light and sound. Additional sensory systems, such as the senses of taste and smell, have evolved for analyzing the chemical environment. These systems have evolved more or less efficiently in diverse species because molecular analysis of the environment allows the animal to respond to the chemical properties of objects, offering an advantage or even a sine qua non of survival.

The role of chemical perception varies greatly depending on species, and supposedly has diminished in higher mammals, like man. However, in human beings, these chemical systems play a far from negligible role, for example, in the detection of pollutants as nuisances in the air we breathe. The first part of this book summarizes our current understanding of odors and olfactory sensory mechanisms, including their influence on behavior. This fundamental knowledge will provide better research tools for solving the problems of nuisance odors. The second part of this book summarizes the modern technologies for odor pollution control that have stemmed from such basic research. The book concludes with a discussion of regulations pertaining to emission control and odors in work areas in Europe, Japan, and the United States.

Many unknowns remain about the physiological mechanisms of olfaction that will be the subject of future research. In the meantime, the knowledge presented in this book on the identification and measurement of olfactory phenomena offers a solid base for the development of odor control technologies.

The Editors

Paris
September 1994

Acknowledgments

The editors wish to thank their colleagues and friends who, by their comments and suggestions, have helped to create this book. Thank you to Jacques Chanel, Professor, University Lyon I, Maurice Chastrette, Professor, University Lyon I, André Duchamp, Associate Professor, University Lyon I, André Holley, Professor, University Lyon I, Jacques Le Magnen, CNRS/Collège de France, Paris, Jacques Sibony, O.T.V. Society, André Tallec, Professor, University Rennes I, and Paule-Andrée Mercerolle.

Guy Martin and Paul Laffort, Editors

Translator's note: This book was translated in close cooperation with the original authors. During the translation from French to English, various texts have been updated.

Contents

Contributors

CHRISTOPHE ANSELME. Scientist, Laboratoire Central (Central Laboratory), Lyonnaise des Eaux, 38 rue du Président Wilson, 78230 Le Pecq, France

GÉRARD BESSON. Scientist, 201 Quai de Valmy, 75010 Paris, France

GÉRARD DAGOIS. Director, PICA Company, 92300 Levallois, France

ANDRÉ DUCHAMP. Associate Professor, Laboratoire de Physiologie Neurosensorielle, (Laboratory of Neuro-sensorial Physiology), Université Lyon I (Lyon University I), 69621 Villeurbanne, France

RÉMI GERVAIS. Research Scientist, Laboratoire de Physiologie Neurosensorielle (Laboratory of Neuro-sensorial Physiology), Université Lyon I (Lyon University I), 69621 Villeurbanne, France

MADELEINE GUEUX. Associate Director, EDIL Company, 243 Avenue Joannes Masset 69009 Lyon, France

MICHEL HAMELIN. Associate Department Head-Pollution Prevention, Agence Pour la Qualité de l'Air (Air Quality Agency), Tour GAN Cedex 13, 92082 Paris La Défense 2, France

FRANÇOIS JOURDAN. Professor, Laboratoire de Physiologie Neurosensorielle (Laboratory of Neuro-sensorial Physiology), Université Lyon I (Lyon University I), 69621 Villeurbanne, France

EGON PETER KÖSTER. Professor, Psychology Laboratory, Utrecht University, 3584-CA Utrecht, Netherlands

PAUL LAFFORT. Research Director, Laboratoire de Physiologie de la Chimioréception (Laboratory of Chemical Receptivity Physiology) CNRS, 91190 Gif-sur-Yvette, France

ALAIN LAPLANCHE. Associate Professor, Laboratoire Chimie des Nuisances et Génie de l'Environnement (Laboratory of Environmental Chemistry and Engineering), ENSCR-Université de Rennes (Rennes University), 35700 Rennes Beaulieu, France

PIERRE LECLOIREC. Department Head, Ecole des Mines d'Alès (Alès School of Mine Engineering), 6, Avenue de Clavières, 30107 Alès Cedex, France

JAQUES LE MAGNEN. Research Director, Laboratoire de Neurophysiologie (Laboratory of Neurophysiology) Collège de France - 75231 Paris Cedex 05, France

JOËL MALLEVIALLE. Director, Laboratoire Central (Central Laboratory), Lyonnaise des Eaux, 38 rue du Président Wilson, 78203 Le Pecq, France

GUY MARTIN. Professor, Laboratoire Chimie des Nuisances et Génie de l'Environnement (Laboratory of Chemistry and Environmental Engineering), ENSCR-Université de Rennes (Rennes University), 35700 Rennes Beaulieu, France

ALAIN MILHAU. Associate Department Head-Pollution Prevention, Agence Pour la Qualité de l'Air (Air Quality Agency), Tour GAN Cedex 13, 92082 Paris La Défense 2, France

JEAN-PHILIPPE OLIER. Previous Director-Pollution Prevention, Agence Pour la Qualité de l'Air (Air Quality Agency) Tour GAN Cedex 13, 92082 Paris La Défense 2, France

HERVÉ PAILLARD. Scientist, Centre de Recherches (Research Center) Anjou Recherche (Anjou Research), Chemin de la Digue BP 76, 78600 Maisons-Laffitte, France

MARIE-LINE PERRIN. Laboratory Head Laboratoire d'Olfactométrie (Olfactometry Laboratory), Institut de Protection et Sûreté Nucléaire (Nuclear Protection and Security Institut), CEA-BP 6-92260 Fontenay-aux-Roses, France

I.H. (MEL) SUFFET. Professor, Environmental Studies Institute, Drexel University, Philadelphia, Pennsylvania 19104, USA

VÉRONIQUE TATRY. Scientist, Pollution Prevention, Agence Pour la Qualité de l'Air (Air Quality Agency) Tour GAN Cedex 13, 92082 Paris La Défense 2, France

Consultants

The authors would like to express their warmest thanks to their colleagues, authors of other works, and outside personalities who have helped make the edition of this work possible with their suggestions and comments, notably:

Jacques Chanel, Professor, Université de Lyon I (Lyon University I)
Maurice Chastrette, Professor, Université de Lyon I (Lyon University I)
André Duchamp, Associate Professor, Université de Lyon I (Lyon University I)
André Holley, Professor, Université de Lyon I (Lyon University I)
Paul Laffort, Research Director, CNRS
Jacques Le Magnen, Research Director, CNRS
Guy Martin, Professor, Université de Rennes I (Rennes University I)
Jacques Roualt, Research Scientist, CNRS
Jacques Sibony, Technical Director, O.T.V. Company
André Tallec, Professor, Université de Rennes I (Rennes University I)

The authors would also like to thank **Madame Paule-Andrée Mercerolle** for her active cooperation on the French manuscript of this book.

Odors
and
Deodorization
in
the Environment

1

Anatomy/Physiology and Mechanisms of Olfaction

A. Duchamp, F. Jourdan, and R. Gervais

1.1 Anatomy of the Olfactory System

1.1.1 The Receptor Organ

In mammals, the olfactory receptor organ is the *olfactory mucosa* situated in the dorsal area of the posterior nasal fossae. In the human species, its surface area is from 2 to 3 cm^2 and covers part of the nasal septum and the middle part of the superior concha. Access to this area by odorous molecules suspended in inspired air is optimal when sniffing occurs and in the reverse direction through the retronasal passage under the impulsion of swallowing followed by expiration.

The structure of olfactory mucosa in mammals is remarkably constant (Figure 1.1).

The olfactory epithelium consists of three cell categories.

- *Olfactory neuroreceptors* (several tens of millions per individual) are bipolar sensory neurons which insure the direct reception of olfactory stimuli by the extremities of their dendrites. The interaction between the odorous molecules and the molecular "receptors" located in the neuronal membrane takes place at the apical pole of the dendrite, which emerges at the surface of the epithelium in the form of an "olfactory vesicle" bearing several dozen sensory cilia (Figure 1.1). The totality of these olfactory vesicles and their cilia are submersed in the mucus layer that covers the epithelium. By the intermediary of its dendritic pole, the primary olfactory neuron is thus in direct contact with the external environment, a singular

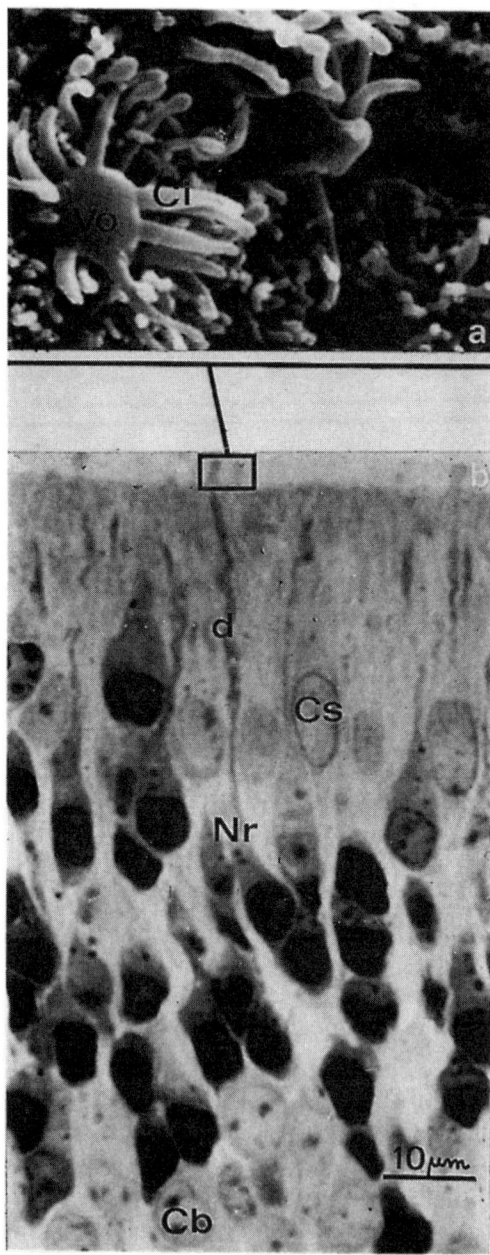

Figure 1.1. Olfactory epithelium of a rat. From each olfactory neuroreceptor extends a dendrite (d) which emerges at the surface of the epithelium in the nasal cavity (a). The olfactory vesicle (Ov) bears a dozen sensory cilia (Ci). The contact between odorous molecules and receptor membranes occurs at this point. The epithelium also contains supporting cells (Sc) and basal cells (Bc) similar to neuronal stem cells.

characteristic which explains the possibilities of viral infestation of the brain by this passage (see Section 1.4.4). The cellular bodies of the neuroreceptors are essentially concentrated in the lower half of the epithelium. From each one of them extends an unmyelinated axon; the olfactory axons gather in small bundles at the base of the epithelium and enter the chorion through the basal lamina.

- *Supporting or sustentacular cells* are column-shaped epithelial cells whose cellular bodies are situated in the top third of the epithelium. They isolate the olfactory neuroreceptors, particularly at the level of their dendrite. The supporting cell surface bears numerous microvillosities which lie in the mucus.

- *Basal cells* are situated in the deepest part of the epithelium, just on top of the basal membrane that separates the epithelial and conjunctive components of the mucosa. At least part of these cells present the remarkable characteristic of being able to multiply and differentiate into olfactory neuroreceptors. Thus, they constitute a stable population of basal stem cells having retained embryonic traits (neuroblasts). This multiplication-differentiation mechanism is restrained in normal olfactory mucosa, where it insures the regular replacement of the neuroreceptor population. It can become explosive, for example, in the case of peripheral lesions of the mucosa or following neuroreceptor loss due to olfactory nerve severage. In this case, it has been shown that the establishment of new olfactory connections is compatible with functional recuperation (Monti Graziadei et al., 1980).

- *The chorion* is characterized by the presence of *Bowman's glands*, mucus-producing glands whose function is essential, since the mucus they secrete onto the epithelium's surface is always between the odorous molecules contained in inhaled or exhaled air and the olfactory receptors. *Olfactory nerve bundles*, which make up the olfactory nerve, gather the unmyelinated axons within Schwann cell processes.

1.1.2 The Olfactory Bulb

The olfactory nerve fibers enter the cranium through the orifices of the ethmoidal bone. Their penetration into the brain occurs at the olfactory bulb level of which they constitute the superficial layer. The *olfactory bulbs* are symmetrical structures that, in the human species, appear in the form of ovoid corpuscles of about 1 cm in length, located in the lower front position with respect to the brain. In all mammals, the bulb presents a remarkably constant laminary structure with spherical symmetry (Figure 1.2). This structure is the result of a well-identified cell neuronal architecture that supports integrative processes of the afferent olfactory information through local microcircuits controlled by the upper brain. The axons of the olfactory neuroreceptors constitute a dense, feltlike coating at the surface of the bulb,

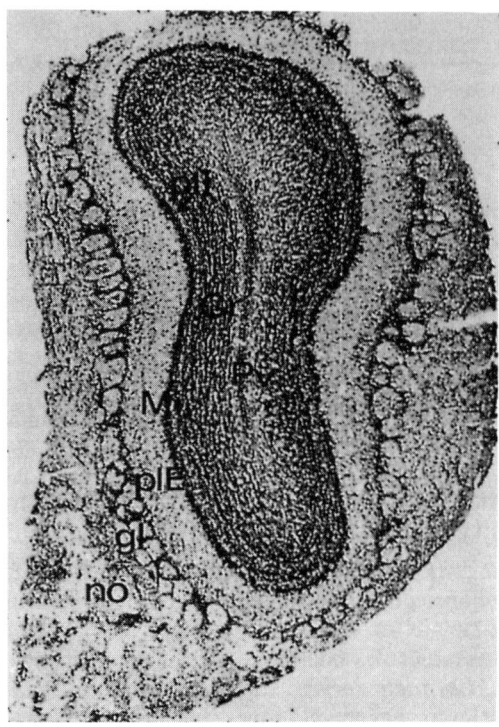

Figure 1.2. Olfactory bulb of a rat, front cross section. The olfactory bulb presents a laminary structure of the cortical type with, from the periphery toward the center, the following layers: glomerular (gl), olfactory nerve (on), external plexiform (Epl), mitral cells (Mi), internal plexiform (Ipl), granular (Gr), and periventricular (Pv).

penetrating a specific location of the bulbar cortex into the olfactory glomeruli. These axons cast synapses onto the dendrites of the olfactory deutoneurons (second-order neurons of the olfactory pathways), the *mitral* and the *tufted cells*. The large pyramidal bodies of these cells are located in the deeper layers of the bulbar cortex. Their myelinized axons gather and go toward the back where they finally form the lateral olfactory tract, which transmits olfactory information toward the higher brain. Several categories of *interneurons* form localized neuronal microcircuits that process the information at different levels of the bulbar cortex. The two principal categories of interneurons are the *periglomerular cells*, located close to the surface and the *granular cells*, which are located in a deeper layer. Only periglomerular cells receive direct synapses from the primary olfactory axons. On the other hand, the two types of interneurons establish dendro-dendritic reciprocal synapses with the mitral and tufted cells. These looped neuronal circuits are most likely the support of self-inhibition and lateral inhibition phenomena observed at the mitral cell level.

Finally, the centrifugal axon fibers come from neurons located in several parts of the brain and on the bulbar interneurons by synapses that are probably responsible for activation.

1.1.3 Accessory Olfactory Systems

Chemiosensory detection, which occurs in the nasal cavity in mammals, is not limited to the olfactory system as described earlier. Other neuronal structures, more or less connected to the principal olfactory system, are equally capable of performing treatment and transmission of chemical stimuli.

1.1.3.1 The Trigeminal Nerve

The peripheral fibers of the *trigeminal nerve,* the fifth cranial nerve, end loosely at the base of the epithelium of the nasal mucosa (Papka and Matulionis, 1983). It has been shown that the trigeminal fibers, besides their usual sensitivity to thermal and mechanical stimuli, are excitable by chemical stimuli (Silver and Moulton, 1982; Keverne et al., 1986). Psychometric studies have shown that anosmic human subjects can perceive many odors used in olfactometry by the trigeminal nerve (Doty et al., 1978). Furthermore, stimulation of the trigeminal fibers in the nasal cavity can influence olfactory perception by modifying the access conditions of the stimuli to the olfactory receptors (Silver and Maruniak, 1981) or by modifying the response threshold of the olfactory neuroreceptors by local release of a neuropeptide, substance P (Bouvet et al., 1987).

1.1.3.2 The Terminal Nerve

The terminal nerve is a ganglionic plexus connected to the olfactory system by its peripheral area and found in most vertebrates, including the human species (Johnston, 1914). The nerve's central projections reach the olfactory tubercle and the supraoptic nucleus of the hypothalamus (Larsell, 1950) while the peripheral fibers go through the olfactory pathway and end in the nasal cavity. It has recently been shown that the terminal nerve is made up of many types of distinct fibers according to ultrastructural and neurochemical criteria (Zheng and Jourdan, 1988b). A neuronal substructure of the terminal nerve is characterized by the presence of luliberine (luteinizing hormone releasing hormone, LHRH) and very likely insures the release of this neuropeptide into the blood vessels of the olfactory mucosa (Zheng et al., 1988). Two other distinct categories of nerve fibers may belong to the autonomic nervous system and to an accessory chemiosensory system, respectively (Zheng and Jourdan, 1988b).

1.1.3.3 The Vomeronasal System

The vomeronasal organ appears as a pair of tubes located in the floor of the nasal fossae, on either side of the septum and, depending on the species, opening into the nasal or buccal cavity. In the human species, the vomero-nasal organ is present during embryonic development, but seems to regress later and then disappears in most adult subjects. The chemical sensitivity of the vomeronasal epithelium is due to the presence of bipolar neuroreceptors whose dendritic extremities usually bear microvillosities (Vaccarezza et al., 1981). Their axons are grouped to form the strands of the vomeronasal nerve that go through the ethmoidal bone along with olfactory fibers and end in the accessory olfactory bulb, located in the dorsal part of the posterior ol-factory bulb. Therefore, the accessory olfactory system is a chemosensory system that is distinct from the principal olfactory system, including the central projection level. The access of the chemical stimuli to the organ receptors can only occur in the liquid phase and involves active pumping processes of stimulus carrying fluids (Wysocki et al., 1980).

1.2 Olfactory Information Coding

1.2.1 Properties of Neuroreceptors and Deutoneurons

The olfactory system is a sensory system specializing in detection, discrimination, and identification of odorous substances. The study of its functioning, first based on the analysis of verbal responses of human subjects, has enjoyed new progress with the discovery of methods that make it possible to record the electrical activity of the first two layers, the olfactory mucosa and olfactory bulb.

The olfactory neuroreceptors are sensory protoneurons and it has been established that they simultaneously fill several functions. Some of them have a specific chemical transducer role. They are located on the olfactory cilia carried by the dendrites of the receptor cells. Others are common to numerous categories of neurons: for example, producing an electrical signal at the cellular body level and transmitting it to the axon extremity where it will trigger a synapse, which is a functional junction with the rest of the olfactory pathways.

1.2.1.1 Transduction

One speaks of transduction when the information changes its support while keeping its meaning. The initial steps of transduction in olfaction have been presented in recent reviews (Lancet, 1986; Chanel, 1987, 1988). The beginning of the olfactory message results from adsorption of odorous molecules occuring in the mucus on *glyco-proteic receptors* of the membrane of the *olfactory cilia*. One could postulate that these receptors are of several types and that they are unevenly distributed among the neuroreceptor population. The receptors are proteins with seven transmembrane domains (Buck and Axel, 1991). As will be seen, a neuroreceptor is only excited by the specific odorous substances that will come into contact with it; some neureceptors, however, do not respond in the same way to the odorous substances to which they are sensitive. The molecule–receptor interaction leads to a cascade of enzymatic reactions that ends with the elaboration of a large quantity of an intracellular messenger, the adenosine cyclic 3', 5'-monophosphate (cAMP) or inositol 1, 4, 5-triphosphate (IP3), that activates *the opening of the transmembrane ionic channels* (Lancet et al., 1986; Lancet and Pace, 1987; Nakamura and Gold, 1987; Breer and Boekoff, 1991). The cumulative effect of this small local depolarization is transmitted by the dendrite to the cellular body whose electric polarization is modified. When this polarization goes from its rest value to the critical value corresponding to the excitability threshold, a train of spikes (during which the cellular body displays short electrical polarity inversion) is generated and transmitted along the axon toward the bulbar relay. Our understanding of information processing in this system is based on the analysis of this message.

1.2.1.2 Electrical Activity of the Olfactory System

At the *bulb level,* Adrian (1950, 1953) noted that presenting an odorous stimulus to the epithelium provokes potential oscillations, which are called *induced waves,* at the surface of the bulbar structure. This means that the presentation of a stimulus induces modifications of the resting activity in the neuronal layers below the recording electrode. Some of the deterministic models explaining the operations of the olfactory system are based on the analysis of these induced waves (Freeman, 1987).

At the *receptor level,* Ottoson (1954) recorded a potential variation on the surface of the epithelium in response to the presentation of an odorous stimulus. This slow negative wave, the *electroolfactogram* (EOG) increases in amplitude when the stimulus concentration increases. Like induced waves, the EOG is hardly suitable for the identification of differentiated actions of odorous molecules on the olfactory system through the analysis of its shape. It represents a complex spatial summation of elementary electrical events related to transduction and located at the level of cilia and dendrites of several thousands of receptor cells. It is more related to transduction than to spikes transmitted in axons deeper in the epithelium. As will be seen, EOGs and induced wave amplitudes are unevenly distributed over the structure they are recorded from. This makes it possible to have access to the concept of spatial coding of information, according to which stimulus identification may be based on a specific spatial distribution of evoked electrical activity (see Section 1.2.2.2).

Since global signals are not defined well enough, one must record receptor cells and bulbar relay cells one by one. This recording of the *unitary elec-*

trical activity is performed by placing a microelectrode, whose dimension is adapted to the size of the biological generator, in the immediate vicinity of the cell body or the axon.

If the neuroreceptor's membrane contains receptor proteins susceptible to binding to a given odorous molecule, a *response by activation* will be recorded in the form of a train of spikes. In the opposite case, the cell does not respond and keeps on sending spikes spontaneously at its specific rate.

At the bulbar level, the mitral and tufted cells send the information that they process in relation with the interneurons to the higher nervous centers. From unitary extracellular recordings, it has been known for a long time that the mitral and tufted cells respond to odors according to modalities described by researchers as more or less complex (Kauer, 1974; review by Holley and Mac Leod, 1977). The deutoneurons have a more or less regular spontaneous activity. The exposure of an animal to an odor may

• Not trigger any modification of the spontaneous activity
• Induce an increase in spike frequency
• Decrease this frequency or even suppress any signal

These three situations are usually referred to as nonresponse or response by activation or by inhibition. In this third modality, one must see the effect of inhibitory interneurons that act momentarily on the recorded cell.

Recent studies of intracellular recordings performed on mitral and tufted cells in salamanders describe response patterns corresponding to exposure to odorous substances (Hamilton and Kauer, 1985, 1989). They make it possible to better understand the meaning of trains of spikes in terms of cellular interactions, specifically with the interneurons.

1.2.1.3 Peripheral Mechanism of the Qualitative Discrimination of Odors

The totality of responses by activation and nonresponses with respect to the odorous substance sample used constitutes the *neuroreceptor response profile*. Gesteland et al. (1963) were the first to underline the fact that each neuroreceptor has its own response profile. This profile provides rather precise information on the functioning of this cell, but brings little information on the functioning of the whole population. It is therefore necessary to probe a large number of receptors. The responses are recorded in matrices where the lines correspond to the probed cells and the columns correspond to the selected stimuli. Pair comparison of the response profiles do not make it possible to point out groups of receptor cells displaying analogous profiles. Pair comparisons of columns show that some stimuli display similar actions on the studied population. Mathematical processing methods, referred to as factor analysis, allows one to assign a position to each stimulus in a multidimensional space. Within this space, the vicinity of points representing the stimuli corresponds to similarity of action on the population of neuroreceptors. On the other hand, a large distance between points means very different actions on the neuroreceptors. This analysis made it possible to identify *qualitative groups,* each containing a set of compounds with similar actions on the neuroreceptors (Duchamp et al., 1974; Revial et al., 1978, 1983). Among some 80 pure substances, these authors pointed out

- An "aromatic" group within which the compounds have the aromatic ring in common
- A "short chain fatty acid" group
- A "terpene" group
- A "camphor" group, made up of camphor and small spherical molecules

Inversely, they could establish that a series of aliphatic primary alcohols could not constitute a group: within the series, successive alcohols are qualitatively close, but the two poles of the series represent very different qualities. This is also true for the cycloketones (at least from C5 to C18). Thus, a common chemical function does not determine a qualitative similarity. Parameters related to the molecular shape seem to play an important role. More specifically, the molecules with a "camphor" odor have been characterized by their spherical shape and their small volume (Eminet and Chastrette, 1983). Weak interactions of the Van der Waals bond type or the hydrogen bond type would explain the interaction between most of the musks and their receptor sites (Chastrette and Zakarya, 1988).

1.2.1.4 Bulb Mechanism of the Qualitative Discrimination of Odors

In order to understand the contribution of the olfactory bulb to qualitative discrimination, the same odors as those used at the peripheral level have been used at the same concentration with the same animal species, the frog (Duchamp, 1982; Duchamp and Sicard, 1984). The deutoneuron responses are sorted like the neuroreceptor responses: the same qualitative groups of odorous substances are found. However, the role of the bulb is not limited to the simple transmisson of the peripheral message to the centers. It contributes to the improvement of the discrimination of compounds barely distinguished at the receptor level, such as enantiomers, especially through the complex involvement of the interneuron network. These studies have also shown that the different types of response do not play the same role in information processing at the bulbar level. *Qualitative discrimination* is based, above all, on responses by activation that result from the direct involvement of deutoneurons through primary connections. Inhibitions are only an indirect consequence—through interneurons—of the activation of nearby deutoneurons. For a given cell, they can be the consequence of the action of different odorous substances on different neighboring cells. Therefore, they must be less specific than activations. Even though anatomical circuits, synaptic mechanisms, and intracellular events of the bulbar inhibition are well established (see the review by Mori, 1987), the role of the olfactory bulb in information processing remains quite unknown. Generally, it is hypothesized that the bulb enhances the contrasts of the resulting activation "image" by the reduction of the noise constituted by the spontaneous activity of the deutoneurons.

1.2.1.5 Intensity Coding

Early works concerning unitary recordings showed that the neuroreceptors (Gesteland et al., 1963; O'Connell and Mozell, 1969), like deutoneurons (Døving, 1964), produce spikes at a frequency that increases with the intensity of the olfactory stimulus. Each neuroreceptor only participates in the recognition of the substances with which the sites of its membrane are likely to bind. In other words, processing by the system of the type and intensity of the stimulus are two aspects that are necessarily related.

At the cellular level, the stimulus intensity is determined by the spike frequency in the initial part of the response by activation (Figure 1.3). How

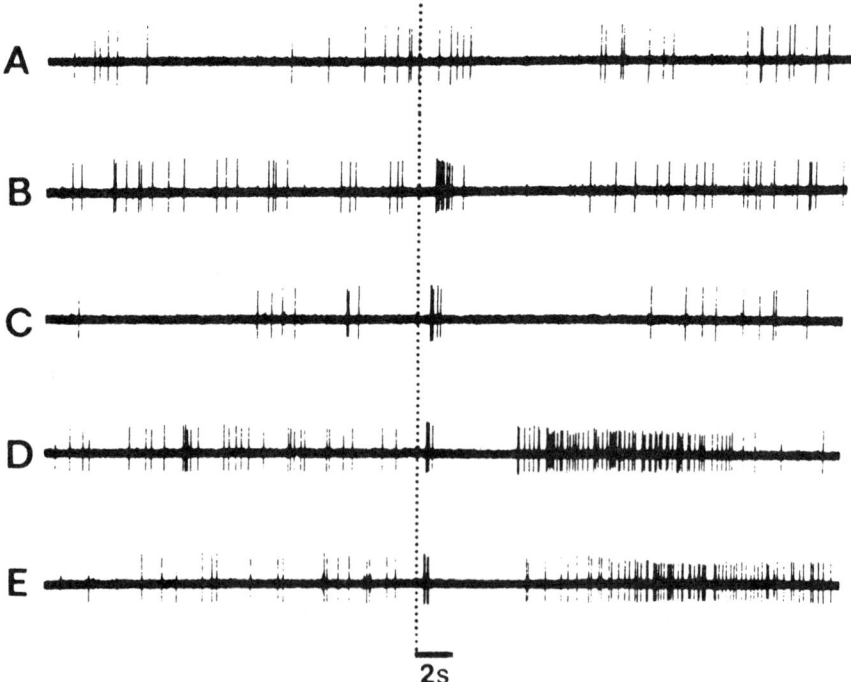

2s

Figure 1.3. Unitary extracellular recording of deutoneurone response for increasing stimulus concentrations. The dotted vertical line shows the beginning of stimulation (presentation of camphor during 2 seconds). Concentrations in Vpm are as follows: A—0.025; B—0.080; C—0.25; D—4.2; E—14.

then is the stimulus intensity determined at the epithelium and bulb level, with each taken as separate entities? To answer this question, one must increase the number of recordings in the mucosa and in the bulb for the same animal species and using the same stimuli. Relying on this condition to insure strict control of the stimuli concentration within a wide range of values (Vigouroux et al., 1988), a study by Duchamp-Viret (1988) led to the following main results.

For a given stimulus, cells of different sensitivities are found. For each of them, the initial spike frequency increases to a maximum. The slope of the relationship between the concentration and the response varies from one to the other as well as the maximum spike frequency. However, from this maximum onward, the response is stable for any higher concentration. Therefore, at the level of cell population, the increasing intensities of a short stimulus (2 seconds) are coded by an increasing number of spikes delivered by an increasing number of active cells. When a summation of the number of spikes is performed for a set of cells responding to a given odor, by successive bins of 200 ms and during the odor presentation time (2 seconds) at different concentrations, the information rate is perturbed during a short time: 3–5 seconds for the neuroreceptors and 1–3 seconds for the deutoneurons. This early perturbation of the information rate, which is synchronized with the odor presentation, probably contains the main part of the information concerning the odor nature and intensity. Late perturbations that are very different from one cell to the other are less tightly correlated to the intensity variations: a few seconds after the odor presentation, two cells that responded by an initial activation can be, on the one hand, in a long silence phase and, on the other hand, in a sustained rebound activity phase. The role of late activities in information processing remains difficult to understand.

The average sensitivity of the deutoneurons is greater than that of the neuroreceptors (Duchamp-Viret et al., 1989). The anatomical convergence of the axons of the latter on the principal cells of the bulbar relay makes it possible to understand this result. As far as the information processing is concerned, it means that at the bulbar level, the activation "image" is established at concentrations for which the peripheral "image" is still undefined and transient.

It has been seen that the odor quality specification relies on the identity of the active neurons. From the better average sensitivity of the bulbar level, it can be inferred that, as far as qualitative discrimination is concerned, this structure should be able to operate at lower concentrations than the neuroreceptors. A study by Duchamp-Viret et al. (1990) confirms this point and shows that the differentiation is more efficiently performed at lower concentrations in the bulb than in the mucosa.

In conclusion, the bulb acts both as a filter and as an integrator of peripheral information containing "noise" and for which it extracts the pertinent part. The bulbar message sent to higher centers keeps its improved quality with respect to the message delivered by the neuroreceptors and the bulb for the whole stimulus concentration range.

1.2.2 Spatial Coding Mechanisms of Olfactory Information

1.2.2.1 Principle

The olfactory system features remarkable detection and discrimination capacities with respect to odorous stimuli. This observation raises the question of the information coding mode used by the system, knowing that each neuroreceptor only displays a relative selectivity with respect to the large variety of stimuli (see Section 1.2.1.2). Thus, the olfactory system in mammals cannot be considered as a set of parallel channels, each presenting a specific sensitivity with respect to a certain number of odors (Holley and Døving, 1977).

The most generally accepted coding model of the nature of odorous substances relies on the spatial dimension of the polyneuronal activation generated by each stimulus at the periphery of the system. If one considers the olfactory mucosa as a sensory surface made of independent and neighboring transducers (the neuroreceptors), each stimulus activates a given number of neuroreceptors whose position on the mucosa is not random. The resulting set of activated receptors constitutes a pattern charateristic of the stimulus. In other words, the chemical sensitivity of the mucosa is spatially organized (chemotopy). The occurence of topological rules in the mucosa projection onto the bulb (Astic et al., 1987) allows the conservation of the chemotopy at the bulbar level. Furthermore, the convergence of the olfactory axons in the glomeruli and the local mechanisms of information processing are capable of improving the quality of the olfactory "image" constituted by the sets of activated glomeruli. The great variety of polyneuronal activation patterns generated at the surface of the receptor organ could explain the discrimination capacities of the overall system.

1.2.2.2 Peripheric Chemotopy: Imposed and Inherent Patterns

The nonhomogenous distribution of sensitivity in the olfactory mucosa in response to different chemical stimuli (chemotopy) has been demonstrated by a study of local EOGs (Daval et al., 1970; Mustaparta, 1971). Several mechanisms can be at the origin of the heterogenous distribution of the response. Some of them are related to the physical and physicochemical characteristics of the carrying media and to the air flow dynamics in the nasal cavity. The patterns "imposed" on the detection system by these constraints concern the variable degree of accessibility of the neuroreceptors as a function of their location in the nasal cavity and the solubility of the odorous molecules into the mucous. Thus, it has been possible to assume the occurence of the adsorption gradient along an anterior–posterior axis. Very hydrosoluble molecules have a tendancy to preferentially activate receptors located in the anterior part of the nasal cavity. This is the principle of the "chromatographic" theory of olfactory coding as proposed by Mozell and Hornung (1981). The configuration of such "imposed" patterns could be influenced by the action of an odorant binding protein (OBP) recently pointed out in mammals (Pevsner et al., 1988). This protein, secreted by a specialized gland, is capable of binding with a certain number of odorous molecules and may facilitate transport into the mucous for the less hydrosoluble molecules. In any case, it is most likely that there exist differential constraints of the receptor organ with regard to stimuli of different chemical natures. However, through the study of EOGs, it has been shown that the peripheral chemotopy persists when stimulation and measurement methods rule out the possibility of imposed constraints (Mackay-Sim and Kubie, 1981).

From this, it should be concluded that the essential determinism of peripheral chemotopy resides in the spatial heterogeneity of the individual sensitivity of neuroreceptors. This spatially organized selectivity generates "inherent" activation patterns and probably explains the bulk of the discriminative capacitites of the system (Holley and Døving, 1977).

1.2.2.3 Bulb Chemotopy

The convergence of a great number of axons of the olfactory nerve in each glomerulus maintains and even enhances the chemotopy at the bulbar level. Works by Leveteau and Mac Leod (1966) showed that each glomerulus gives an integrated electrophysiological response of the "all or nothing" type for each olfactory stimulus delivered to an animal. The functional unit constituted by the olfactory glomerulus along with the capacity of the local interneuronal mechanisms to improve the definition of the spatial patterns were confirmed and illustrated by the method of metabolic mapping by 2-Desoxyglucose (2DG) (Jourdan et al., 1980) (Figure 1.4). This method uses a radioactive tracer, the (^{14}C) 2DG, whose accumulation in neurons is proportional to their activity level during the experimental olfactory stimulation time imposed on the animal. This method allowed the visualization and the comparative study of activation patterns triggered by different odors (Jourdan, 1982; Royet et al., 1987). Thus, it has been demonstrated that the characteristics of the glomerular activation pattern (location, shape, and extension) are relative to the stimulus nature, reproducible from one animal to another for a given odor, and symmetrical in the left and right olfactory bulbs.

A spatially organized bulbar response has also been pointed out by the "voltage dependant" dye method (Kauer et al., 1987) characterized by its excellent time resolution. This confirms the compatibility between spatial coding with the short time separating the occurence of a stimulus with its identification.

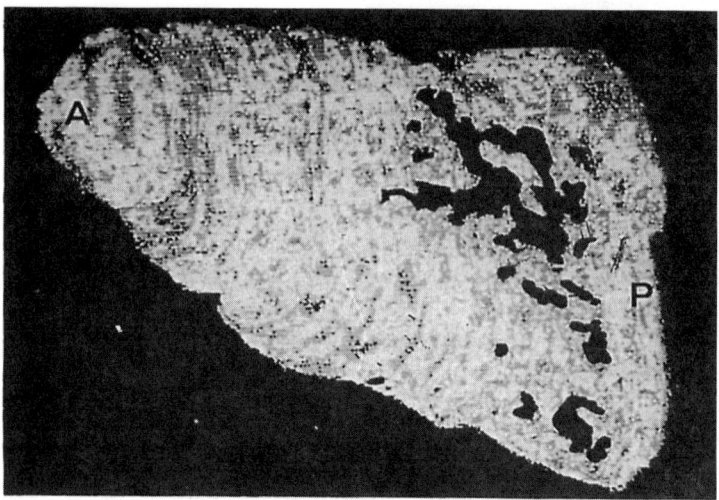

Figure 1.4. Glomerular activation pattern revealed by 2-Desoxyglucose. This figure represents the reconstituted olfactory bulb of a rat stimulated by a pure odor: cyclohexane. The bulb is seen from its median side (A: anterior pole; P: posterior pole). The 2-Desoxyglucose has selectively accumulated in the cortex bulb areas that are the most activated by the odorous stimulation, illustrated by the dark areas situated in the posterior portion. In this way, each odor induces a spatial activation pattern that is specific and reproducible from one animal to another.

The olfactory information spatial coding model is also illustrated by the fact that rats are capable of distinguishing between the activation spatial patterns obtained by electrical stimulation through a network of microelectrodes installed within the bulb (Mouly et al., 1985). Such imposed stimulation patterns, akin to "pseudo odors" and differing from each other by very subtle spatial characteristics are very easily identified by the animal when they are used as operational conditioning stimulus (see the definition of these terms in Chapter 5). Therefore, these results demonstrate that the central brain structure is capable of discriminating olfactory "inputs" that differ only by the bulbar spatial pattern of the concerned deutoneurons. This does not influence the information processing modes subsequently involved in further information integration and processing (Mouly and Holley, 1986).

1.3 Specific Chemosensory Subsets

There is much experimental evidence that points to the existence of specialized chemosensory processes in the perception of specific stimuli and especially in the chemical signals involved in interindividual communication, either within or outside a given species. The occurence of such signals that can trigger either neurovegetative or behavioral endocrinian reactions (pheromones, allomones, and kairomones) is no longer doubtful in mammals, including the most evolved (Keverne, 1983). Processing of these biologically very important signals could be performed by chemosensory systems included in the olfactory system or by neuronal subsets located within the main olfactory system.

Even though it is not in contradiction with the olfactory information spatial coding theory, this hypothesis increases its complexity by introducing the idea that the "status" of all olfactory stimuli is neither the same nor constant with respect to the olfactory neuroreceptors. Taking this variable into account in functional models of the olfactory system becomes a necessity.

1.3.1 Accessory Chemosensory Systems

The implication of the *vomeronasal system* in olfactory signal processing controlling endocrinian (Keverne, 1983) and behavioral (Halpern, 1987) aspects in mammalian reproduction has been well illustrated. Thus, spontaneous abortion in mice provoked by the odor of an unfamiliar male (Bruce effect) depends on the functional integrity of the vomeronasal system (Rajendren and Dominic, 1984). The accessory olfactory system could be a determining factor as well in the pheromonal relationships of maternal behavior (Brouette, 1984; Lepri et al., 1985). The *terminal nerve* also seems to be involved in the chemosensory control of reproductive behavior, since its selective destruction supresses this behavior in male hamsters (Wirsig and Leonard, 1987). However, it still remains to be proven whether this effect is due to the supression of a specific chemosensitivity or to an alteration of the neuroendocrinian type subsequent to the destruction of the part of this nerve susceptible to the secretion of the LHRH neuropeptide in the nasal cavity (Zheng et al., 1988). The trigeminal nerve probably participates in the general chemosensitivity (common chemical sense) and therefore interacts with "pure" olfactory perception. Contrary to the other accessory systems described earlier in this chapter, no specific chemosensory function can be associated to the vomeronasal system (see Keverne et al., 1986).

1.3.2 Neuronal Subsets of the Main Olfactory System

The hypothesis that specific subsets exist in the olfactory system relies on anatomical and neurochemical data. The main olfactory system may be less homogenous than has been suspected for a long time. This heterogeneity is indicated not only by the presence of a low but significant percentage of neuroreceptors that have microvillosities instead of cilia (Jourdan, 1975), but also by "atypical" glomeruli in the olfactory bulb (Zheng et al., 1987). These glomeruli are characterized by an original localization, a high acetylcholinesterase concentration, and an ultrastructure that indicates the presence of primary olfactory afferents stemming from atypical neuroreceptors (Zheng and Jourdan, 1988a). The functional specificity of such subsets remains to be demonstrated.

However, the fact that more and more attention is being paid to neuronal subsets, which can be characterized in particular by very specific reactivity to certain monoclonal antibodies (Hempstead and Morgan, 1985; Schwob and Gottlieb, 1986), obliges one to consider the existence of neurochemically and functionally specialized subsets in the mucosa and olfactory bulb in mammals.

1.4 The Olfactory Brain

1.4.1 The Central Olfactory Tracts

Organization of central processing of olfactory information is beginning to be documented in rodents and in primates (Holley and Mac Leod, 1977; Tagaki, 1984). Data provided by electrophysiological and anatomical studies make it possible, for a given structure, to identify the location of its projections (the efferents) and the structure that receives these projections (the afferents). This type of experimental approach is difficult to apply to humans, where one is limited to postmortem examinations of tissue cross sections that reveal the histological organization of the structures, but not the connections that they establish.

The axons of the deutoneurons reach many of the telencephaon structures which generally present a simple cortical organization in three layers (the paleocortex). This mainly concerns the piriform cortex, the entorhinal cortex and the amygdala area. The piriform cortex and the amygdala are at the origin of projections that reach the neocortex (six-layered cortex) directly or indirectly (after relay at the thalamus level). In primates, the olfactory neocortex corresponds to the orbito-frontal cortex. Furthermore, the lateral part of the entorhinal cortex is at the origin of massive projection toward the hippocampus (Figure 1.5).

The organization of the central olfactory pathways is original. Unlike other sensory modalities (vision, audition, . . .), *olfactory information accesses limbic system structures very rapidly,* like the amygdala and the hippocampus, which play an important role in memory and emotional manifestations. Processing by the cerebral areas that are the most recent phylogenetically, which are the centers of the most complex cognitive processes, does not seem to be dominant. These few anatomical characteristics make it possible to see why olfactory perception has as much to do with emotion as with reason.

As already emphasized, the olfactory bulb does not only receive the neuroreceptor axons as afferents. Indeed, all the central structures that it reaches directly (piriform cortex, entorhinal cortex, . . .) project, in turn, onto the bulb. Furthermore, the bulb is the site of axon endings whose cellular bodies belong to the cholinergic complex of the basal telencephalus and

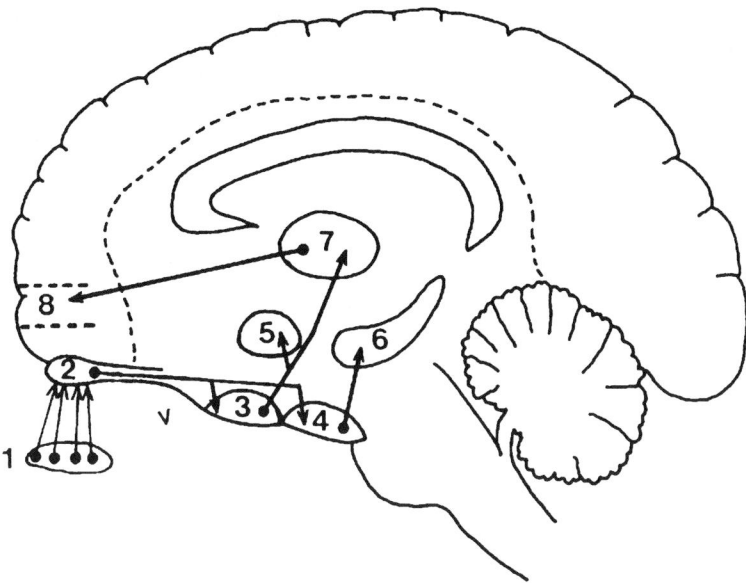

Figure 1.5. Representation of the central projection pathways of the principal olfactory system in humans: **1.** neuroreceptors; **2.** olfactory bulb; **3.** piriform cortex; **4.** entorhinal cortex; **5.** amygdala; **6.** hippocampus; **7.** thalamus; **8.** orbito-frontal neocortex.

to neurons belonging to the locus caeruleus and the "raphe." The locus caeruleus and the "raphe" are structures of the brain trunk whose neurons respectively use noradrenaline and serotonine as neurotransmitters.

1.4.2 Central Perception Modulation

As suggested by anatomical data, information does not circulate in only one direction in the olfactory system. After the message has been sent by the olfactory bulb and then by the piriform cortex to the higher centers, in return, these centers exercise an action at the entry level. This phenomenon is particularly well studied at the *first relay level of the system*, which is the olfactory bulb. Electrophysiological and behavioral studies have shown that the olfactory bulb's response to odors largely depends on commands from the central brain and not simply on the message transmitted by the neuroreceptors (Pager, 1986; Gervais et al., 1988).

In rats able to move about freely, it is possible to record the activity of a small mitral cell population. It could be noted that cell reactivity to odors that represent a behavioral interest for the animal is particularly high: for example, the odor of a usual type of food for a fasting subject, the odor of a receptive female for a sexually experienced male rat, or the odor of a male for a female in œstrus. On the other hand, the high response rate to these odors disappears after sectioning the fibers that transmit information from the higher centers to the olfactory bulb. This points to the fact that the result of olfactory information analysis by the higher centers is reflected as far as the olfactory bulb by way of centrifugal fibers.

Behavioral studies provide information on this central command function. In the first few hours that follow giving birth, a ewe memorizes the odor of its lamb. This olfactory imprint allows the ewe to establish a selective bond with its young: this particular lamb will be the only one in the herd that the ewe will feed. The others will be actively rejected. In rats that give birth for the first time, the sense of smell greatly contributes to the organization of maternal behavior. Interestingly, normal behavioral response is considerably disturbed in both ewes and rats whose olfactory bulbs are deprived of their noradrenergic afferents issued from the locus caerulus (Pissonier et al., 1985; Dickinson and Keverne, 1988; Levy et al., 1990). According to this data, the central noradrenergic command makes it possible *for the bulbar neuronal network to keep a trace of the olfactory experience.* The formation of this trace could be necessary for odor memorization. Proving this hypothesis is presently the object of particular attention because its confirmation would point out an important central nervous system property: the experience not only modifies the functioning of higher structures (limbic system and neocortex) but also that of structures that transmit sensorial information toward the centers.

1.4.3 Olfactory Memory

The study of olfactory memory is of increasing interest for neurobiologists. Indeed, understanding the mechanisms underlying the most evolved forms of memory requires experimental studies using adequate animal models. It so happens that of all the types of memory studied in rodents, olfactory memory is the one that presents the most similarity with human memory. It is in the area of olfaction that the rat remembers best and establishes associations between a large number of elements, as we know how to do so well in the visual and language domain (Slotnick and Katz, 1974; Staubli et al., 1984, 1987). Furthermore, in mammals (including primates, which include humans), the hippocampus and the thalamus play a key role in storing new information (Miskin and Appenzeller, 1987). Importantly, lesions in these structures in rats reproduce the type of amnesia observed in humans in which these structures are also affected. In fact, the animal rapidly forgets olfactory information that it has just learned, but remembers those learned before the operation (Staubli et al., 1984, 1986, 1987; Eichenbaum et al., 1980, 1986, 1988). Humans exhibit this same type of amnesia whichever sensory modality is solicited. These observations are interesting because one can hope that careful analysis of the mechanisms brought into play in rodents to learn odors may provide information on those underlying certain aspects of memory in humans. Moreover, since the animal's olfactory experience seems to modify the functioning of the olfactory structures themselves (see Section 1.4.2), one can look for mechanisms supporting memory at this level. This approach is facilitated by the fact that the olfactory structures present a relatively simple neuronal organization, whose histological, neurochemical, and electrophysiological characteristics are now well established. These studies are promising in the sense that they are carried out in animal species that are well studied, inexpensive, and in which olfactory information plays a key role in behavioral organization.

In humans, the sense of smell does not represent the dominant sensory modality. For this reason, a rather restrained number of studies have been carried out on the physiology of the human olfactory function in general and olfactory memory in particular. Some researchers, perhaps inspired by Marcel Proust, have tested the tenacity and evocative power of human olfactory memory (see Engen's synthesis, 1989). The results confirm Proust's intuition in the sense that, in a *nonverbal task,* the subjects easily recognize—out of twenty different odors—those presented several months earlier. The scores are close to 100% for 2-month delays and about 70% after 12 months. This good long-term olfactory retention contrasts with visual information retention, where the scores are very high on the scale of several days or weeks, but close to zero for 4-month delays. On the other hand, short-term olfactorymemory turns out to be very poor if the subjects must *verbally identify* the odors that they have learned to name. This last observation illustrates the difficulties most people have every day when they talk about odors. The reasons why vocabulary is so limited when dealing with olfaction remain obscure. This characteristic could be due to particular central brain processing of olfactory information related to the nature of the stimulus itself. Indeed, the physical dimensions of odor are much more difficult to define than those of a visual stimulus, for example. Current data reveals that the numerous dimensions of olfactory stimulus do not seem to be totally dissected within the first two structures of the olfactory system: the neuroreceptors and the olfactory bulb. This difficulty that the peripheral structures have in classifying the information may even result in problems at the verbal expression level. Furthermore, narrow anatomical relationships between the olfactory structures and the limbic system may have something to do with strong hedonic dimension of odors (agreeable and disagreeable odors). In summary, if odor memory is tenacious in spite of the difficulties surrounding its verbalization, the essential function of the human sense of smell might be to provoke an emotional response that is largely determined by individual experience rather than leading to an elaborate cognitive analysis.

1.4.4 Olfaction and Pathology

Careful evaluation of the sensory disfunction that accompanies the most common neurological illnesses has been the object of several investigations.

These research projects were carried out in order to know if a lesion or degeneration of a given cerebral structure could be associated with specific and reproducible problems of perception or memorization of an information category. Data along these lines would ultimately make it possible for clinicians to identify the type of cerebral illness their patients were suffering from by using analyses of response performances to sensory stimulation.

Subjects presenting Korsakoff's syndrome are generally suffering from a lesion of the prefrontal cortex and the medial part of the thalamus. They have great difficulty recalling events previous to and after the lesion. Parkinson's disease is manifested by motor troubles due to degeneration of the "black substance" that uses dopamine as neuromediator in neurons. Finally, Alzheimer's disease is characterized by premature senility (subjects from 50 to 60 years of age) associated with abnormal cell death of cholinergic neurons in the front part of the brain. One of the symptoms of this illness is shown by forgetting recent events. In these three syndromes, odor detection and discrimination capacities are invariably decreased (Porter and Butter, 1980; Ward et al., 1983; Serby, 1987; Warner et al., 1986; Doty et al., 1987, 1988). This data emphasizes the multiplicity of cerebral areas and neurotransmitter systems brought into play in the central processing of olfactory information, without, however, any one of them having a specific olfactory role. Under these conditions, it is not likely that a pathological cerebral state would have olfactory function alteration as its only sign, unless the lesion were to selectively affect one of the first information processing steps, which involve the olfactory mucosa, the olfactory bulb, and the piriform cortex.

The major spin-off of these studies is that they may help clinicians in the early discovery of certain cerebral affections. In fact, it turns out that odor detection and discrimination disturbances observed in patients suffering from Alzheimer's disease appear during the first stages of the disease (Roberts, 1986; Serby, 1987; Warner et al., 1986). This phenomenon might be explained within the framework of a controversial hypothesis according to which Alzheimer's disease is caused by pathogenic agents present in the environment which gain access to the brain by the nasal tract (Roberts, 1986). The olfactory system has long been known as a rapid access to the brain; for example, by cocaine "sniffers." This empirical observation is objectivized by the fact that in rats, a nervous tract marker placed on the olfactory epithelium can be found later in the olfactory bulb and piriform cortex and also in more central structures whose cellular bodies address the projections to the olfactory bulb. In particular, it is the cholinergic neurons in the basal telencephalus and the noradrenergic neurons in the locus caeruleus that innervate the majority of the telencephalic structures besides the olfactory bulb (Shipley, 1985). Furthermore, certain viruses, such as the herpes virus, can use these same routes (Tomlinson and Esiri, 1983; Mc Lean and Shipley, 1987). If this is so, it is easy to understand how a pathogenic agent can disturb olfactory function before causing more general problems. In this context, one can understand the interest of developing a simple, reliable, and inexpensive olfactory test that could be used routinely by clinicians.

Bibliography

Adrian, E. D., 1950. The electrical activity of the mammalian olfactory bulb. *Electroencephalog. Clin. Neurophysiol.* **2**, 377–388.

Adrian, E. D., 1953. Sensory messages and sensations. The response of the olfactory organ to different smells. *Acta Physiol. Scand.* **29**, 4–14.

Astic, L., Saucier, D., Holley, A., 1987. Topographical relationships between olfactory receptor cells and glomerular foci in the rat olfactory bulb. *Brain Res.* **424**, 144–152.

Bouvet, J. F., Delaleu, J. C., Holley, A., 1987. Olfactory receptor cell function is affected by trigeminal nerve activity. *Neurosc. Lett.* **77**, 181–186.

Breer, H. and Boekhoff, I., 1991. Odorants of the same odor class activate different second messenger pathways. *Chem. Senses* **16**, 19–29.

Brouette, I., 1984. Une phéromone du tout jeune rat: le propionate de dodécyle; caractérisation et identification. Fonction dans le comportement maternel. Thèse de 3ème cycle, Université Claude Bernard, Lyon.

Buck, L. and Axel, R., 1991. A novel multigene family may encode odorant receptors: a molecular bassi for odor reception. *Cell* **65**, 175–187.

Chanel, J., 1987. The olfactory system as a molecular descriptor. *NIPS* **2**, 203–208.

Chanel, J., 1988. L'odeur:un portrait chromatographique de la molécule. In Audition et olfaction, éléments de la communication animale, Comité pour l'association des médecines humaine et animale, collection Fondation Mérieux, 105–147.

Chastrette, M., Zakarya, D., 1988. Sur le rôle de la liaison hydrogène dans l'interaction entre les récepteurs olfactifs et les molécules à odeur de musc. *C.R. Acad. Sci.* **307**, série 2, 1185–1188.

Daval, G., Leveteau, J., MacLeod, P., 1970. Electroolfactogramme local et discrimination olfactive chez la grenouille. *J. Physiol. Paris* **62**, 477–488.

Dickinson, C., Keverne, B., 1988. Importance of noradrenergic mechanisms in the olfactory bulb for the maternal behavior of mice. *Physiol. Behav.* **43**, 313–316.

Doty, R. L., Brugger, W. E., Jurs, P. C., Orndorff, M. A., Snyder, P. J., Lowry, L. D., 1978. Intranasal trigeminal stimulation from odorous volatiles: psychometric responses from anosmic and normal humans. *Physiol. Behav.* **20**, 175–185.

Doty, R. L., Reyes, P., Gregor, T., 1987. Presence of both odor identification and detection deficits in Alzheimer's disease. *Brain Res. Bull.* **18**, 597–600.

Doty, R. L., Deems, D. A., Stellar, S., 1988. Olfactory dysfunction in parkinsonism. *Neurology* **38**, 1237–1244.

Døving, K. B., 1964. Studies of the relation between the frog's electroolfactogramm (EOG) and single unit activity in the olfactory bulb. *Acta Physiol. Scand.* **60**, 150–163.

Duchamp, A., 1982. Electrophysiological responses of olfactory bulb neurons to odour stimuli in the frog. A comparison with receptor cells. *Chem. Senses* **7**, 191–210.

Duchamp, A., Revial, M. F., Holley, A., MacLeod, P., 1974. Odor discrimination by frog olfactory receptors. *Chem. Senses* **1**, 213–233.

Duchamp, A., Sicard, G., 1984. Odour discrimination by olfactory bulb neurons: statistical analysis of electrophysiological responses and comparison with odour discrimination by receptor cells. *Chem. Senses* **9**, 1–14.

Duchamp-Viret, P., 1988. Le codage intensitif du stimulus olfactif. Rôle des neurorécepteurs et des deutoneurones. Etude électrophysiologique chez la grenouille. Thèse de Doctorat, mention Neurosciences, Université Claude Bernard, Lyon, 136 pages.

Duchamp-Viret, P., Duchamp, A., Vigouroux, M., 1989. Amplifying role of convergence in olfactory system. A comparative study of receptor cell and second order neuron sensitivities. *J Neurophysiol.* **61**, 1085–1094.

Duchamp-Viret, P., Duchamp, A., Sicard, G., 1990. Olfactory discrimination over a wide concentration range. Comparison of receptor cell and bulb neuron abilities. *Brain Res.* **517**, 256–262.

Eichenbaum, H., Shedlack, K. J., Heckman, K., 1980. Thalamo-cortical mechanisms in olfaction I-Effects of lesion of the medio-dorsal thalamic nucleus and frontal cortex on olfactory discrimination in the rat. *Brain Behav. Evol.* **17**, 255–275.

Eichenbaum, H., Fagan, A., Mathews, P., Cohen, J., 1986. Normal olfactory discrimination learning set and facilitation of reversal learning after medio-temporal damage in rats: implications for an account of preserved learning abilities in amnesia. *J. Neurosci.* **7**, 1876–1884.

Eichenbaum, H., Fagan, A., Mathews, P., Cohen, J., 1988. Hippocampal system dysfunction and odor discrimination learning in rats: Impairment or facilitation depending on representational demands. *Behav. Neurosci.* **102**, 331–339.

Eminet, B. P., Chastrette, M., 1983. Discrimination of camphoraceous substances using physicochemical parameters. *Chem. Senses* **7**, 293–300.

Engen, T., 1989. La mémoire des odeurs. *Recherche* **207**, 170–177.

Freeman, W. J., 1987. Simulation of chaotic EEG patterns with a dynamic model of the olfactory system. *Biol. Cybern.* **56**, 139–150.

Gervais, R., Holley, A., Keverne, B., 1988. The importance of central noradrenergic influence on the olfactory bulb in the processing of learned olfactory cues. *Chem. Senses* **13**, 3–12.

Gesteland, R. C., Lettvin, J. Y., Pitts, V. S., Rojas, H., 1963. Odor specificities of the frog's olfactory receptor. In *Olfaction and Taste I*, edited by Y. Zotterman. Pergamon Press, Oxford, 19–34.

Halpern, M., 1987. The organization and function of the vomeronasal system. *Annu. Rev. Neurosci.* **10**, 325–362.

Hamilton, K. A., Kauer, J. S., 1985. Intracellular potentials of Salamander mitral/tufted neurons in response to odor stimulation. *Brain Res.* **338**, 181—185.

Hamilton, K. A., Kauer, J. S., 1989. Patterns of intracellular potentials in Salamander mitral/tufted cells in response to odor stimulation. *J. Neurophysiol.* **62**, 609–625.

Hempstead, J. L., Morgan, J. I., 1985. Monoclonal antibodies reveal novel aspects of the biochemistry and organization of olfactory neurons following unilateral olfactory bulbectomy. *J. Neurosci.* **5**, 2382–2387.

Holley, A., Døving, K. B., 1977. Receptor sensitivity, acceptor distribution, convergence and neural coding in the olfactory system. In *Olfaction and Taste VI*, edited by J. Le Magnen and P. Mac Leod, Paris, 113–123.

Holley, A., MaccLeod, P., 1977. Transduction et codage des informations olfactives chez les Vertébrés. *J. Physiol. Paris* **73**, 725–828.

Johnston, J. B., 1914. The nervus terminalis in man and mammals. *Anat. Rec.* **8**, 185–198.

Jourdan, F., 1975. Ultrastructure de l'épithémium olfactif du rat: polymorphisme des récepteurs. *C.R. Acad. Sci. Paris* **280**, 443–446.

Jourdan, F., 1982. Spatial dimension in olfactory coding: a representation of the 2-deoxy-glucose patterns of glomerular labelling in the olfactory bulb. *Brain Res.* **240**, 341–344.

Jourdan, F., Duveau, A., Astic, L., Holley, A., 1980. Spatial distribution of (14C)2-deoxy-glucose uptake in the olfactory bulbs of rats stimulated with two different odours. *Brain Res.* **188**, 139–154.

Kauer, J. S., 1974. Response patterns of amphibian olfactory bulb neurons to odour stimulation. *J. Physiol.* **243**, 695–715.

Kauer, J. S., Senseman, D. M., Cohen, L. M., 1987. Odor elicited activity monitored simultaneously from 124 regions of the salamander olfactory bulb using a voltage sensitive dye. *Brain Res.* **418**, 255–261.

Keverne, E. B., 1983. Pheromonal influences on the endocrine regulation of reproduction. *Trends Neurosci.* **6**, 381–384.

Keverne, E. G., Murphy, C. L., Silver, W. L., Wysocki, C. J., Meredith, M., 1986. Non-olfactory chernoreceptors of the nose: recent advances in understanding the vomeronasal and trigeminal systems. *Chem. Senses* **11**, 119–133.

Lancet, D., 1986. Vertebrate olfactory reception. *Annu. Rev. Neurosci.* **2**, 329–355.

Lancet, D., Chen, Z., Heldman, J., Pace, U., 1986. Cyclic nucleotide cascade in olfactory transduction. *Biophys. J.* **49**, 183a.

Lancet, D., Pace, U., 1987. The molecular basis of odor recognition. *TIBS* **12**, 63–66.

Larsell, O., 1950. The nervus terminalis: mammals. *J. Comp. Neurol.* **30**, 3–68.

Lepri, J. J., Wysocki, C. J., Vandenbergh, J. G., 1985. The mouse vomeronasal organ: effects of chemosignal production and maternal behavior. *Physiol. Behav.* **35**, 809–814.

Leveteau, J., MacLeod, P., 1966. Olfactory discrimination in the rabbit olfactory glomerulus. *Science* **153**, 175–176.

Levy, F., Gervais, R., Kindermann, U., Orgeur, P., Pikety, V., 1990. Importance of beta-noradrenergic receptors in the olfactory bulb of sheep for recognition of lambs. *Behav. Neurosci.* **104**, 464–469.

Mackay-Sim, A., Kurie, J. L., 1981. The salamander nose: a model system for the study of spatial coding of olfactory quality. *Chem. Senses* **6**, 249–257.

McLean, J. H., Shipley, M. T., 1987. Transnasal transneuronal transport of Herpes simplex virus type I into the rat brain. *Abstr. Soc. Neurosci.* **13**, 387.2.

Miskin, M., Appenzeller, T., 1987. L'anatomie de la mémoire. *Pour la Science,* août, 26–36.

Monti Graziadei, G. A., Karlan, M. S., Bernstein, J. J., Graziadei, P. P. C., 1980. Reinnervation of the olfactory bulb after section of the olfactory nerve in monkey (Saimiri sciureus). *Brain Res.* **189**, 343–354.

Mori, K., 1987. Membrane and synaptic properties of identified neurons in the olfactory bulb. *Prog. Neurobiol.* **29**, 275–320.

Mouly, A. M., Holley, A., 1986. Perceptive properties of the multi-site electrical microstimulation of the olfactory bulb in the rat. *Behav. Brain Res.* **21**, 1–12.

Mouly, A. M., Vigouroux, M., Holley, A., 1985. On the ability of rats to discriminate between microstimulations on the olfactory bulb in different locations. *Behav. Brain Res.* **17**, 45–58.

Mozell, M. M., Hornung, D. E., 1981. Imposed and inherent olfactory mucosal activity patterns and experimental design prompted by the work of David Moulton. *Chem. Senses* **6**, 267–276.

Mustaparta, H., 1971. Spatial distribution of receptor responses to stimulations with different odours. *Acta Physiol. Scand.* **32**, 154–166

Nakamura, T. and Gold, G. 1987. A cyclic nucleotide-gated conductance in olfactory receptor cilia. *Nature* **325**, 442–444.

O'Connell, R. J., Mozell, M. M., 1969. Quantitative stimulation of frog olfactory receptors. *J. Neurophysiol.* **32**, 51–63.

Ottoson, D., 1954. Sustained potentials evoked by olfactory stimulation. *Acta Physiol. Scand.* **32**, 384–386.

Pager, J., 1986. Neural correlates of odor-guided behaviors. *Experientia* **42**, 250–256.

Papka, R. E., Matulionis, D. M., 1983. Association of substance P immunoreactive nerves with the murine olfactory mucosa. *Cell Tissue Res.* **230**, 517–526.

Pevsner, J., Hwang, P. M., Skalar, P. B., Venable, J. C., Snyder, S. H., 1988. Odorant-binding protein and its mRNA are localized to lateral nasal gland implying a carrier function. *Proc. Nat. Acad. Sci. USA* **85**, 2383–2387.

Pissonnier, D., Thierry, J. C., Fabre-Nys, C., Poindron, P., Keverne, B., 1985. The importance of olfactory bulb noradrenalin for maternal recognition in sheep. *Physiol. Behav.* **35**, 361–363.

Porter, H., Butter, N., 1980. An assessment of olfactory deficits in patients with damage to prefrontal cortex. *Neurophyschol.* **18**, 621–628.

Rajendren, G., Dominic, C. J., 1984. Role of the vomeronasal organ in the male-induced implantation failure (Bruce effect) in mice. *Arch. Biol. (Bruxelles)* **36**, 587–590.

Revial, M. F., Duchamp, A., Holley, A., MacLeod, P., 1978. Frog olfaction: odour group, acceptor distribution and receptor categories. *Chem. Senses* **36**, 23–33.

Revial, M. F., Duchamp, A., Holley, A., 1983. New studies on odour discrimination in the frog's olfactory receptor cells. II: Mathematical analysis of electrophysiological responses. *Chem. Senses* **3**, 179–194.

Roberts, E., 1986. Alzheimer's disease may begin in the nose and may be caused by aluminosilicates. *Neurobiol. Aging* **7**, 661–567.

Royet, J. P., Sicard, G., Soucher, C., Jourdan, F., 1987. Specificity of spatial patterns of glomerular activation in the mouse olfactory bulb: computer-assisted image analysis of 2-Deoxyglucose autoradiograms. *Brain Res.* **417**, 1–11.

Schwob, J. E., Gottlieb, D. I., 1986. The primary olfactory projection has two chemically distinct zones. *J. Neurosci.* **6**, 3393–3404.

Serby, M., 1987. Olfactory deficits in Alzheimer's disease. *J. Neural. Transm. suppl.* **24**, 69–77.

Shipley, M. T., 1985. Transport of molecules from nose to brain: Transneuronal anterograde and retrograde labeling in the rat olfactory system by wheat germ agglutinin-horseradish peroxidase applied to the nasal epithelium. *Brain Res. Bull.* **15**, 129–142.

Silver, W. L., Maruniak, D. A., 1981. Trigeminal chemoreception in the nasal and oral cavities. *Chem. Senses* **6**, 295–306.

Silver, W. L., Moulton, D. G., 1982. Chemosensitivity of rat nasal trigeminal receptors, *Physiol. Behav.* **28**, 927–931.

Slotnick, B. M., Katz, H. M., 1974. Olfactory learning set formation in rats. *Science* **185**, 796–798.

Staubli, U., Ivy, G., Lynch, G., 1984. Hippocampal denervation causes rapid forgetting of olfactory information in rats. *Proc. Natl. Acad. Sci. USA* **81**, 5885–5887.

Staubli, U., Fraser, D., Kessler, M., Lynch, G., 1986. Studies on retrograde and anterograde amnesia of olfactory memory after denervation of the hippocampus by entorhinal cortex lesions. *Behav. Neural Biol.* **46**, 432–414.

Staubli, U., Fraser, D., Faraday, R., Lynch, G., 1987. Olfaction and the data memory system in rats. *Behav. Neurosci*, **6**, 757–765.

Takagi, S. F., 1984. The olfactory nervous system of the old world monkey. *Jpn. J. Physiol.* **34**, 561–573.

Tomlinson, A. H., Esiri, M. M., 1983. Herpes simplex encephalitis: immunohistochemical demonstration of spread of virus via olfactory pathways in mice. *J. Neurol. Sci.* **60**, 473–484.

Vaccarezza, O. L., Sepich, L. N., Tramezzani, J. H., 1981. The vomeronasal organ of the rat, *J. Anat.* **132**, 167–185.

Vigouroux, M., Viret, P., Duchamp, A., 1988. A wide concentration range olfactometer for delivery of short reproducible odor pulses. *J. Neurosci. Methods* **24**, 57–63.

Ward, C. D., Hess, W. A., Calne, D. B., 1983. Olfactory impairment in Parkinson's disease. *Neurology* **33**, 943–946.

Warner, M. D., Peabody, C. A., Flattery, J. J., Tinklenberg, J. R., 1986. Olfactory deficits and Alzheimer's disease. *Biol. Psychiatry* **21**, 116–118.

Wirsig, C. R., Leonard, C. M., 1987. Terminal nerve damage impairs the mating behavior in the male hamster. *Brain Res.* **417**, 293–303.

Wysocki, C. T., Wellington, T. L., Beauchamp, G. K., 1980. Access of urinary non volatiles to the mammalian vomeronasal organ. *Science* **207**, 781–783.

Zheng, L. M., Caldani, M., Jourdan, F., 1988. Immunocytochemical identification of luteinizing hormone-releasing hormone positive fibres and terminals in the olfactory system of the rat. *Neurosci.* **24**, 567–578.

Zheng, L. M., Jourdan, F., 1988a. Atypical olfactory glomeruli contain original olfactory axon terminals: an ultrastructural horseradish peroxidase study in the rat. *Neurosci.* **26**, 367–378.

Zheng, L. M., Jourdan, F., 1988b. Ultrastructural of the rat terminal nerve: organization of neuronal subsets and accetylcholinesterase cytochemistry. *Chem. Senses* **13**, 473–485.

Zheng, L. M., Ravel, N., Jourdan, F., 1987. Topography of centrifugal accetylcholinesterase-positive fibres in the olfactory bulb of the rat: evidence for original projections in atypical glomeruli. *Neurosci.* **23**, 1083–1093.

2

Psychophysical Methods of Evaluation in Environmental Studies

E.P. Köster

2.1 Introduction

Since 1860, when Fechner defined *psychophysics* as being "the exact science of the functional relationship between the physical and psychic (mental) worlds," psychophysicists have tried to establish laws giving prominence to relationships that exist between physical stimuli coming from the environment and human sensations.

In those fields of investigation where the nature of the physical stimulus can be clearly defined, as in vision and audition, psychophysicists' contributions have led to the establishment of important theories on how the sensory system works and to the comprehension of complex phenomena such as color mixing. In other research fields, like olfaction, where the exact nature of stimulus is still unknown and where, consequently, the experimenters cannot vary the functional properties of the stimulus as they wish, the results are much more limited. Given the fact that the only olfactory stimulus property that can be varied systematically is the concentration of the odorous substance that reach the nasal fossae (not even those which reaches the olfactory epithelium) and that this property is linked to the intensity of the olfactory sensation, most psychophysicists have turned to odorous intensity measurement. The methods they used will be discussed in detail in the present chapter.

Some researchers have equally used psychophysical methods to estimate odorous quality. Even though the value of their results is limited because of the arbitrary choice of odorous substances, their methods are interesting and will be described briefly.

The object of this chapter is not to give details on all the experimental procedures used in psychophysics. It seems more important to give an overview of all the existing methods, to discuss the advantages and disadvantages of their use in olfaction, in particular in odor pollution assessment, and to establish how they can make it possible to respond to the questions we are all asking ourselves.

2.2 Psychophysical Methods for the Evaluation of Odorous Intensity

2.2.1 General Points

In psychophysics, discussions of intensity are traditionally centered on three questions:

1. At what concentration does a stimulus become perceptible? (*detection problem*)
2. How much do two concentrations of the same nature have to differ for a difference to be perceived? (*discrimination problem*)
3. How is the perceived intensity of a stimulus a function of its concentration? (*scaling problem*)

These three problems will be treated in the following paragraphs.

2.2.2 Detection Problem

2.2.2.1 General Points

What must the concentration of a stimulus be for it to become perceptible? This question leads directly to the "absolute" or "detection" threshold concept and to the problems related to threshold measurements. All psychophysicists are generally in agreement in admitting that the "absolute" threshold is more of a statistical concept than "absolute" and that, for a given sensory modality, there is no physical intensity that constitutes the breaking point under which no stimulus at all is perceived and above which all stimuli are perceived. In fact, the probability that a stimulus will be perceived increases in a continuous manner over a range of increasing concentrations above a certain point. Figure 2.1 gives theoretical curves expressing the relationship between detection probability (percentage of positive responses) and the concentration of an odorous substance for two subjects (S and S') with different sensitivities (S being more sensitive than S'). The *detection threshold* is generally defined as the concentration at which a subject detects an odor with a probability of 0.50. Therefore, in Figure 2.1, the concentrations C and C' respectively, are the threshold concentrations for S and S'.

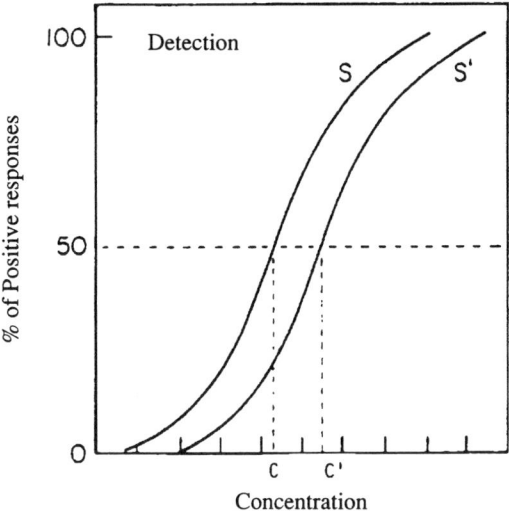

Figure 2.1. Relationships between the odorous concentration of the stimulus and the percentage of positive responses given by two subjects S and S'. The detection threshold is the concentration that provokes 50% of the positive responses.

2.2.2.2 Factors that Modify the Detection Threshold: Description

The detection threshold can be affected by a certain number of factors that will be described briefly here and that must be taken into account during threshold determination.

Adaptation

The sensitivity of a continuously stimulated olfactory system decreases rapidly at the beginning and then generally reaches a plateau that is maintained throughout the duration of the stimulation. The point on the concentration scale that corresponds to this new sensitivity level depends on the stimulus concentration to which the sensory organ is exposed. The higher the stimulus concentration is, the more sensitivity decreases. Given the fact that the threshold concentration (or barely perceptible concentration) corresponding to the new sensitivity level is sometimes situated above the stimulus concentration being used, complete adaptation (i.e., reaching the new level) can result in a complete loss of stimulus perception. The necessary time for complete adaptation or loss of stimulus perception depends on the concentration and duration of the stimulus. The higher the stimulus concentration is, the longer the olfactory system takes to adapt completely.

After the stimulation itself stops, the original sensitivity of the sensory system is progressively restored. This phenomenon is called *recovery*; its time-dependent curve has an exponential form analogous to that of adaptation. The necessary time to return to the original threshold depends on the preliminary adaptation level.

Figure 2.2 gives a schematic example of adaptation and recovery curves for two stimuli with different concentrations. The curves in this figure represent variations in the detection threshold as a function of time. Köster (1971) studied adaptation at the threshold level and showed that adaptation in olfaction is relatively rapid, but that recovery after adaptation is a slow process. Even if the subjects take only one sniff of a threshold concentration, it takes them at least 90 s to recover their initial olfactory sensitivity. He equally concluded that autoadaptation (loss of sensitivity to the same stimulus as the one the subject was exposed to) is always stronger than crossadaptation (loss of sensitivity to a stimulus after exposure to another stimulus) and that certain odorous substances raise the autoadaptation and crossadaptation levels much more than others.

In consequence, for practical applications, it is advisable to assure oneself, first, that the threshold measurements are carried out in a nonodorous environment so as to avoid previous adaptation of the subjects and, second, that the interval between successive measurements with the same subject is at least 120 s.

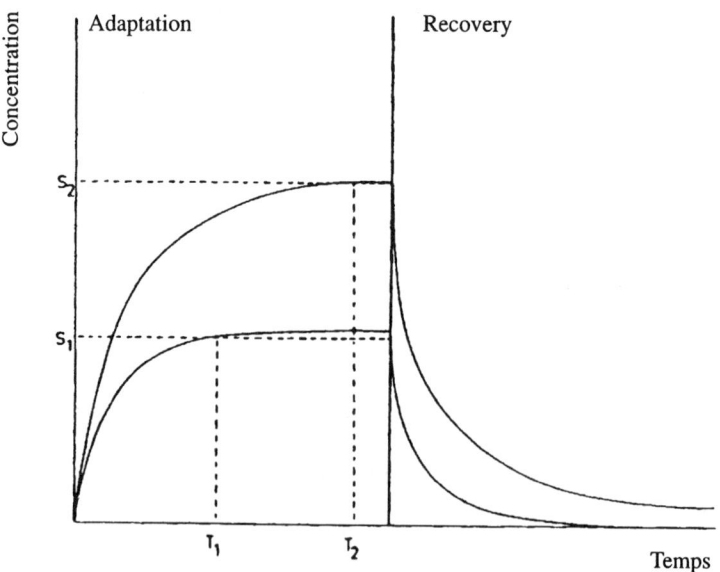

Figure 2.2.   Diagram examples of adaptation and recovery curves for two adapting stimuli with different concentrations (S1 and S2). The points indicate the moments (T1 and T2) where the threshold exceeds the adaptation stimulus (complete adaptation).

Response Bias

The signal detection theory has focused attention on two factors likely to modify the willingness of subjects to give a positive response when they encounter a barely detectable signal.

Signal Anticipation. If subjects have the impression that a signal occurs frequently and is in doubt as to whether or not they smell anything, they will be inclined to give a positive answer more readily than when they suspect that the signal occurs only infrequently.

Motivation. If subjects have the impression that it is very important to detect a signal when it is present, they will be inclined to give a positive answer more readily in situations in which they is in doubt. The best way to get around this response bias problem is to use forced choice procedures. In this case, the subject is obliged to give a response, even in the absense of a detectable signal, by choosing one of the proposed options.

Illnesses

Ordinary head colds, influenzas, and a certain number of illnesses related to the ORL tract and the nervous system can influence the threshold of a subject. Subjects in good health should be used for threshold determination.

Diurnal Rhythms

Even though diurnal rhythms influence the olfactory threshold to an extent, the resulting variation is not important enough to be taken into account in threshold determination. In practice, measurements are avoided between midnight and 8 A.M. and at those moments during the day when attention is lowered, in particular at the beginning of the afternoon (digestion, sleepiness).

Satiation

The olfactory threshold is slightly lower when one has a sensation of hunger than when one is satisfied. It is for this reason that experiments preferably take place before meals.

Fatigue

General fatigue can equally influence the threshold. Morning experimental sessions are preferable to those carried out at the end of the afternoon.

Fatigue can also play a role during an experimental session. In principle, the session must last less than one hour. Sessions longer than 2 hours should be avoided unless sufficient time and possibilities for recreation are provided.

Hormonal Influences

Olfactory sensitivity is affected by hormonal factors. On the average, women are more sensitive than men, but this sensitivity is subject to greater variations due to the ovarian cycle (Le Magnen, 1952; Koelega and Köster, 1974; Doty, 1976). If there is a sufficient number of subjects, there is no need to take particular measurements to neutralize these influences.

Odorous Contamination

Olfactory thresholds must be measured in an environment free from odorous contamination. The subjects are asked not to apply any perfume the day of the test and to not smoke 15 min. before the test and during the test itself.

Age

On the average, olfactory sensitivity decreases with age, but individual sensitivity differences within an age group are large. Generally, subjects from 18 to 45 years old are used for the average threshold determination of a group.

2.2.2.3 Methods of Detection Threshold Determination: Description

There are a number of classical methods to measure the detection threshold. What differentiates them is, on the one hand, the order in which the stimuli are presented to the subject and, on the other hand, the nature of the responses asked of the subjects. They are not all suitable for studies on olfaction and/or environmental odors and it is advisable to take particular precautions with most of them.

Method of Limits

With the method of limits, a number of stimuli, differing in small discrete steps on a logarithmic concentration scale (usually ^{2}log), chosen in such a way that the lowest one is not perceptible and the highest one is clearly perceived, are presented alternatively in ascending and descending series. An ascending series is continued until the subject's response changes from "no" to "yes" and a descending series is stopped after a change in the response from "yes" to "no," as illustrated in Table 2.1.

In order to avoid the possibility of the subject building up an expectation about the position of the transition point in the series, merely on the basis of the number of trials or the time elapsed since the beginning of the series, the concentration which serves as a starting point is varied at random in successive ascending and descending series. The concentrations of the stim-

Table 2.1. Data Obtained With the Method of Limits

Concentration	↓	↑	↓	↑	↓	↑	↓	↑
24	+							
23	+							
21	+		+					
20	+		+					
19	+		+				+	
18	+		+		+		+	
17	+		+		+		+	
16	+	+	+		+		+	
15	+	−	+	+	+		+	+
14	−	−	−	Σ	−	+	−	−
13		−	−		−	−		−
12		−	−		−	−		−
11		−	−		−	−		
10		−	−					
9			−					
8			−					
7			−					
Threshold by series	14.5	15.5	14.5	14.5	14.5	13.5	14.5	14.5
Average threshold = 14.5								

uli at the transition points are averaged over a number of at least 10 ascending and 10 descending series and this average is considered to be the threshold for a given subject (see Table 2.1).

Usually, transition points in ascending series are situated at a slightly higher concentration than that of descending series. This phenomenon is understandable if one realizes that, in descending series, the subjects have a better knowledge of what they are looking for in the next attempt because they experienced the stimulus in the previous attempt. On the other hand, in ascending series their memory of the quality can die down, not having perceived the stimulus for several attempts. In addition, the subjects can have a tendency to persevere in their responses, which can cause them to continue to say "no" in ascending series and "yes" in descending series.

However, these overlapping transition points in ascending and descending series are rarely or never encountered in olfaction. On the contrary, one often finds an interval between transition points, those in descending series being higher than those in ascending series. This phenomenon can be illustrated by Pangborn et al.'s (1964) results, which represent a positive response frequency at different concentrations of heptanone-2 obtained from three different stimulus presentations: sequential-up, sequential-down, and random. As can be seen in Figure 2.3, the lowest detection threshold (equivalent to the lowest transition point) is obtained by the sequential-up method (su), the highest by the sequential-down method (sd), and the intermediate threshold is obtained by the random method.

This phenomenon can be perfectly explained in terms of olfactory adaptation (see Section on Adaptation earlier in this Chapter.) As shown by Köster (1971), strong stimuli have a greater adaptation effect than weaker stimuli, but even short weak stimuli reduce olfactory sensitivity for a considerable period (more than 2 min.). In Pangborn's study, the interval between stimuli was only 15 s. Given the fact that with the sequential-down, method the first stimulus is always stronger than the following one and that with the sequential-up method the first stimulus is always weaker than the following one, differences in adaptation effects exerted by the preceding stimuli could very well explain the differences in thresholds obtained by these two methods. The sequential-up method is preferable, unless sufficient time is allowed for the recovery of olfactory sensitivity between stimuli, adaptation affects above all thresholds obtained with the sequential-down method and, to some extent, those obtained with the randomized order method. It is both more economical (shorter interstimulus interval permit-

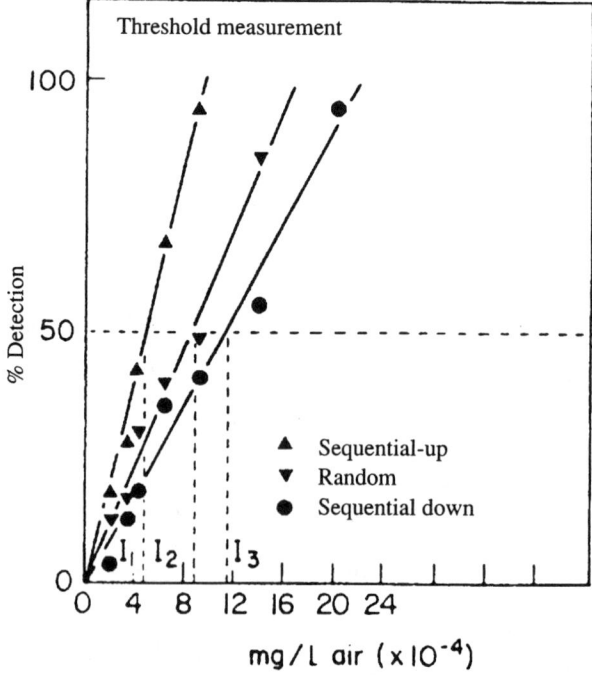

Figure 2.3. Results obtained by Pangborn et al. (1964) in threshold determination experiments with three different stimulation presentation protocols. The odorous substance is heptanone-2.

ted) and more sensitive (lower threshold) than the sequential-down method, which is usually avoided.

"Staircase Method"

A modified method of limits, the so-called staircase method is recommended by some authors (Wetherill and Levitt, 1965). With this method, the experimenter begins stimulation with a very weak concentration and increases it each time the subject does not smell it; this is the same procedure as for the sequential-up method. However, in the staircase method the experimenter does not begin a new series of stimuli when the subjects change their responses and give positive responses. Instead, the experimenter now presents a stimulus at a concentration one step weaker than the previous stimulus. If this stimulus is perceived again the experimenter will still go a step lower on the next trial and, if this is not perceived, will take a step higher on the next trial. The experimenter will continue to do so until the subject has reached a certain criterion which is usually set at three or four successive changes of response in the series. The detection threshold is then set at midpoint between the odor concentrations of the stimuli that provoke this series of alternating responses.

Although this method seems very economical, since it avoids presenting a large number of stimuli which are not in the vicinity of the threshold, it presents serious disadvantages. In particular, it may take a very long time before the subject meets the criterion. This is due to the fact that the sensitivity of the subject varies over time and may do so to an extent that transcends the difference between successive stimuli in the staircase method. In that case, it may take a very long time before subjects reach the criterion, if in fact they ever do. In addition, with this method, the presented stimulus concentration depends only on the subject's response to the previous stimulus. Not only does this make automated stimulus presentation more difficult, but it also makes simultaneous experimentation on several subjects with the same olfactometer practically impossible.

Method of Constant Stimuli

Compared to the random method, the sequential-up method presents the inconvenience of requiring a certain number of additional stimuli to avoid the previously mentioned expectation effects. Since these supplementary stimuli are always too weak to be perceived, they do not contribute to the actual measurement.

For this reason and because is is easier to automate experiments in which stimulus presentation is independent of the subject's response (particularly when the same stimulus is given to a group of subjects at the same time) many experimenters choose the random method or, as it is often called, the method of constant stimuli. Five or seven stimuli at different concentrations are selected on the basis of very fast evaluation by the ascending method, so that the weakest is perceived in nearly 10% of the cases and the strongest in 90%. These stimuli are then presented to the subjects many times in a random order or in an order in which each stimulus follows each of the other stimuli an equal number of times, the order of successive stimuli pairs being determined at random as much as possible. The interval between two successive stimuli must be at least 1 min. to reduce adaptation effects (see Section 2.2.2.2) and, in order to avoid general subject fatigue (see Section 2.2.2.2), the experimentation is generally carried out in sessions during which a maximum of 35 stimuli are presented. At the end of the experiment the positive response frequencies to each of the five or seven stimuli are calculated for each subject and plotted as a function of concentration (Figure 2.4, left). When positive response frequencies are expressed in standard deviation units (probit analysis), a straight line can be adjusted and the threshold can be determined by interpolation to the point corresponding to 0.5 probability of positive responses (number of standard deviations = 0) (Figure 2.4, right). This way, all measurements have contributed to the threshold determination.

Use of Blank Stimuli (Nonodorous) and the Forced Choice Method

In the two methods previously described, the subjects are asked to indicate whether or not they perceive the presented stimulus by simply saying "yes " or "no"; this method presents a certain number of difficulties. The subjects may be biased to say either "yes " or "no" for reasons unrelated to their actual sensitivity. For example, if threshold measurement is carried out as part of the procedure of panel member selection in sensory evaluation and being chosen as a "subject" is desired, the candidates may lie and say "yes " much more often than they actually perceive the stimulus. Even if

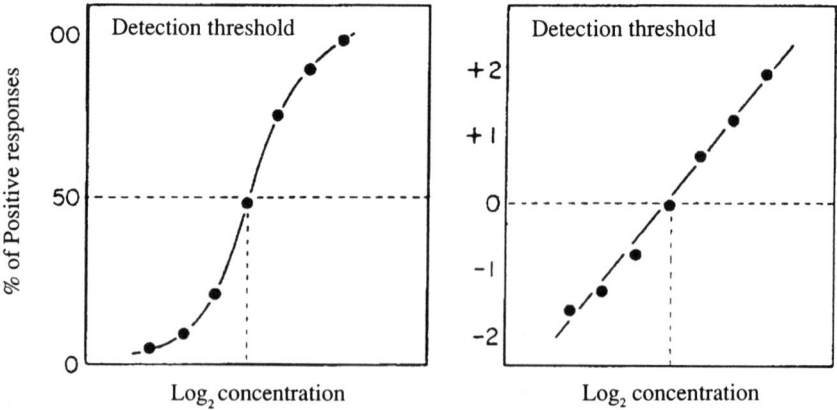

Figure 2.4. Detection threshold functions expressed in percentages of positive response (left) and in the number of corresponding standard deviations (right).

they do not intend to lie, their responses may be biased in one way or another without their knowledge.

In order to be able to check the influences of this type of bias, experimenters have introduced blank stimuli (i.e., containing no odorous substance) into their sequences, which consists of only the carrier fluid. If the subject still responds positively to some of these blank stimuli, they can be told that they are wrong, since they perceive stimuli that do not exist. Such a procedure carries the risk of making a subject wary with regard to their own responses and consequently lead to a tendency to respond negatively at the least doubt. The threshold value obtained under these conditions is too high.

For this reason, a certain number of experimenters have used a method that forces the subject to make guesses even when they have a doubt. With this method, known as "forced choice," the subject receives a defined number of stimuli (generally 3–5) at each presentation. These stimuli may be presented either simultaneously in different places (vision) or at different moments (audition, olfaction). All the stimuli except for one are blanks and the subject is asked to find the signal that is different from the others. Even if the subject cannot specify a difference, a choice is required by trying to guess. After the experiment, the results are corrected by the following formula:

$$P_{(corr)} = \frac{P_{(obs)} - P_{(random)}}{100 - P_{(random)}} \times 100$$

$P_{(corr)}$ = percentage of positive responses to this stimulus after correction
$P_{(obs)}$ = percentage of observed positive responses to a stimulus
$P_{(random)}$ = percentage of anticipated positive responses uniquely due to each presentation)

Figure 2.5 (right) illustrates the results of such a correction procedure. The percentages in Figure 2.5 (right) can, of course, be tranformed into the number of standard deviations so as to obtain a straight line as in Figure 2.4 (right). In olfaction, the forced choice technique is often used to measure detection thresholds. It is usually used in combination with the method of constant stimuli. Blank stimuli presented together with odorous stimuli contain only the carrier liquid (generally benzyl benzoate, diethyl phthalate or water) used to reduce the odorous stimulus concentration. One must be very careful not to give the subject other clues except the olfactory information for correct response determination. Since the subject has access to the blank stimuli to make a direct comparison, very slight differences in the color or the amount of substance in the test bottles can give this type of information, which would remain unnoticed in a procedure with successive presentations of the stimuli.

It is also very important to use the same batch of diluent to prepare odorous concentrations and blank stimuli. Very slight differences in the purity

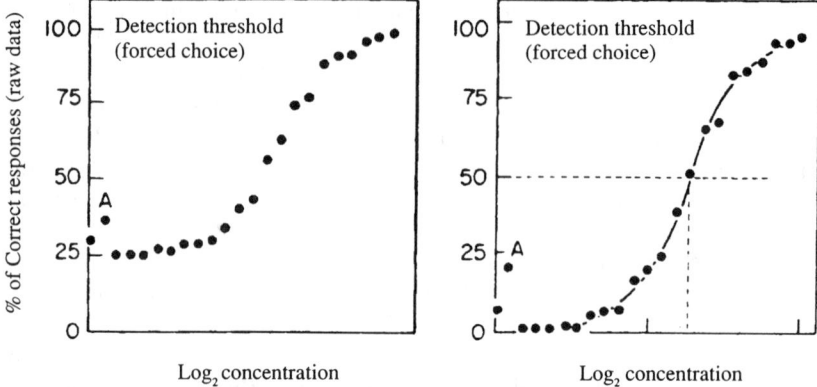

Figure 2.5. Results of a forced choice threshold experiment before (left) and after (right) correction of responses due to random choice (here, random choice corresponds to one good choice out of four).

of different batches can distort the experiment by giving irrelevant information. Sometimes, when a nearly but not completely odorless dilution liquid is used, subjects can quite easily discriminate concentrations well below the concentration at which they reach by mere chance (Figure 2.4, left). The explanation of this strange phenomenon is based on the fact that minute quantities of odorous substances that are not themselves perceptible any longer, "block" the odor of the carrier liquid according to the well-known mechanism of odor compensation (supression of perceived odorous intensity) which tends to occur in mixtures with unequal concentrations (Köster, 1968; Mac Leod, 1968). In these cases, the subject detects the right stimulus because its smell is weaker than that of the blank stimuli simultaneously presented.

Signal Detection Method

In 1961, Swets, Tanner, and Birdsall showed that mathematical decision theory provides a rather elegant theoretical framework for the description of the influence of response biases on psychophysical measurements. Even though psychophysicists had become aware of response biases in their measurements, they had not done anything up until this point to evaluate them sytematically. For years, the influence of these response biases has been studied with great attention. Green and Swets (1966) and, more recently, Engen (1971) have extensively described this research.

 Since measurements by signal detection require extended experimentation, the method had never been used in environmental research and will not be discussed here. Nevertheless, the theory has attracted attention to the two sources of response bias in normal threshold measurement: expectancy and motivation (see Section 2.2.2.2).

2.2.2.4 Application of Detection Methods to Environmental Studies

In environmental studies, the method of limits is often used to obtain a first rough estimate of the detection threshold of a given pollutant odor. Generally, only the sequential-up form is used so as to avoid adaptation. After this approximative threshold determination, the method of constant stimuli (see Section 2.2.2.3) is often used to obtain a more precise detection threshold. In some cases, only the method of limits is used in the sequential-up or staircase form. Even though this last procedure seems quite efficient, it sometimes poses problems not only because of variations of subject sensitivity, but also because the odor source may vary during the procedure. As a result, many subjects only reach the criterion after very long ascendant and descendant sessions and some never reach it. Because of this fact, it is not recommended to use this method and certainly not in combination with the forced choice procedure. (Although such a combination is completely irrational, because the guessing in the forced choice procedure will produce positive answers to nonperceived stimuli in a number of cases, a warning is given here in view of the fact that such a procedure has been advocated recently by some investigators in France.) The best method to use for environmental studies is the method of constant stimuli combined with the forced choice method, provided that the number of stimuli repetitions per subject or the number of subjects (see discussion, Section 2.4.2) is sufficiently important (not less than 36 judgements per stimulus).

It was possible to prove that this method gave the lowest and most reliable thresholds in a number of environmental studies. Detection threshold measurement plays an important role in environmental studies because odor strengths are often being expressed in odor units (1 odor unit is equal to the quantity of odor necessary to reach the detection threshold). The number of times an odor source must be diluted to reach a detection threshold gives an indication of the odorous strength of a source expressed in odor units.

Even though it has become standard to express odor strength this way, it should be clear from the beginning that the odor unit is merely a concentration measurement and not directly related to the perceived intensity of the odor. Perceived intensities of two different odorous sources each giving off 100 odor units can easily be very different as a function of their respective psychophysical function slopes (see Figure 2.12). Perceived intensities of two different odors would only be equal by definition at the detection threshold level (1 odor unit). At this level, perceived intensity of odors is so weak that it is only detected in half the cases where it is presented. In fact, there are still two remaining uses for odor strength characterization in terms of odor units.

Determination of Odor Reduction Efficiency

To judge the efficiency of certain devices tending to reduce the odorous strength of a source (for example, by scrubbing or burning smokestack effluents), comparison between the number of odor units emitted by the source before and after installation of an odor reducing apparatus may be taken as a proof of its efficiency. However, it should be understood that the perceived intensity of an odor may not have been proportionally changed.

Calculation of Odor Perception Contours Around a Source

Odor dispersion models have often been used to calculate the distance from a source at which the odor will be perceptible for a given percentage of the time. Since such calculations are based on the dilution of the odor concentration needed to find the point at which the threshold is reached under the prevailing climatological circumstances around the source, it is logical to express the odor strength in odor units.

2.2.3 Discrimination Problem

How much difference must there be between two concentrations of the same type for a difference to be perceived?

There is no fundamental difference between the detection problem previously described and the discrimination problem that will now be discussed. In both cases, the subject is asked to say what the difference is between a stimulus and something else, which could be another stimulus or a "background noise". In fact, the detection problem is a special case of the more general discrimination problem. Given the fact that discrimination problems are rarely encountered in environmental studies, the differential threshold will only be discussed briefly here.

2.2.3.1 Discrimination Threshold Determination

When considering detection, a discrimination threshold can be determined. This threshold, which is often called the differential threshold or "just noticeable difference" (JND), is defined as being the difference in concentration between two stimuli detected by the subject with a 0.5 probability.

Generally, the JND is determined by a paired comparison method in which a standard stimulus (S) is presented simultaneously with a comparison stimulus (C). The subject is asked to indicate if the comparison stimulus is stronger than the standard stimulus or not. A series of comparison stimuli is selected (generally from 7 to 9) on the basis of preliminary tests, so that the logarithms of concentration C_i are distributed as evenly as possible around the concentration S, the geometric average of the concentration series C being equal to S.

On the logarithmic concentration scale, the distances are equal between the different C_i of a series. The largest C must be perceived as being stronger than S in about 90% of the cases and the smallest must be perceived as being weaker than S in about 90% of the cases. Each C is compared to S many times (at least 50) and the pairs are presented to the subject in a random order. After experimentation, the judgement frequencies that designate C as "stronger than S" are calculated for each pair and these percentages are plotted against the concentration scale as indicated in Figure 2.6. The same method is used for responses "weaker than S". In olfaction, this precaution is particularly important due to the very important influence that adaptation can have.

The adaptation influence is partially nullified by a tendency (that many subjects seem to have) to overestimate the intensity of the second in a pair of stimuli. This general phenomenon, which is often called "time error," is usually attributed to the fact that the memory trace of the first stimulus may already be fading when the second stimulus is presented. There exists a vast and somewhat confusing literature on time error effects which one need not be concerned with here, because the effects of time errors can be largely reduced by counterbalancing the stimulus orders as was shown for adaptation.

Generally, when the point of subjective equality (PSE) is shifted, all the points of the curve are displaced with it. This is especially true when a response bias or a time error causes the displacement. In these cases, the form of the curve may remain unchanged or be only very slightly changed. Ad-

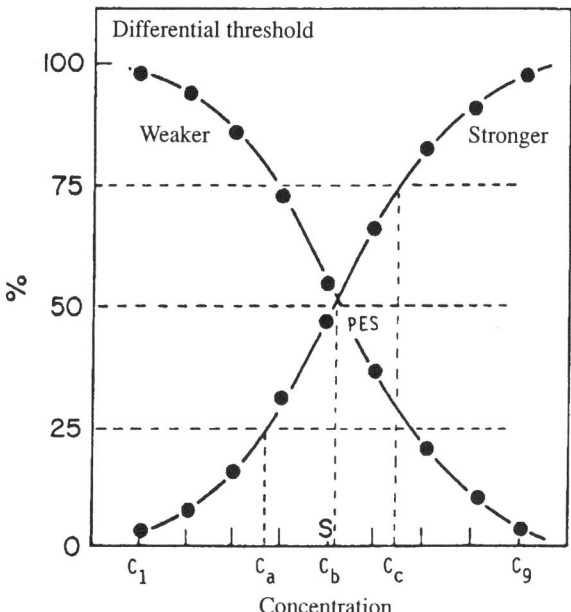

Figure 2.6. Curves indicating the number of times (in %) that a certain comparison stimulus C is judged to be weaker or stronger than the standard stimulus S. SEP = subjective equality point.

aptation, which can more strongly affect perceived intensities at the lower end of the scale than those situated at the high end, may be the cause of a slight distortion. However, if all the precautions previously mentioned are taken, changes in the curve are so slight that it can be used to calculate differential thresholds even if the PSE is considerably shifted. This is possible because the JND is a differential concentration measurement and because modifying the two concentrations in the same direction and in an equal way does not change the difference between them. In calculating the difference threshold (JND), one should depart from the observed concentration which belongs to point of subjective equality (PSE) rather than from the concentration of S. The problem of finding the JND can then be stated as follows: how much does one have to add or subtract from the intensity corresponding with the PSE to reach an intensity which the subject perceives as being different from it in 50% of the cases? Given the fact that experimentation by pairs is typically experimentation by forced choice with a probability level of 0.5, "stronger" or "weaker" percentages must be corrected with the formula described above as the "forced choice method." This means that the "stronger" and "weaker" corrected responses become zero at the PSE and that the "stronger" and "weaker" 0.75 responses (i.e., 25% "stronger") of the initial data are transformed into 0.5 corrected values. The JND is the concentration difference of the logarithmic scale corresponding to 0 and 0.5 corrected responses. By converting responses to the number of standard deviations (probit analysis), it becomes possible to fit a straight line to the data and make all the points contribute to threshold determination (see right-hand side of Figure 2.4).

An easier and more direct method, though less precise, for JND determination for a given PSE is to establish the difference between the 0.25 and 0.75 probabilities of the initial data on the logarithmic concentration scale and divide by two (Figure 2.6).

Weber's Law. If the value of S is changed, one quickly finds that the JND depends on the concentration of S. The higher the concentration of S, the larger the differential threshold is. This relationship has been formalized in Weber's law, which reads: JND $= K \cdot C$, where C is the concentration of S to which the JND is measured and K is a constant, called the Weber fraction.

Weber's law holds well for most sensory modalities in the middle part of the concentration scale. Stone and Bosley (1965) showed that this is also true for olfaction. The Weber fractions found for different substances vary from 0.20 to 0.36.

2.2.3.2 Discrimination Method Application
to Environmental Studies

Discrimination methods are rarely used in the assessment of odor sources in the environment. The effects of measures taken to reduce the output of an odor source production are usually far too large to make such measurements useful.

In some cases they are used to verify the functioning of measuring equipment such as olfactometers. A good example is given by a method making it possible to verify that preliminary dust filtration of odorous effluents entering the olfactometer does not change the effluent's odor. This method is described in the appendix of the French Standard Method AFNOR X43–104 norm (1990).

In addition, Weber's law is important for those relying on Fechner's law to describe the perceived intensity of odors. In this law, perceived intensity is expressed as the number of JNDs above the threshold (see below).

2.2.4 Scaling Problem: How is the Perceived Intensity of a Stimulus Related to its Physical Concentration?

2.2.4.1 General Points

Up until this point, only absolute or differential sensitivity measurement methods have been described. The smallest differences in concentration intensity that bring about a change in sensation experienced by the subjects are determined by these methods. Even though these methods constitute useful information for a number of practical and theoretical applications, it is clear that the sensations experienced by subjects may vary over much larger ranges than the JND or the absolute threshold. Thus, the question arises as to how these larger differences are related to the differences in the concentrations of the stimuli provoking them. If the concentration of a stimulus is doubled, will the corresponding sensation be doubled as well? Or may one assume (as is still often done in environmental esearch) that, by setting up a scale of physical intensities in which the absolute threshold is chosen as the scale unit (the odor unit, see Section 2.2.2.4), one ensures that equal intensities on this scale will give rise to equal intensities of sensation, regardless of qualitative or even cross-modal differences in the stimuli employed?

In other words, is there a direct linear relationship between the concentration of the stimulus (C) and the strength of the sensation (S) experienced by the subject? Even though there are still several controversies on the exact nature of this relationship, a consensus that this relationship is not linear can be put forward.

Fechner assumed that Weber's law was correct and that the JND, the minimum detectable amount of change, should be considered the unit on the physical scale corresponding to one unit of sensation. On this basis, he suggested a logarithmic relationship $S = k \log C$, in which k is a constant which may be different depending on the nature of the stimuli or the different sensory modality. This can be easily deduced from the fact that the concentration steps (C) needed to let the perceived intensity (S) grow in an arithmetical order form a geometrical series:

$$C_{(i+1)} = C_i + \text{JND}_i = C_i + mC_i = C_i(1 + m)$$

where C_i is the required concentration to give a perceived intensity S_i and C_{i+1} is the concentration at only one JND (JND_i) higher corresponding to the immediately following intensity ($S_{(i+1)}$). The relationship between C and S for the Weber fraction $m = 0.50$ is illustrated in Figure 2.7.

This law was contested by Stevens (1936), who suggests a different law: $S = C^n$, in which n is yet another constant which can vary with the stimulus quality or the sensory modality being studied. This power law, which can be transformed into log $S = n$ log C, establishes that there is a linear relationship between the logarithm of the physical intensity and the sensation or the logarithm of the perceived intensity. The constant n is the slope of the straight line traced in log $*$ log coordinates. One of Stevens' objections to Fechner's law was based on the observation that doubling the physical intensity steps did not reflect doubled perceived intensity. Given the fact that the objective of this chapter is not to discuss the respective merits of these laws and other proposed variants, only a short description of the most currently used methods in olfaction for determining sensory values of perceived intensities will be given. For an exhaustive description of the theoretical and practical implications of the use of scaling methods, readers are encouraged to consult Engen's (1971) excellent work, which gives numerous examples drawn from the field of olfaction.

It is important to remember that, even if some of the scaling methods described below do not give a direct look at the exact nature of the sensation

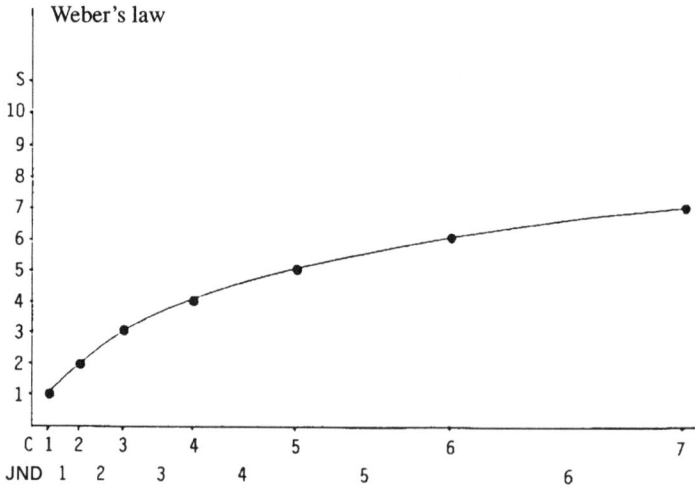

Figure 2.7. Relationship between the steps of the physical scale (abscissa) and those of the sensation scale (ordered), after Fechner.

scale, they may provide a means for matching stimuli at fixed levels of sensation provided that the subject can handle them in a reliable way.

2.2.4.2 Category Scaling

One of the first methods that comes to mind when one is faced with the problem of determining the strength of sensation in relation to the physical intensity of the stimulus is to let the subjects rate the strength of their sensation on a given rating scale. Such a scale might have the following form (adapted after Engen, 1971):

Description	Evaluation
Extremely weak	1
Very weak	2
Weak	3
Medium weak	4
Medium	5
Medium strong	6
Strong	7
Very strong	8
Extremely strong	9

The subject is asked to classify each stimulus presented in one of the categories mentioned on the scale. Generally, at the beginning of the experiment, a "very strong" and a "very weak" stimulus are presented to the subject to indicate the range of values. The scale is often reduced to seven or even five categories because many experimenters believe that the subject cannot process too many categories in a satisfactory way. In these cases, the "extremely weak (strong)" and "medium weak (strong)" categories are usually suppressed.

For each of the presented stimuli (presented at least 10 times), the mean of the category judgements is calculated and then plotted against the physical intensity. Figure 2.8 shows the result of two experiments with the same stimuli but with different numbers of category values on the sensation scale.

As can be seen from this figure, the two curves are similar in shape, both leveling off at their extremes. This leveling off is one of the disadvantages of the use of category scaling. Subjects are confronted with sets of categories that are forced on them and they seem to use these categories more freely in the middle ranges than at the extremes. The subjects' reluctance to mention the extreme categories may also explain why the curve found with the nine categories is somewhat steeper than the curve found with the five categories. In both cases the extremes are avoided, but this affects the five-category scale more than the nine-category scale.

The fact that the subjects behave this way has a theoretical importance. It signifies that the categories do not represent equal steps of sensation for them; in other words, the separation between their "very strong" and

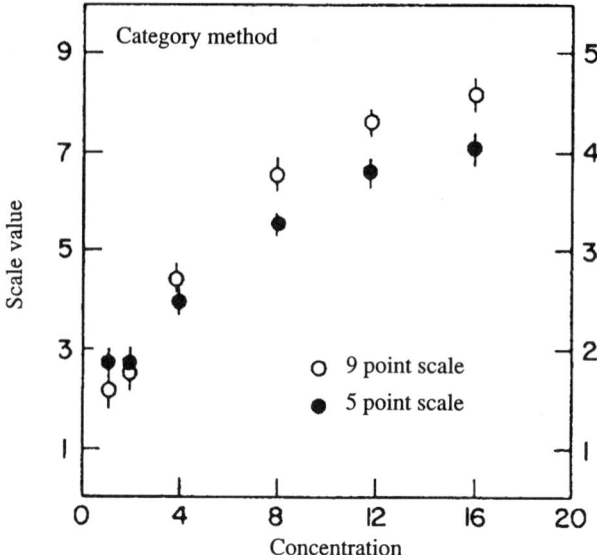

Figure 2.8. Results of perceived intensity measurements with scales of five and nine categories for the same concentrations.

"strong" judgements differs from the separation between their "strong" and "medium strong" judgements. Because of this fact, the category method does not seem to respond in a very satisfactory way to the following question: Which of the two laws mentioned above can best be applied to the relationship between sensational intensity and physical intensity? To respond to this question in an adequate way, a method that allows subjects to choose their own sensation scale is needed rather than forcing them to think in terms of predetermined categories. Methods such as magnitude estimation and cross-modality matching (see Section 2.2.4.5) seem to provide a better response to this requirement. However, category scaling also presents certain advantages. Subjects, generally familiarized with this type of evaluation procedure, can easily accomplish their task. For this reason, they use the categories in a rather reliable way as can be seen from the standard deviations found for the different points in Figure 2.8. In olfaction, category scaling has been used on numerous occasions (Mitchell, 1971).

2.2.4.3 Intensity Matching

Another method for the assessment of perceived intensity is intensity matching. The external odorous stimuli to be measured can be compared to a series of preestablished concentrations of a reference substance. The subjects are asked to indicate at which or between which of the steps of the fixed odor intensity scale the intensity of the external odor is located. After a series of preparatory experiments carried out by Dravnieks, the ASTM

(1975) adopted this procedure as a standard method and recommended the use of a butanol scale at eight concentrations as reference material. Dravnieks himself (1974) developed an olfactometer with eight preestablished concentrations for reference stimuli presentation.

Even though this method has been used frequently, it does have some disadvantages. In a detailed study, using three odor reference scales, Punter et al. (1984) showed that the results do not fulfill the criteria of symmetry, transivity, and reflexivity that are found with methods bringing one substance into play at a time. This occurs even when butanol stimuli at the same concentrations as those of the reference scale are studied. In fact, one then observes that weak stimuli are estimated to be stronger and the strong stimuli are estimated to be weaker than the equivalent reference concentrations.

The paired comparison method can also be used to match the perceived intensities of different odorous substances. In experiments on the odorous intensity of odor mixtures (Köster, 1968, 1969), a fixed standard stimulus (a molar concentration of m-xylene) and a number of concentration series of different pure odorous substances and their mixtures in a number of fixed mixing ratios used as the comparison stimuli were used. For each pure substance and each mixing ratio, the concentration corresponding to the point of subjective equality with the standard stimulus was determined. In this way, it could be shown that odorous substances, when mixed in certain ratios, will supress each other's odorous intensity.

Matching of odorous intensities is also important if one wants to prepare stimuli for quality assessment. Intensity differences should be avoided whenever the odorous qualities of odorous substances are compared, because otherwise the subjects are often unable to single out the quality aspects.

2.2.4.4 Magnitude Estimation

Magnitude estimation is the most direct of all scaling methods because it leaves the subjects completely free to establish their own sensation scale. The subjects receive a random order series of concentrations and are asked to assign numbers to each of them in such a way that they feel that the ratios between the assigned numbers correspond with the ratios between the perceived intensities. The subjects are told that they may use any numbers they wants, including decimal numbers. This method, which was devised by Stevens, is generally believed to provide the best demonstration of Stevens' power law.

Two variations of this method can be used. In the first one (fixed modulus method), a standard concentration (usually the middle one from the series) is given to the subjects, who are asked to call the sensation value corresponding to that concentration 10. In the second variation (free modulus method) no standard is given and the subjects are given complete freedom to choose their own scale.

For each stimulus concentration the geometrical mean of the numbers assigned by the subject during different presentations is calculated by converting the numbers into their logarithms and by taking the exponential of the mean. This last step is often omitted and the logarithmic means are directly plotted as a function of the concentration logarithms. Figure 2.9 gives an example of the results of two different subjects who received each of nine concentrations twice (Cain, 1968; Engen, 1971).

As can be seen in this figure, the subjects' results differ considerably in two ways: on one hand, the slope of the straight line that can be drawn through the points for each subject and on the other hand, the quality of fit of the points to each of these straight lines. Subject one requires about one full order of magnitude of concentration in order for the perceived odorous intensity estimate to differ equally from an order of magnitude, whereas Subject 2 needs two orders of magnitude of concentration to reach the same difference of perceived intensity. Does the difference in exponent of Stevens' power function, which is directly reflected by the difference in the slope of the two lines, mean that the two subjects experience very different sensations when they judge the same stimuli, or does it mean that the subjects simply use the numerical scales in a different way? This is a difficult question to answer. It is known that many inexperienced subjects have difficulties using a scale based on ratios. Some have a tendency to produce a category scale and others start using numbers in an almost haphazard way.

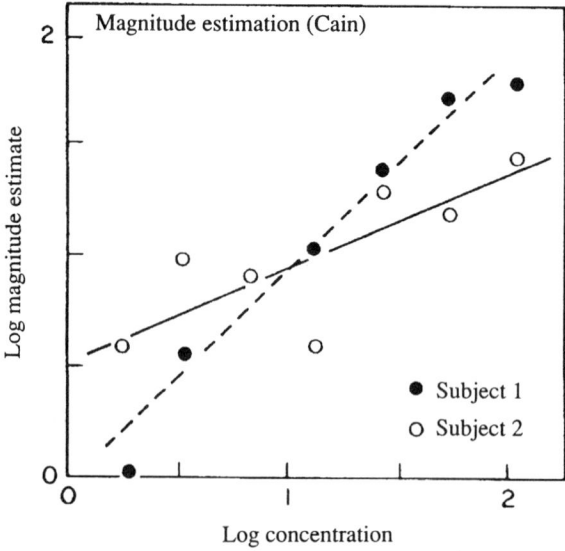

Figure 2.9. Results obtained with two different subjects by magnitude estimation with the same stimuli (after Cain, 1968).

The large dispersion of points around the line drawn for Subject 2 makes one wonder if the subject has used the ratio scale effectively. Even the ordering of stimuli according to their concentration is not correct. This throws a doubt on the precision of the magnitude estimation method to establish sensation intensity. Out of the 15 subjects studied by Cain, who were given each of the nine concentrations to evaluate twice, only one (Subject 1, Figure 2.9) produced perceived intensities in an order that correlated perfectly with the order of concentrations. Subject 2 does not by any means represent the only or even the worst case of nonlinearity in the set of data in Cain's experiments.

These results might be improved considerably by taking more measurements per point, but even then the large differences in the slopes of the lines remain. In his discussion of the individual results of Cain's experiment, Engen gives 0.56 as the mean slope for the whole group of subjects. However, the standard deviation is 0.22, which represents a rather large variability.

Although individual functions, based on a sufficient number of measures per point and obtained with well-trained subjects, usually support Stevens' law and groups of such subjects will also produce stable results over time, the large variability around individual points and slopes of lines obtained with untrained subjects throws some doubt on the meaningfulness and practical value of the method. A simple matching of intensities on the basis of equal sensation values appears to be more reliable when carried with either category scaling or paired comparison than with magnitude estimation.

2.2.4.5 Cross-Modality Matching

In magnitude estimation, the subjects are asked to indicate the strength that they experience by giving a numerical value. With cross-modality matching, the subjects are asked to express the intensity of what they experience by matching it with an equally strong sensation in another sensory modality. This method presents two advantages: it is independent of the subject's "number behavior" and it is based on comparative rather than absolute judgement.

Normally, a stimulus of one of the two modalities is presented to the subject at a fixed physical intensity. The subject varies the other modality until the two sensations are equal. This procedure is then repeated at another physical intensity level of the fixed stimulus and so on. However, it is difficult to perform such a measurement with the olfactory stimulus as the variable one, because even if one uses an olfactometer that allows continuous variation of odorous concentration over a large range, the adaptation that occurs during this long experimentation will distort the results. Consequently, preference is given to using the method with a variable stimulus, chosen from another sensory modality, which is easier to handle and that does not present as rapid an adaptation phenomenon. In some cases, auditory stimuli (1,000 Hz or white noise) are used for this purpose.

The odorous concentrations are presented a number of times in random order and the sound level at which the subjects start their matching is varied systematically to make sure that such variables, such as time or the number of turns on the potentiometer used to vary the sound intensity, do not systematically influence the measurement. After the experiment, the sound intensity mean (in dB) judged to be equal to a given odorous intensity is calculated for each odorous concentration. Since the dB scale is logarithmic, no conversion to logarithms is necessary, as is the case with the numbers in magnitude estimation. The results of such an experiment with two odorous subtances are given in Figure 2.10.

The results obtained for the two substances agree with Stevens' power law and the slopes of the straight lines drawn through the points are very similar. This slope similarity is, however, mere coincidence. In general, it seems that cross-modality matching is an easier and more natural task for the subject than magnitude estimation. As a result, both intra- and interindividual variability is smaller. Ekman et al. (1967) have also used finger span as a cross-modal indicator of olfactory strength in many experiments. (The subject must exert pressure with a finger or a hand, depending on the procedure, on a dynamometer in proportion with perceived olfactory intensity.)

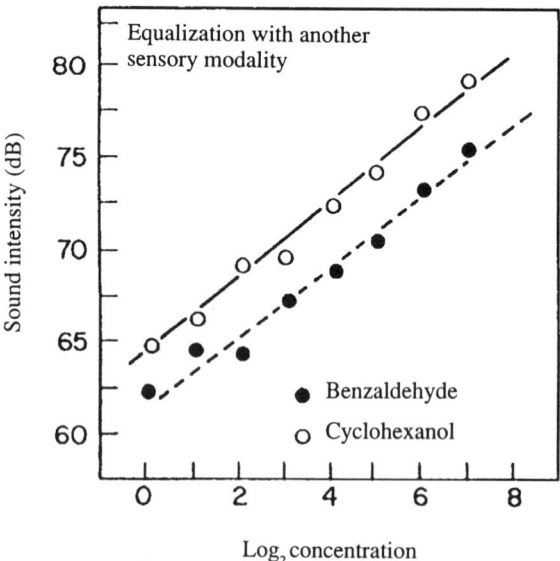

Figure 2.10. Results obtained by the equalization with another sensory modality method in which the perceived intensities of 8 concentrations of 2 odorous substances were compared with the perceived intensities of white noise.

2.2.4.6 Factors Affecting the Perceived Intensity of Stimuli

The perceived intensity of an odor may change under the influence of a number of factors. Some of these factors depend on the subject's state and others on external factors. The most important of these factors are as follows.

Adaptation

According to Cain and Engen (1969), who measured adaptation effects using magnitude estimation, adaptation influences the slope of psychophysical function for weaker concentrations than those to which the subject has been adapted. This part of the psychophysical function becomes steeper whereas the part above the adaptation concentration remains practically unchanged. This is illustrated in Figure 2.11.

The data in this figure were obtained after complete or nearly complete adaptation to two different concentrations of the same odorous substance. It is easy to understand that the lower part of the curve is almost vertical in view of the fact that under these conditions, the detection threshold must have shifted almost to the adaptation stimulus concentration. The fact that psychophysical function of higher concentrations is hardly affected by previous adaptation is somewhat surprising.

It is clear that perceived intensity at these very high levels is not very strongly affected by a loss of sensitivity for lower concentrations. However, it should be remembered that magnitude estimation is a method based on ratio judgements and that the resulting slope may not reflect shifts in the

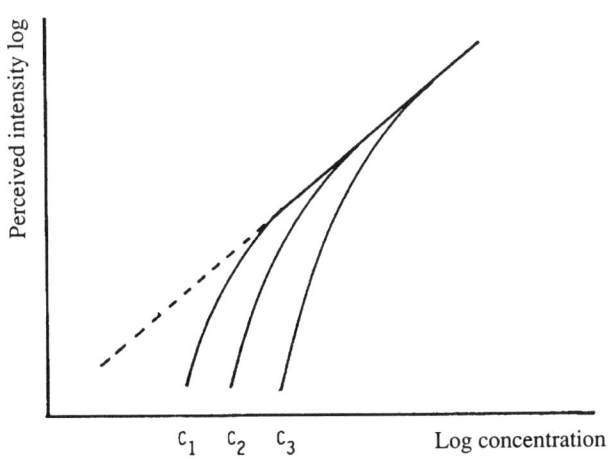

Figure 2.11. Psychophysical functions obtained with a nonadapted subject (dotted line) and with the same subject after adaptation to three different concentrations C1, C2, and C3.

absolute perceived intensities of the concentrations presented. In this case, the result would be an artifact of the method. Otherwise, if the results reflect true absolute judgements of unchanged perceived intensities, the result might only be explained if one supposes that, at different concentration levels, different groups of receptors are brought into play. Electrophysical findings by Duchamp et al. (1974) and Revial et al. (1978, 1983) are in agreement with this point of view.

Habituation

Unfortunately, there is no systematic research on the influence of habituation on the psychophysical function, even though it has been generally assumed that humans get used to monotonous stimulation. The only proof of this habituation phenomenon that can be cited is the fact that the perceived intensity of a new stimulus after a series of identical stimuli is generally overestimated.

Range Effects and Other Artifacts

It is well established that the range of stimuli used to determine psychophysical function may strongly influence its slope and form (Poulton, 1989). Great care must be taken to choose stimuli that are well balanced around a concentration that corresponds with a perceived intensity that is truly "middle of the road" and to choose the concentration steps so that both the upper and lower limits of the human sensitivity range are not exceeded.

2.2.4.7 Use of Perceived Intensity Measurements and Psychophysical Function in Environmental Research

Even though perceived intensity is certainly a more direct indication of the subject's experience when exposed to odorous substances than odor concentration expressed in odor units, it is a much less important tool in environmental research. This is due to the fact that nowhere in the world does perceived intensity provide the basis for a legal norm on pollution. Norms are either based on emissions expressed in odor units or on measurements of odorous annoyances. Nevertheless, the study of perceived intensity has two important practical applications.

The measurement of perceived intensity can be used to indicate odor emission effects of measures taken to reduce the odor emission of a source. If no olfactometric equipment is available or if it is difficult to apply such techniques, it is possible to demonstrate such emission effects by comparing results obtained in perceived intensity measurements made before and after the installation of the odor reduction device. In such cases, category scales or intensity matching procedures are recommended, notably when the measurements have to be carried out with untrained subjects. Whenever possible, it is recommended to use the same group of about ± 10 people to carry out the two measurements. If this is not possible, the panels used in the two different measurements should be considerably larger (at least 24 people) in order to make a valid comparison. With such methods, the existence of odor emission effects can be demonstrated, but not much can be said with great precision about the extent of the effects.

It is very important to have information about the slope of the power function (n in $S = C^n$) for given odorous emissions in deciding the priority and extent of the measures to be taken to reduce the odorous pollution caused by them. This is easily understandable with the help of a schematic example. If the odors coming from two different sources A and B have psychophysical functions as different as those in Figure 2.12, it is necessary to reduce the concentration of B much more than that of A to reduce their perceived intensities by half. It also means that when both odors are of equal strength at the source, perceived intensities odor B will be perceived over a much larger distance than odor A. Finally, in the extreme case where odor A is much stronger at the source than B, what may occur is that odor A is not perceived outside the limits of an industrial complex, whereas odor B is still perceived in an urban neighborhood 10 km away at almost the same intensity as at the source.

In view of all this, it is obvious that concentration measures such as odor units do not necessarily provide an indication of perceived intensity of the corresponding odor. It is essential to know the psychophysical function slope as well. It is probably easier to obtain such information by using magnitude estimation or cross-modality matching methods.

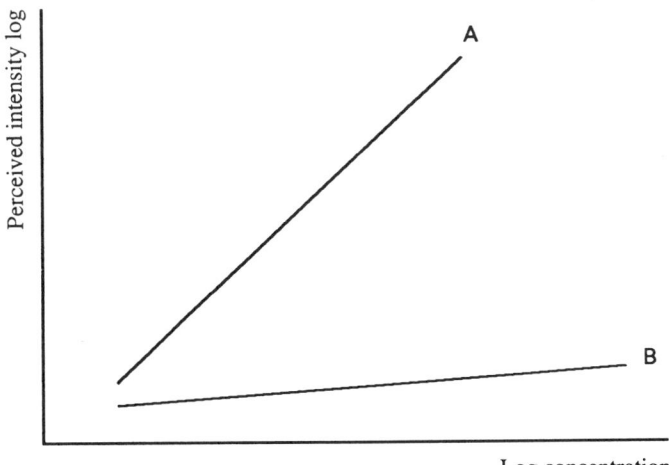

Figure 2.12.   Psychophysical functions of two odors (diagram example).

2.3 Psychophysical Methods to Assess Odorous Quality

2.3.1 General Points

As was pointed out in the introduction, the history of psychophysics shows that a great deal more attention has been paid to the systematic measurement of odorous intensity than to the assessment of odorous quality. Pioneers such as Linné, Zwaardemaker, and Hennings tried to classify odors on a purely intuitive basis and, until recent years, only occasional research has been carried out to verify their conclusions. In most of this recent research, the methods that have been used do not make it possible to establish odor similarities directly, but rely on measurements linked to intensity, such as crossadaptation (Ohma, 1922; Cheesman and Mayne, 1953; Moncrieff, 1956, 1957; Köster, 1971), or on anosmia (Guillot, 1948a, 1948b; Amoore et al., 1968) to draw indirect conclusions about odor similarity. These methods have not been very conclusive.

The question arises as to what methods are likely to solve the problem of odorous quality and qualitative similarity more directly. The question can be divided into two separate questions:

1. How can similarities between odors be measured?
2. How can similarites or differences in odorous quality be specified?

These two questions will be treated here briefly and several of the most currently used methods will be described.

2.3.2 How are Odor Similarities Measured?

A number of methods have been used to measure relative similarity in a given set of odors. The results provided by some of these methods have been used to to study the concept of similarity itself. What common characteristics must these odors have in order to be qualified as similar by a subject or, in other words, what qualifiers or criterion must the subject use to judge the similarity of odors? To respond to this question, Torgerson (1958) and Kruskal (1964a, 1964b) developed multidimensional scaling methods. In these methods, similarity measurements obtained by one of the methods that will be described here are considered to be indications of the distances of stimuli in a multidimensional similarity space. The methods make it possible to estimate the minimal number of dimensions needed without noticeable loss of information. The complexity of this data treatment will not be discussed here; only the methods by which similarity measures are obtained will be described.

2.3.2.1 Scaling with Paired Comparison

The simplest way to find out how similar two things are to a subject is to give them both to the subject, asking that their similarity be rated on a scale.

Generally, a category scale is used and all the possible pairs of stimuli in a given set of stimuli are presented to the subject. It is important to vary the order of presentation of the substances within each pair, as results from each of the two presentation orders may differ. For example, if one of the odors is an unpleasant odor of rotten fruit and the other is a typical fresh fruit odor, the subject may use the agreeable-disagreeable character as the principle criterion to judge the similarity in one of the presentation orders and the fruity character in the other order. The result may be the same, but it does not have to be so. After the experiment, the mean of the category scale values is calculated and these mean scale values are converted into relative distances of the stimuli in a multidimensional similarity space. Woskow (1964) used this method to assess similarites between 25 odorous substances and found that the data could well be fitted into a three-dimensional space. This means that his subjects essentially used three independant criteria to judge similarities. The first and most significant dimension found by Woskow correlated very strongly with a separate evaluation of substances on an agreeable-disagreeable scale, thus suggesting that this hedonic dimension plays a very important role in judging the similarity between odors.

2.3.2.2 *Ordinal Scaling Method*

Although similarities can be expressed directly in scale values, some experimenters prefer to ask the subject to make relative rather than absolute judgements of similarites. In this case, ordinal scaling methods are often used in which the subject simply indicates which of the two or more different pairs of stimuli presents the greatest similarity. A number of these ordinal scaling methods were developed by Coombs (1964). The simplest of these methods is the method of propellers.

Starting with a given set of stimuli (A, B,. . ., N) all possible combinations of three stimuli are presented to the subject in all possible orders (A-BC, A-CB, B-AC, B-CA, etc.); the subject is then asked to indicate which of the two stimuli last presented resembles the first one most. Thus, the subject compares two similarites (between A and B and between A and C, for example) and makes a relative judgement about them. From the data obtained in this way, an estimate of the relative distances between the stimuli can be deduced and then further analyzed by multidimensional analysis.

Given the fact that presenting series of triads of stimuli in the three possible orders takes a considerable amount of time, other experimenters prefer to present the three stimuli in a random order and ask the subject which two stimuli are the most similar and which two are the most different. With this method, known as the "complete method of triads," the subject must compare and rank order three similarities (A-B, A-C, and C-B) per presentation. Obviously, this task is more difficult for the subject than the method of propellers where only two similarities are compared. The author attempted the complete method of triads, but the task seemed to be too difficult for the subjects, who tend to forget their judgement of the first similarity before judging the third one. On the other hand, if the subjects are allowed as much time as they like to study the three stimuli, thus giving them a better chance to remember the three similarities, they crossadapt themselves too much, which considerably distorts the results.

Other ordinal scaling methods described by Coombs require even larger sets of stimuli (4 or 5) or more similarity judgements per presentation. Since the complete triad method seems to be unsatisfactory in olfaction, these methods will not be described here.

2.3.2.3 Confusion Matrix

A third and more indirect way of evaluating similarities between stimuli is based on measuring the degree to which they are confused with each other by the subjects. The more readily two stimuli are confused, the more similar they must be.

In a typical experiment based on this principle, subjects will first receive each stimulus of the set together with a special code by which they will have to name the stimulus throughout the rest of the experiment. After this first step, the stimuli are presented to the subjects in a random order and an equal number of times; the subjects must identify each stimulus presentation by giving its code name. In all cases, the subjects are given the correct code name and they are presented with the next stimulus. Thus, the subjects are confronted with a paired associate learning task. They are asked to continue until they have reached a certain learning level (for example, no mistakes in two successive presentations of the set of stimuli). Meanwhile, all of their responses are recorded and a confusion matrix is drawn giving the number of times for each stimulus that each different code was mentioned. Separate confusion matrices may be drawn for each group of 5 or 10 presentations of the whole set of stimuli in order to check learning speed. The data in these confusion matrices can be converted into distances between stimuli and a multidimensional analysis can be carried out.

In order to use this method correctly, great care must be taken in choosing the code names and making sure that the subject is equally familiar with them at the beginning of the experiment. If the code names are not well chosen, the subjects may be led to respond according to similarities between the code names rather than sensory similarities, or they may prefer to name a code that is easier to remember when they are in doubt.

Consequently, codes that are too simple, such as the numbers from 1 to n or the first letters of the alphabet should be avoided. Two- or three-digit numbers or two-letter codes are preferable, provided that one makes sure that there are no striking resemblances between certain individual codes. Also, special codes such as multiples of 100 or 111 should also be avoided.

Before starting the actual experiments with the odors, the subjects must know the codes by heart and in any order. To reach this criterion, the subjects are trained in a recognition experiment in which the codes are presented to them at random among other codes of the same type. The subjects are asked to indicate "yes" or "no" if the codes that are presented to them belong to the set to be studied. This training continues until they correctly recognize the codes at least five times in succession.

After this criterion has been reached, the experiment in which the codes are associated with the odorous stimuli can begin. Pairing of codes and odors is systematically varied over the different subjects so that each code is paired with each odor an equal number of times.

As shown by Köster (1972), results obtained with this procedure agree with those obtained with experiments in which paired comparison is used in combination with category scaling, but the confusion method is much more time-consuming.

2.3.2.4 Descriptive Methods

An even more indirect method for assessing qualitative similarities is to ask the subjects to scale each of the stimuli on a number of well defined descriptive attributes such as camphoraceous, minty, etc. The degree to which these attributes are judged to be relevant in the description of the quality of two odorous stimuli is considered to be an indication of their similarity. This method has the disadvantage of being very indirect as far as the similarity measurement itself is concerned, but it has the advantage of providing a more direct indication of the nature of the qualitative similarity. Consequently, supplementary experiments intended to specify the nature of these similarities can be avoided.

The danger of this method is that it relies to a large extent on semantics. There may exist—and this is perhaps especially true in olfaction, where there seems to be a lack of fundamental descriptive terminology—dimensions of similarity for which neither the subjects nor the experimenters have appropriate terms and which may not be taken into account in the description as a result of this lack.

2.3.3 How are Similarities or Differences in Odorous Quality Characterized?

There are two approaches to the problem of stimulus quality characterization. In the first, mentioned in the preceding paragraph, one attempts to specify the quality directly in descriptive terms thus relying on semantics. In the second, one tries to avoid the difficulties related to semantics by using standard substances as references. In the latter case, odorous qualities are specified in terms of similarity with a set of well-known odorous compounds.

Both methods have been used extensively and with varying degrees of success. Harper, Bate Smith, and Land (1968) have given a very good overview of the existing methods in their book, *Odour Description and Odour Classification*. Rather than going into a lengthy description of all the possible techniques and the numerous difficulties encountered in using them, this book is recommended to the reader.

2.4 The Use of Psychophysical Data

2.4.1 General Points

Before concluding this chapter, several remarks should be made about the use of psychophysical data for theoretical or practical purposes.

In psychophysics, the subject is used as a measuring instrument. As in all measurement problems, thorough knowledge of the measuring instrument's properties is necessary to evaluate the validity of the results obtained with it. This is the reason why a large number of psychophysicists are mainly interested in the form of psychophysical functions and in the possible sources of experimental bias such as presentation order, training, anticipation, and motivational effects. Therefore, it can often seem as if psychophysics has become a goal in itself; its objective seems to be unraveling the measuring instrument's complexities. External stimuli only provide a simple and adequate means to study the decision processes, which are supposed to take place in the subject in a reduced and well-controlled experimental situation.

Most of what is presented as psychophysics may consequently seem uselessly complicated to those who simply wish to use psychophysical methods to evaluate external stimuli in a pratical perspective. This is true up to a certain point. If one wishes to assess the intensity difference between two stimuli, it is not necessary to go into all the details of each individual's response bias or personal psychophysical function. In such a case, it is sufficient to obtain judgements from a large group of subjects, provided one takes into account such factors as presentation order (by balancing them carefully), possible coding effects (by choosing codes that are equally attractive and by systematically varying stimuli codes for different subjects), as well as expectancy and motivational effects (by keeping the stimulus probability constant during all experimental phases and making sure that subject motivation does not influence them to say "yes" rather than "no" or the reverse).

Thus, even though it is by no means necessary to study all these effects for their own sake, one should recognize their possible influence on experimental results and control them accordingly.

2.4.2 Individual Versus Group Data

In this context, another problem should be discussed. Some psychophysicists are vehemently opposed to the use of group data because they contain two sources of variability. The exactness of this objection cannot be denied, but its validity depends greatly on the use one wants to make of the data. This can be illustrated by a simple example.

If one wants to determine the average threshold for a given substance in a large population, one may rigorously determine the probability curve for individual detection within a group of subjects by carrying out a constant stimuli experiment in which all the stimuli are presented to each subject a large number of times. From each of these curves, one can obtain the individual thresholds and then the group threshold by taking the mean of the individual threshold.

The tested subjects may show the same sigmoidal response curves, as suggested by the two subjects in Figure 2.1. In this case, one can generalize and conclude that the probability curve for this substance is characterized by said form. This is a good method to gain insight on the measuring instument's properties.

This form similarity, however, may be related to a totally different sensitivity, as can be noted from the variability around the average threshold. In addition, this variability may not be reliable because of the small number of subjects taking part and the possiblility that they are not truly representative of the population. If, on the other hand, the same stimulus as that used in threshold determination is presented to a large group of subjects a small number of times (or even only one time for reasons of economy) and the positive response curve for the entire group is drawn, one would also obtain a sigmoid curve. This curve, from which the average group threshold may be determined in the same way as for individual thresholds in the preceding experiment, does not, however, only represent the detection probability function. Its form reflects sensitivity variations between the subjects more than growth in detection probability as a function of concentration for each of the subjects. In fact, individual probability curves generally present a more rapid growth (steeper slope) than the average group function obtained in the same way. Even though the group function does not provide any insight into the functioning of single individuals, it is of considerable practical value because it makes it possible to predict what fraction of the population will detect a stimulus at a certain concentration, provided that the sample of tested subjects is sufficiently representative.

Therefore, both methods are valid, but they respond to different questions. The decision to use one method or the other depends on the question to be answered and the necessity to economize an experimental procedure because it is always possible to combine them at the expense of greater experimental effort.

2.4.3 Data Banks

Several compilations of psychophysical data on olfaction have been published: Laffort (1963), Van Gemert and Nettenbreijer (1977), Fazzalari (1978), Van Gemert (1984), and Devos et al. (1990).

2.4.3.1 Numerical Threshold Values

Since threshold values obtained for the same odorous substances vary wildly between the authors, it is necessary to consider such data with caution. These variations are mainly due to to the following differences.

1. *Purity of odorous substances studied.*
2. *Dilution methods.* Some authors use liquid solutions in solvents that are considered to be odorless, whereas others dilute odorous gas current lines in pure air with a dynamic flow olfactometer.
3. *Presentation techniques.* When the concentrations are sniffed from bottles, wide openings should be used to avoid supplementary dilutions by inhalation of air outside the bottle. Olfactometers should have a flow of at least 20 L/mn to supply a sufficient amount of odorized air and avoid this supplementary dilution. Many olfactometers do not meet this criterion.
4. *Groups of subjects.* Interindividual sensitivity differences are important. For this reason, restricted groups of subjects may differ in their average threshold values. To obtain a sufficiently representative threshold of the population, groups of at least 24 subjects are recommended.
5. *Subject motivation and remuneration modalities.* Some authors reward their subjects regardless of their results, while others add supplementary rewards (financial or other) depending on the quality of their responses.

Certain authors of compilations have taken several of these factors into account and attempted to estimate their influence by calculating the differences between authors for thresholds obtained for the same odorous substances. They have also sometimes attempted to estimate standard detection thresholds by using weighing coefficients proper to each author.

2.4.3.2 Psychophysical Functions

A compilation of psychophysical function exponents has also been published (Patte et al., 1975) which reproduces values obtained by different authors for a certain number of odorous substances. Again, large differences are found, which are due, for the most part, to the same reasons as the threshold differences.

2.4.4 Recommendations and Norms

There are recommandations (Nielsen et al., 1986) and an AFNOR norm (Norm X43–101) (1986) for olfactometer use in atmospheric pollution research.

Bibliography

AFNOR, 1986. Qualité de l'air. Méthode de mesurage de l'odeur d'un effluent gazeux. Détermination du facteur de dilution au seuil de perception. Norme X43-101. AFNOR, Tour Europe Cedex 7, 92080 Paris La Défense, France, 13 pp.

AFNOR, 1990. Qualité de l'air. Atmosphères odorantes. Méthodes de prélèvement. Norme expérimentale X43-104, AFNOR, Tour Europe Cedex 7, 92080 Paris La Défense, France, 14 pp.

Amoore, J. E., Venstrom, D., Davis, A. R., 1968. Measurement of specific anosmia. *Percept. Mot. Skills* **26**, 143–164.

ASTM (American Society for Testing Materials), 1975. Standard recommended practice for referencing suprathreshold odor intensity, Annual book of ASTM Standards, part 31, ASTM, Philadelphia, PA, 13 pp.

Cain, W. S., Engen, T., 1969. Olfacctory adaptation and the scaling of odor intensity. In *Olfaction and Taste III*, edited by C. Pfaffman. Rockefeller University Press, New York, 127–141.

Cheesman, G. H., Mayne, S., 1953. The influence of adaptation on absolute threshold measurements for olfactory stimuli. *Quart. J. Exp. Psychol.* **5**, 22–30.

Cheesman, G. H., Townsend, M. J., 1945. Further experiments on the olfactory thresholds of pure chemical substances, using the "sniff-bottle methods." *Quart. J. Exp. Psychol.* **8**, 8–14.

Coombs, C. H., 1964. *A theory of data.* Wiley, New York.

Devos, M., Patte, F., Rouault, J., Laffort, P., VanGemert, L. J., 1990. *Standardized Human Olfactory Thresholds*, Oxford Univerity Press, Oxford, 165 pp.

Doty, R. L., (ed.) 1976. *Mammalian Olfaction, Reproductive Processes and Behavior.* Academic Press, New York.

Dravnieks, A., 1974. A building-block model for the characterization of odorant molecules and their odors. *Ann N.Y. Acad. Sci.* **237**, 144–163.

Duchamp, A., Revial, M. F., Holley, A. MacLeod, P., 1974. Odor discrimination by frog olfactory receptors. *Chem. Senses* **1**, 213–233.

Ekman, G., Berglund, B., Berlund, U., Lindvall, T., 1967. Perceived intensity of odor as a function of time of adaptation. *Scand. J. Psychol.* **8**, 177–186.

Engen, T., 1971. Psychophysics. In *Experimental Psychology*, edited by J.W. Kling and L.A. Riggs. Holt, Rinehart and Winston, New York.

Fazzalari, F. A., 1978. Compilation of odor and taste threshold values data. ASTM Data series D54 8A p. 4598, Philadelphia, PA.

Fechner, G. T., 1860. *Element der Psychophysik*, Leipzig, Breikopf und Härtel, 907 pp.

Green, D. M., Swets, J. A., 1966. *Signal Detection Theory and Psychophysics.* Wiley, New York.

Guillot, M., 1948a. Anosmies partielles et odeurs fondamentales. *C.R. Hebd. Acad. Sci. Paris* **226**, 1307–1309.

Guillot, M., 1948b. Sur quelques caractères des phénomènes d'anosmie partielle. *C.R. Soc. Biol. Paris* **142**, 161–162.

Harper, R., Bate Smith, E. L., Land, D. G., 1968. *Odour description and odour classification.* Churchill, London.

Koelega, H. S., Köster, E. P. 1974. Some experiments on sex differences in odor perception, *Ann. N.Y. Acad. Sci.* **237**, 234–246.

Köster, E. P., 1968. Relative intensity of odour mixtures at suprathreshold level. *Olfactologia* **1**, 29–41 (Suppl. Cahier d'Oto-Rhino-Laryngologie 3, no. 5).

Köster, E. P., 1969. Intensity in mixtures of odorous substances. In *Olfaction and Taste III*, edited by C. Pfaffmann. Rockefeller University Press, New York.

Köster, E. P., 1971. Adaptation and cross-adaptations in olfaction. Thesis, Utrecht, The Netherlands.

Köster, E. P., 1972. Odour similarities in nine odorous compounds: a methodological study. Report no. 72-3. Psychological laboratory, Utrecht University.

Kruskall, J. B., 1964a. Multidimensional scaling by optimizing goodness of fit to a nonmetric hypothesis. *Psychometrika* **29**, 1–27.

Kruskall, J. B., 1964b. Non-metric multidimensional scaling: A numerical method. *Psychometrika* **29**, 115–129.

Laffort, P., 1963. Essai de standardisation des seuils olfactifs humains pour 192 corps purs. *Arch. Sci. Physicol.* **17**, 75–105.

LeMagnen, J., 1952. Les phénomènes olfacto-sexuels chez l'homme, *Arch. Sci. Physiol.* **6**, 125–160.

MacLeod, P., 1968. Interactions dans un mélange d'odeurs—Étude électrophysiologique, *Olfactologia* **1**, 25–27.

Mitchell, M. J., 1971. Investigations of olfactory similarity scaling. Thesis, University of Canterbury, Christchurch, New Zealand.

Moncrieff, R. W., 1956. Olfactory adaptation and odour likeness. *J. Physiol. (London)* **133**, 301–316.

Moncrieff, R. W., 1957. Olfactory adaptation and odor-intensity. *Am. J. Psychol.* **70**, 1–20.

Nielsen, V. C., Voorburg, J. H., L'Hermite, P., 1986. *Odour Prevention and Control of Organic Sludge and Livestock Farming*. Elsevier Applied Science Publishers, London.

Ohma, S., 1922. La classification des odeurs aromatiques en sous-classes. *Arch. Néer. Physiol.* **6**, 567–590.

Pangborn, R. M., Bergh, H. W., Roessler, E. B., Webb, A. D., 1964. Influence of methodology on olfactory response. *Percept. Mot. Skills* **18**, 91–103.

Patte, F., Etcheto, M., Laffort, P., 1975. Selected and standardized values of suprathreshold odor intensities for 110 substances. *Chem. Senses Flavor* **1**, 283–305.

Poulton, E. C., 1989. *Bias in Quantifying Judgments*. Lawrence Erlbaum Associates, London.

Punter, P. H., Verhelst, N. D., Verbeek, A., 1984. *Odor Intensity Measurement Using a Reference Scale*, Symposium "Characterization and control of odoriferous pollutants in process industries." Louvain la Neuve, Belgium.

Revial, M. F., Duchamp, A., Holley, A., MacLeod, P., 1978. Frog olfaction odour group, acceptor distribution and receptor categories. *Chem. Senses* **3**, 23–33.

Revial, M. F., Duchamp, A., Holley, A., 1983. New studies on odour discrimination in the frog's olfactory receptor cells. II: Mathematical analysis of electrophysiological responses. *Chem. Senses* **8**, 179–194.

Stevens, S. S., 1936. A scale for the measurement of a psychological magnitude loudness. *Psychol. Rev.* **43**, 405–416.

Stone, H., Bosley, J., 1965. Olfactory discrimination and Weber's Law. *Percept. Mot. Skill* **20**, 657–665.

Swets, J. A., Tanner, W. P., Birsdall, T. G., 1961. Decision processes in perception. *Psychol. Rev.* **68**, 301–340.

Torgerson, W. S., 1958. *Theory and Methods of Scaling* (7th ed., 1967). Wiley, New York.

VanGemert, L. J., Nettenbriejer, A. H., 1977. *Compilation of Odour Threshold Values in Air and Water.* Central Institute for Nutrition and Food Research TNO, Zeist, Netherlands and National Institute of Water Supply, Voorburg, Netherlands, 79 pp.

VanGemert, L. J., 1984. *Compilation of Odour Threshold Values in Air.* Supplement V. Report No. 84, 220, p. 49. TNO-CIVO Food Analysis Institute.

Wetherill, G. B., Levitt, J. H., 1965. Sequential estimation of points on a psychometric function. *Br. J. Math. Statist. Psychol.* **18**, 1–10.

Woskow, M. H., 1964. Multidimensional scaling of odors, Thesis, University of California, Los Angeles.

3

Hedonic Aspects of Odors and Odor Pollution Control

E.P. Köster

3.1 General Points

3.1.1 Nature and Origin of the Agreeable and Disagreeable Character of Odors

The affective or hedonic tonality of odorous perception is one of its principle characteristics. Almost all odors are immediately liked or disliked; very few seem completely neutral. However, one can note large interindividual differences in affective response, appreciation diversity being more pronounced for "agreeable" odors than for "disagreeable" odors. This observation leads one to formulate the hypothesis according to which the olfactory apparatus would serve as an innate alarm system with regard to potentially dangerous sources, such as those resulting from decay.

In fact, Zoeteman (1978), who carried out an extensive study on sensory evaluation and the chemical composition of drinking water in the Netherlands, devoted part of his work to the relationship that may exist between olfactogustative detectability and the acute toxic effects of organic compounds. He found that the 93 substances that he had studied could be detected by odor at concentrations at least ten times weaker than those which might be dangerous. This author also showed that drinking water appreciation was clearly related to daily consumption of pure tap water.

Nevertheless, one could formulate two objections to the idea of the olfactory sense as an innate alarm system. First of all, there are a certain number of odorless substances, carbon monoxide for example, that are known to be extremely toxic. There are also numerous dangerous substances such as cyanide that do not have a disagreeable odor at all. For this reason and also because in olfaction, unlike gustation (Steiner, 1979), the existence of innate disgust or pleasure reflexes has not been shown, it is generally thought that the vast majority of our affective reactions (if not the totality) are acquired by experience. It has also been shown that newborn humans (apart from a slight initial preference that could be attributed to intrauterine learning) acquire a preference for the odor of their mother's breast during the first feeding sessions (Schaal, 1988). Furthermore, it is well established that the aversion reactions of most humans in the presence of urine or fecal odors do not exist in young infants and are acquired during toilet training before the third year of life (Stein et al., 1958).

3.1.2 Several General Rules

If it is true that preferences and aversions are acquired during an individual's past history, it is difficult to establish general rules about the pleasantness or unpleasantness of odorous sensations. The only general rules that can be formulated are as following.

3.1.2.1 All Agreeable Odors Become Disagreeable at Very Strong Concentrations

Although the intensity level at which the phenomenon occurs differs according to the substance, the rule (in its generality) holds true for all odors.

3.1.2.2 The Agreeable or Disagreeable Character of an Odor Depends Largely on its Context

Perfumes may be pleasant in a concert hall, but are annoying in a private garden located near a cosmetics plant. In a more general way, all odors become disagreeable outside their "natural" context. Consequently, for example, olfactory hedonic properties determined in laboratories have little, if any, validity in situations in the outside world.

3.1.2.3 Only Odors which are Associated with Very Common Experiences Shared by People in the Same Culture can be Trusted to Have the Same Effects on People

This is why the odor of methyl salicylate associated with a drink such as root beer is appreciated by Americans but not Europeans. In many countries, the odor of pine trees is associated with Christmas celebrations or walks in the forest; thus, it evokes peace and serenity. It is unlikely that it holds the same attraction for Polynesians, for example. Of course, above and beyond cultural differences, there is a certain common affective value with regard to body odors, but in humans this might well be attributed more to a similar mode of acquisition than to an innate character, unlike that often observed in other mammalian species.

3.1.2.4 Odor Aversions are More Persistent than Odor Preferences

Whenever an aversion to an odor occurs following an association with a disagreeable experience, subjects will have a tendency to avoid that odor in the future in order to avoid the associated experience. By doing so, they deprive themselves of the possibility of discovering that the odor itself is harmless and that its association with the disagreeable experience was accidental. On the other hand, when an odor is associated with an agreeable event, avoidance behavior does not occur and the subject, on the contrary, will have a tendency to search for the odor in order to repeat the agreeable experience. In doing so, the next event associated with the odor will very likely be less favorable than the first time and the subject's preference for the odor will decrease. This process may explain why, in general, one notices more diversity in odor preferences than in aversions.

3.1.2.5 Tolerance with Regard to Disagreeable Odors Varies Considerably Depending on the Person

For some people, a slight disagreeable odor may already be a considerable annoyance, whereas it is not a source of discomfort for others.

All of these considerations make odorous annoyance level assessment on the basis of simple estimations on a hedonic scale very difficult. Nevertheless, methods for the assessment of odor acceptability have been extensively compared (Köster et al., 1987) and the use of a hedonic reference scale has been recommended in the Netherlands.

3.1.3 Annoyance Measurements

At the same time that methods making it possible to estimate odor annoyance via the indirect way of assessing acceptablility of the odor in the laboratory were being used, another method was developed and tested on a large scale with the help of population panels (Köster et al., 1984; Köster et al., 1985; Maiwald et al., 1989). This latter method has also been applied in a study on the odorous annoyance provoked by the industrial complex at l'Etang de Berre in the south of France (Perrin et al., 1989).

Also, a number of questionnaire methods have been used with independent groups of observers in areas of the Netherlands where studies were carried out at the same time either on the basis of emission measurements (Ham et al., 1987) or with the help of population panels. In the latter case, where a questionnaire based on the one developed by Kastka (1982) and modified by Janse was used, it was possible to compare directly the relationships and advantages of two annoyance determination methods (Punter and Blaauwbroek, 1989). The so-called "questionnaire" method differs from that used with population panels in two respects.

1. The questions asked in them are numerous and complex; for example, on habitation pleasantness, the incidence of disagreeable odors during the last three months, etc. In other words, the gathered responses rely on the memory over a certain period of time in the lives of the people questioned.
2. People participating in the inquiry are only questioned one time.

The totality of this research has shown that establishing an odor acceptability level or perceived odorous intensity has little importance if one wants to evaluate the annoyance experienced by the population located in the vicinity of an odor source. Furthermore, one should note that in most countries, legislation on pollution is entirely based on the principle of an absence of annoyance caused to others. In view of this fact, the rest of this chapter will be entirely devoted to annoyance assessment.

3.2 Method Comparison

3.2.1 Methods Based on Odor Emissions

In most research on odor pollution, samples of the emitted odor are taken at the source and the concentration is expressed in terms of odor units (equivalence to the dilution factor necessary to reach the threshold). Once the concentration is known, dispersion models are applied to estimate the emission concentrations at different distances from the source and in different directions, taking into account prevailing climatic conditions and geographical particularities. On the basis of emission concentrations calculated in this way and knowledge about the sensitivity of the population, the perceived odor strength at different distances from the source is estimated.

Although this approach is most certainly useful, notably when the odorous source is well known and localized (plant smokestacks, for instance), it presents several disadvantages:

1. Emissions from a source may present important variations as a function of time, necessitating an extensive sampling program;
2. Climatic conditions can vary so much that they pose a problem for the application of the dispersion models;
3. Geographical particularities can modify dispersion in a noticable way, also bringing the application of the dispersion models into question;
4. Knowledge of odor strength does not, in any case, provide a direct indication of the degree of annoyance provoked by the odor; a strong manure odor in a rural area may be less disagreeable than a weak industrial odor.

The first three disadvantages are relatively minor and can be overcome by the means of appropriate sampling techniques and improved dispersion models, which have been tested as to their validity by the introduction of molecular tracers into the source. The last disadvantage is more important and of a different nature. Large-scale research programs carried out in the Netherlands (Köster et al., 1983) did not make it possible to establish any universal rules about the relationships between strength and odorous nuisances.

3.2.2 Methods Based on Complaints

In some cases, attempts have been made to relate the calculated emission strength in the environment to indirect indications of the annoyance experienced by the general public. Thus, many studies relate the emission concentration to the number of complaints received. Even though the relationships found are often quite convincing (Clarenburg, 1987), there are several objections that can be made against the use of complaints as an indication of the annoyance.

- Complaints are ungraded all-or-nothing responses. Thus, they are not sensitive enough for the evaluation of low-level nuisances. They are only useful above a certain threshold of dissatisfaction.
- Complaints are human responses with a "refractory period" (in the sense of the term as used in nerve physiology): the same person will only complain again after a certain time lapse.
- Interindividual differences in dissatisfaction thresholds and refractory periods are quite high. A large part of the population never makes complaints, whereas for some people this is typical behavior. Therefore, complaint behavior is not representative of an annoyance experienced by the total population.
- Complaint frequency can be modified by factors other than the annoyance itself; such as the case of the degree of accessibility to authorities and the cost of the complaints themselves (both financial and in terms of human effort). This is also the case when people have little confidence in the efficiency of making complaints. This is particularly true of odorous pollution, where solutions—which are never immediate—may take months or even years.
- Complaint frequency is likely to be affected by factors that, even if they change the effects of the annoyance, are extraneous to the odorous pollution itself. Thus, press publications may modify tolerance for an annoyance or temporarily lower the complaint threshold. Pressure groups may also have a strong influence on the number of complaints registered over a given time period. Even if these groups only represent a small part of the population, they can considerably modify the number of complaints by inciting all their members to double the number of their complaints, for example.

In general, complaints may be considered good indicators of sudden appearances of odorous pollution, such as those resulting from industrial accidents, but they are a poor indication of the general dissatisfaction level of the population.

3.2.3 Methods Based on Population Panels

Nuisances due to odorous pollution may also be evaluated in a direct way by using population panels, which are groups of volunteers who have committed themselves to give judgments on nuisances of odorous origin at regular intervals over an extended period of time. With this type of judgment, many of the difficulties encountered with complaint use can be overcome.

- Population panels work with precisely constructed scales to indicate the degree of annoyance experienced (such scales allow graded responses).
- Intervals between responses are fixed by the experimenter and are thus independent from refractory periods or outside influences. Population panel members receive a post card on which the scale is printed and the date and time at which they are asked to make their judgment is indicated.
- Panel members are carefully selected to form a group that is as representative as possible of the total population with respect to sex, age, socio-professional background, dependence on industry, time spent in the area, and attitude with regard to environmental problems. They are all known to the experimenters by means of a code, which makes it possible to monitor the regularity of their participation, follow the evolution of their individual responses over time and correlate their responses with any of the characteristics mentioned earlier (age, dependence on industry, etc.).

Since the time and frequency of their judgments are known, it is also possible to relate their responses to other types of measurements of the pollution carried out at the same time, such as in situ emission concentration measurements or meteorological conditions (wind velocity and direction, for example).

- Since the panel members have committed themselves to participate regularly, their response frequency is not affected by outside factors; appropriate measures are taken to stimulate and maintain their motivation.
- Although population panels are also sensitive to outside factors that are likely to change their annoyance tolerance, the impact of these factors on results obtained this way is smaller than on the number of complaints and it is possible to detect such changes almost immediately. The effects are smaller because the panel members constitute a constant group that cannot grow overnight following an article in the press or under the influence of a pressure group and also because the panel cannot increase the frequency of its responses as is the case with complaints. The only possible consequence is a change in response grading; the panel members may develop a tendency to give higher scale values. However, in a well-estab-

lished panel, this change will be detected almost immediately because of the concomitant modification of the mean and standard deviation of the response, independent of the influence of meteorological changes. Data analysis at an individual level makes it possible to clarify such response changes later, since it is highly unlikely that all the panel members would be influenced in the same way.

Nevertheless, other measures can be taken to detect modifications in scale grading use. In the first place, emissions can be measured at the same intervals and at the same time that the panel operates. In the second place, the panel's use of the scale can be regularly calibrated by asking them to judge samples prepared in the laboratory and sent to them by mail, according to the same criteria as they judge ambient air.

Taking all of these measures into account, one can be assured that if changes of the tolerance for odor annoyance do occur as the result of extraneous factors, they represent a true change in the attitude of the population for which the panel is representative. They cannot be the effect of manipulation by a part of that population.

On the basis of these considerations, it was decided to test the use of population panels for the direct measurement of annoyance in the project described here.

3.3 Description of the Population Panel Method

3.3.1 Annoyance Scale Development

A carefully established annoyance scale must satisfy the following criteria:

1. The concepts used must be easily understood by all population strata;
2. The concepts must be unequivocally interpretable by all the panel members;
3. The scale must be sensitive and reliable;
4. The scale must be stable over time.

Twelve different scales were tested. Fifty subjects, who were selected to be as representative of the population as possible, participated. Great care, in particular, was taken that special groups were represented (notably, elderly people and those with a low educational level) who might have difficulties using this type of scale.

The twelve scales differed in two respects:

1. The qualifying terms used: annoying , disagreeable, irritating, strong
2. The type of scale: verbal five-item scale, preceded or not by a "filter" question ("Do you smell anything? If so, is it annoying?"), graphic scale, nine-point numerical scale

These scales were each presented twice on different days. Five odorous substances were judged this way: mercapto-ethanol, isobutyric acid, diacetyl,

International Flavors and Fragrances "Standard Malodor," and dipropyl sulfide. Each of the odorous substances were presented at five different concentrations in such a way that the weakest ones would be slightly under the level detectable by half of the subjects and the strongest ones would be clearly perceptible.

The results of this preliminary study are reported in Table 3.1, which clearly shows that scale no. 1 was selected for use with population panels. This scale is reproduced in Figure 3.2, as it was presented to panel members in post card form.

3.3.2 Site Choice

The sites that were under investigation with population panels were chosen in such a way that a certain amount of odor pollution would have a good chance of being pointed out. The selected areas were located around pollution sources and included villages or neighborhoods that were sufficiently populated so that recruitement of a hundred people around the source (and located according to different wind directions) presented no problem. An example is given in Figure 3.1.

3.3.3 Panel Recruitment

To recruit the volunteers, letters are distributed to the population in the concerned areas, informing people about the purpose of the investigation and requesting their participation. The letters are distributed in the mailboxes of every fifth house on a number of preselected streets inside each area. This procedure is adopted to avoid too much direct contact between the members of a panel during the measurement period. In fact, it is thought that imme-

Table 3.1.   Comparison of Different Annoyance Scales

	Type of scale	Reliability	Sensitivity	Univocal character	Simplicity of use
1	Filter-annoying	+ +	+ +	+ +	+
2	Filter-disagreeable	+ +	+	0	+ +
3	Filter-irritating	+ +	+	−	0
4	Filter-strong	+	+ +	0	+ +
5	Graphic-annoying	+ +	+	0	0
6	Graphic-disagreeable	+ +	+	−	+
7	Graphic-irritating	+ +	+	−	+
8	Graphic-strong	0	+ +	+ +	+
9	Numerical-annoying	−	+ +	− −	+
10	Numerical-disagreeable	−	+ +	−	+ +
11	Numerical-irritating	−	+ +	− −	0
12	Numerical-strong	+	+ +	−	+ +

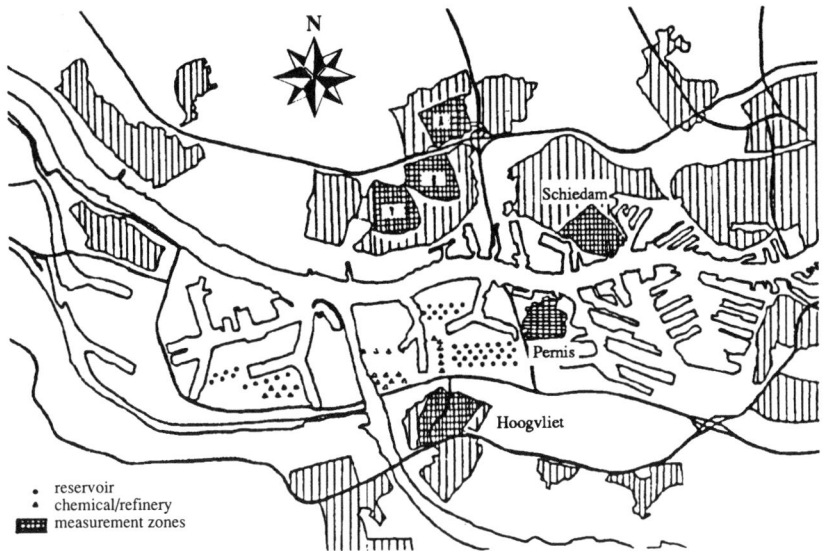

N

Schiedam

Pernis

Hoogvliet

• reservoir
▲ chemical/refinery
▦ measurement zones

Figure 3.1. Localization of five zones around industrial sources (● and ▲) where odor annoyance index (O.A.I.) measurements were carried out in 1985–1987 (Maiwald et al., 1989). The vector origins are projected at the center of the zones. The vectors indicate the O.A.I. in the prevailing wind direction during the measurements.

diate neighbors are likely to influence each other. A questionnaire concerning age, sex, civil status, level of education, time spent at the same address, profession, and attitude with regard to environmental problems are included with the letter, as well as a prepaid postage envelope. If the response level is too low, advertisements are placed in local newspapers encouraging the solicited people to participate in the study. If, in spite of these efforts, too few responses are obtained, the remainder of the panel is recruited by telephone.

3.3.4 Panel Instruction

Once they are selected, the panel members receive a letter including instructions and an invitation to attend a meeting (one for each area), during which complementary information is given to them and possible questions are answered. The object of the investigation and the experimental procedures are explained to them.

On this occasion, a film is shown on pollution problems and actions undertaken to attempt to solve them. The panel members are also informed that they will receive a post card every week. The scale shown in Figure 3.2 is printed on the back of these cards, on the top of which there is a label glued on indicating the code number of the person concerned as well as the

Please smell attentively:

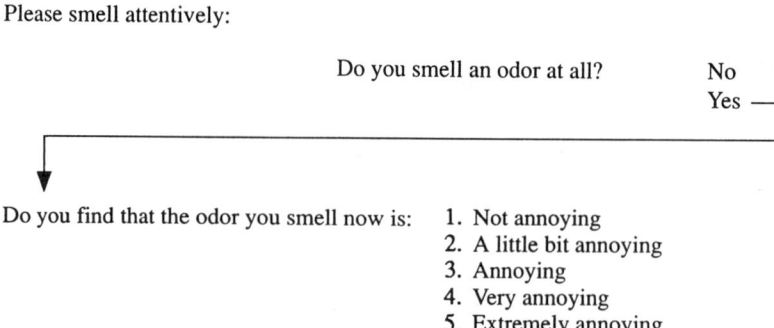

Figure 3.2.   Scale chosen for use with population panels.

measurement date. The subjects are asked to keep the card in a safe place until the indicated date. On this date, they must go ouside to a preselected place on the street or on their balcony at a specific time, for instance, from 7:00 P.M. to 8:00 P.M. They are then supposed to sniff the air attentively and respond to the questions asked. They are also asked to mail the prepaid post cards as soon as possible after having made a judgment. Great care is taken to explain to the subjects that their judgments must only reflect their impression of the odor quality at the moment in question. If they notice other odor phenomena during the course of the same day, they are free to provide separate comments on the back of the card, but their judgment on the intensity scale must not reflect the latter. They are also asked not to take notice of kitchen odors and to retard their judgment by half an hour if it seems impossible to them to give a judgment independent of such odors.

After the information meetings, which are generally attended by only part of the panel, all the members receive a summary of the questions asked during the meetings and answers provided by the project leader.

3.3.5 Panel Motivation

Panel members participate on a purely volunteer basis; therefore, it is necessary to take measures with regard to them in order to keep their motivation at a high level. The general principles used are as follows.

1. *Approaching panel members on a personal basis as much as possible*; For example, by signing letters by hand, speaking to them individually during panel meetings, responding to all correspondence including questions and remarks on the investigation post cards as well as by sending cards or flowers to those who fall ill.
2. *Providing material and information that make the panel members stand out from the rest of the population*; For example, stickers to be placed in

the house to remind them of their weekly duty and/or quarterly bulletins with general information on pollution problems, about nasal anatomy and the way sense of smell works, and on response rate or other aspects of the experiment in progress.

3. *Sending a gift at the end of the year*; For example, a bottle of wine with a special label.

4. *Organizing meetings to inform panel members about the progress of the experiment.*

3.3.6 Data Treatment and Odor Annoyance Index Calculation

From the obtained responses, an Odor Annoyance Index (O.A.I.) is calculated with the following formula:

$$O.A.I. = \frac{ON_0 + ON_1 + 25N_2 + 50N_3 + 75N_4 + 100N}{N_{total}}$$

where

N_0 = number of people having responded "no odor"
N_1 =   "   "not annoyed"
N_2 =   "   "a little bit annoyed"
N_3 =   "   'annoyed"
N_4 =   "   "very annoyed"
N_5 =   "   "extremely annoyed"

According to this formula, the O.A.I. varies from 0 to 100 and only responses indicating an annoying odor are taken into account. Coefficients attributed to each response are arbitrary, but have been chosen after careful consideration on the basis of the response behavior of panel members in many field studies. They present the advantage of taking response grading into account and being easy to manipulate. The indexes obtained this way are very highly correlated ($r > 0.98$) with indexes calculated on the basis of "psychological distances" between the different types of response.

Starting from weekly O.A.I. collected over a year's time or more, the annoyance experienced by the total population can be estimated by calculating the O.A.I.$_{50}$ (the median) and the O.A.I.$_{90}$ (the 90th percentile, i.e., the annoyance index equal to or immediately above 90% of the observed values) The 95% confidence interval for these two values can also be calculated. As an experimental norm, the Dutch Ministry responsible for the environment has proposed that the upper limit of this 95% for the confidence interval for the O.A.I.$_{50}$ should not exceed 5 and that the upper limit of the O.A.I.$_{90}$ should not exceed 10 on a yearly basis. In practice, this is equivalent to saying that the O.A.I.$_{50}$ (O.A.I.$_{90}$) should not be higher than 5 (10) in more than 63.5% (19.2%) of the cases or in 33 (10) out of 52 weekly measurements

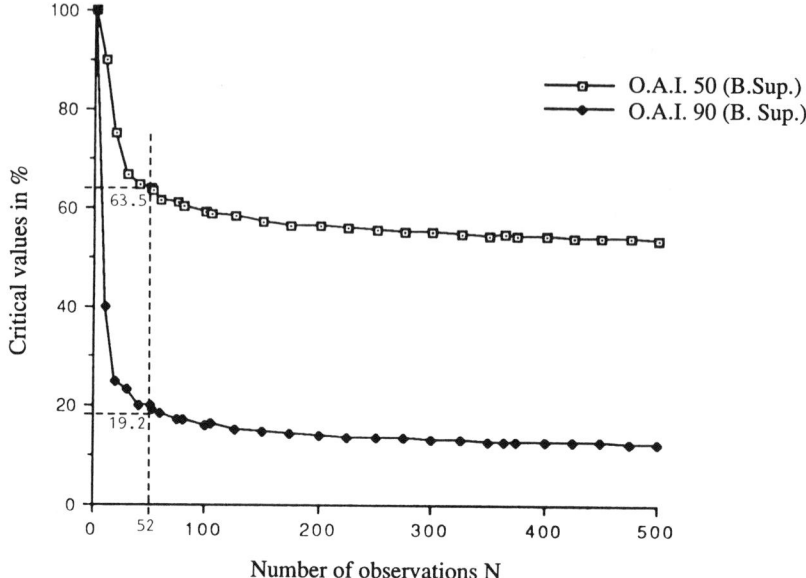

Figure 3.3. Critical values of the average odor annoyance index (O.A.I.$_{50}$) and of the 90th percentile of the odor annoyance index (O.A.I.$_{90}$) (upper limits of the confidence interval) as a function of the number of observations.

per year. When a large number of measurements are carried out, these critical values are lowered as can be seen in Figure 3.3.

3.4 Some Results Obtained with the Population Panel Method

3.4.1 Panel Member Participation

In order to succeed, the method relies on panel member participation. Two important questions on this subject can be asked.

1. *How many members of the original panel will be lost during the measurement period (panel mortality).* Loss of panel members may be due to all sorts of different causes, notably loss of interest, illness, death, or moving out of the area under study.
2. *How often do remaining panel members respond?*

Response frequency in the first study mentioned is illustrated in Figure 4 which shows percentages of subjects in the towns of Zeist, Hoogvliet, Pernis, and Schiedam who sent in their weekly post cards more than 20% of the time and less than 20% of the time. Results from the last three sites, which

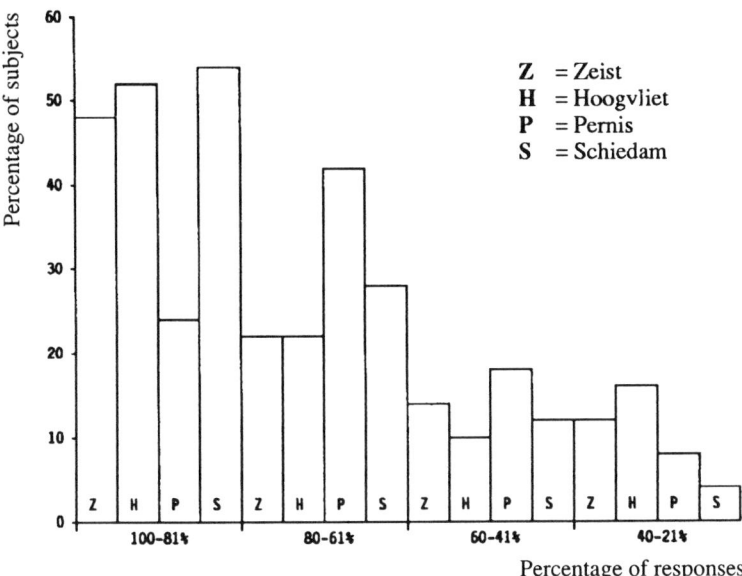

Figure 3.4. Percentage of subjects in the four cities (Z, H, P, S), and according to four categories of response percentages, that sent in their post cards (Köster et al., 1986).

are located in the same polluted area, were also combined in a fifth category in the figure. As can be noted, in three of the sites with responses of more than 50%, the panel members sent in more than 80% of the cards received by each of them and, out of the totality of the sites, less than 20% responded less than 40% of the number of weeks. Furthermore, it seems clear that in Zeist, which is a town without industrial pollution, the number of responses is as high as in the three towns with high pollution. In the second study (Maiwald et al., 1989), it has been shown that, during the measurement period, the total response rate (including those from panel mortality) decreases exponentially as follows:

$$R_t = R_0^{-at}$$

where R is the number of responses during the time t counting from the beginning of the measurements and a is the annual decrease rate appropriate for each panel.

Parameter a varies from 0.12 to 0.55 in the 13 sites studied; its median value was 0.30. In a supplementary study in which the subjects are asked to make six daily judgments, the values found for a were between 0.31 and 1.13 (median = 0.69), indicating that the volunteers were probably asked to accomplish too heavy a task.

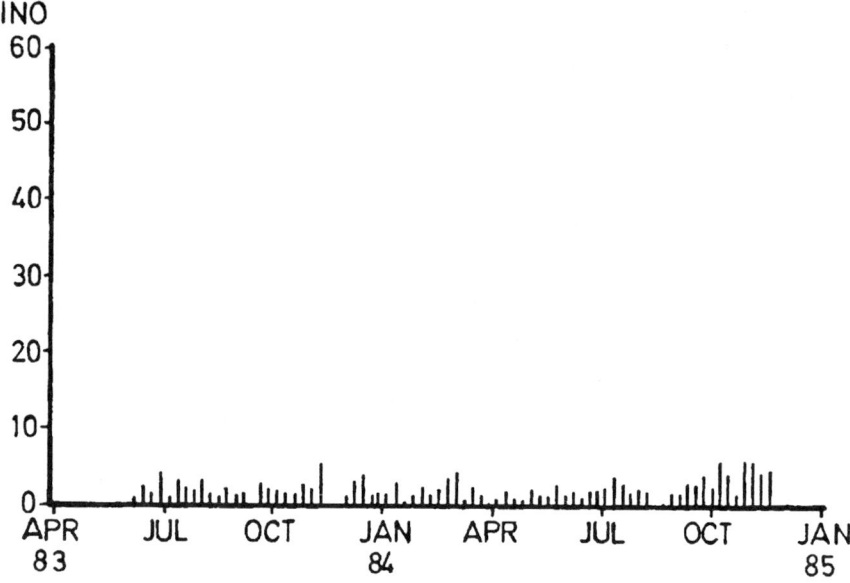

Figure 3.5. Odor annoyance indexes during 75 weeks in Zeist (the two missing values between October 1983 and January 1984 are not zero values, but observations that were not sent because of a postal strike) (Köster et al., 1985).

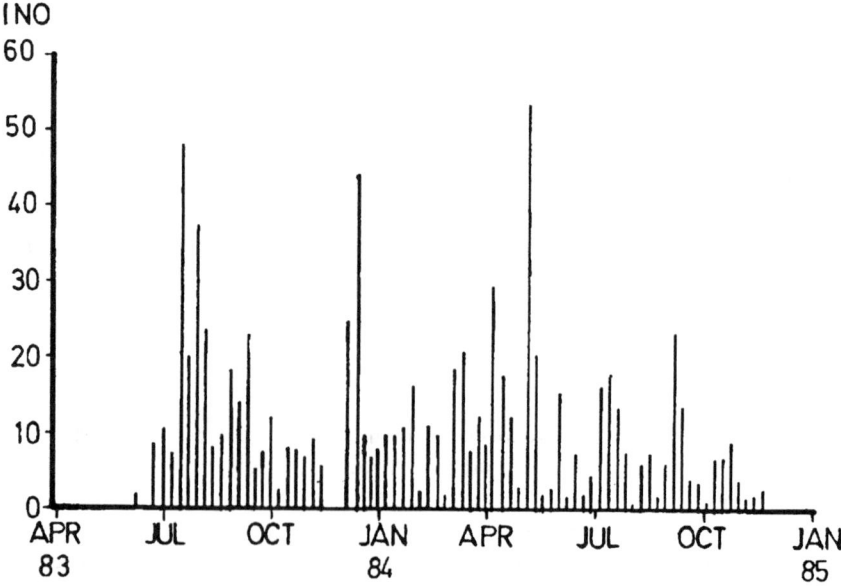

Figure 3.6. Odor annoyance index values during 75 weeks in Hoogvliet (see remark in Fig. 3.5) (Köster et al., 1985).

3.4.2 Analysis of the Estimated Annoyance

The annoyance experienced by the population can be found by drawing weekly odor annoyance indexes, as reported in Figures 3.5 and 3.6 for Zeist and Hoogvliet during the first study mentioned. According to this data, it is obvious that there is no odor pollution in Zeist (the index rarely exceeds value 5 and never reaches 10), though there is a high pollution level in Hoogvliet [where an index as high as 50 was reached and O.A.I. values equal to 5 (10) were exceeded in 55 (29) out of 75 measurements]. It has also been noticed that O.A.I. values found at Hoogvliet on pollution-free days are the same as average values found at Zeist. This shows that panels evaluate O.A.I. values in a similar way even though they are exposed to different levels of pollution. This result has been confirmed through the use of a calibration test of both panels (for further details, see Punter et al., 1986).

Another way of representing data is to plot O.A.I. values on a compass card in the form of vectors whose orientations are determined by prevailing winds during the measurements. This is illustrated by figure 3.7. According to this double representation it can be seen that the main source of odor pollution perceived in Hoogvliet is located north of the town and that odor nuisances are not to be feared when winds are blowing from the south. Combining the results obtained in different geographical locations, a convergence can be drawn on the location of the polluted areas where the main sources of odorous nuisances occur (see Figure 3.1).

The fact that, for a given location, the perceived annoyance depends, for the most part, on prevailing winds raises another problem. It is not certain that prevailing winds occuring during the 52 annual hours when the measurements were carried out are representative of the annual average. There-

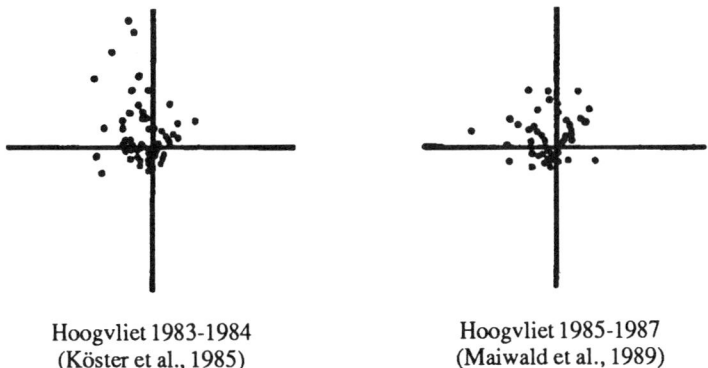

Hoogvliet 1983-1984
(Köster et al., 1985)

Hoogvliet 1985-1987
(Maiwald et al., 1989)

Figure 3.7.   O.A.I. values obtained in Hoogvliet for the periods 1983–1984 (left) and 1985–1987 (right) expressed as a function of prevailing winds during the panel's work (polar coordinates).

$$\text{O.A.I.}_{90} = 31 \, e^{-0.18D} \, (D = \text{distance in km})$$

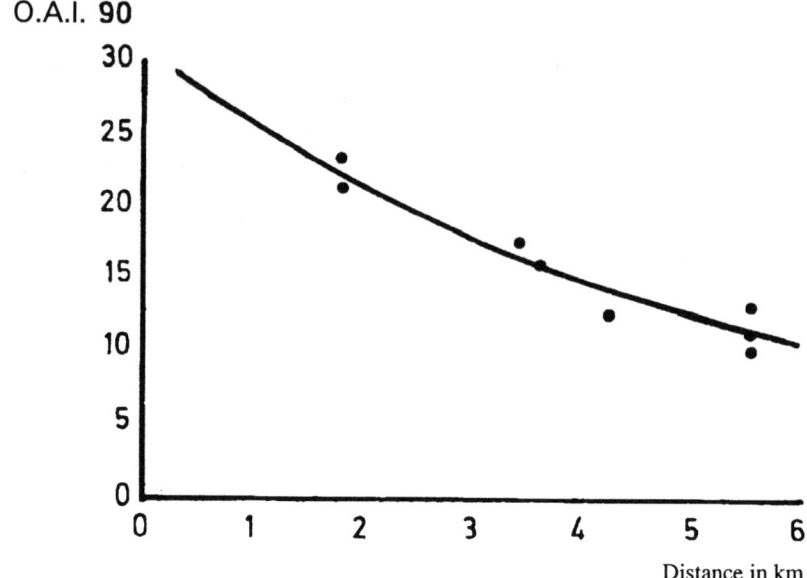

Figure 3.8. Relationship between the O.A.I. values and the distance from the odor source in km (Maiwald et al., 1989).

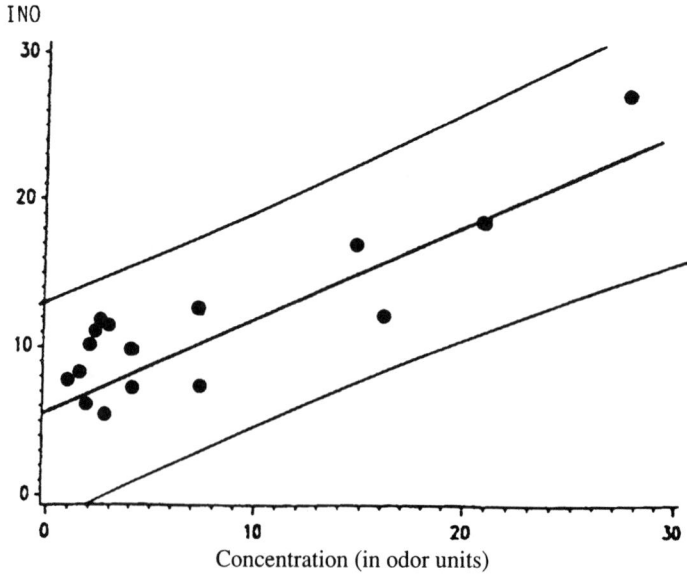

Figure 3.9. Relationship between the O.A.I. values and the odorous concentration expressed in odor units (Maiwald et al., 1989).

fore, a procedure has been designed to overcome this possible nonrepresentativity (for a detailed desription, see Maiwald et al., 1989).

In the most recent study, Maiwald et al. (1989) studied the relationship between $O.A.I._{90}$ and the distance between the sources and the studied sites. As plotted in Figure 3.8, the results can be expressed by the equation

$$O.A.I._{90} = 31 \exp(-0.18\ D)$$

where D is the distance in km.

Furthermore, in the same study, these authors correlated O.A.I. values to the odor concentration as determined on the same sites where the panels operated. Odor concentrations were determined by olfactometry using samples drawn windward of the odorous source as reference air. The results (Figure 3.9) show a good correlation between the odorous strength thus calculated and the level of generated annoyance. In the first study where pure air was used as a reference in the olfactometric measurements, Köster et al. (1985) could not point out such a correlation.

3.5 Final Discussion

Punter (1989) compared results obtained on the same sites with population panels and with questionnaire methods. He found that there was a good agreement between the two as well as between odor intensity (measured in odor units) and the degree of annoyance from a source. However, the latter relationship may vary when the odors are qualitatively different (Hangartner, 1990).

On the whole, one may conclude that direct odor annoyance measurement constitutes a very satisfactory tool in odorous pollution evaluation.

Bibliography

Clarenburg, A., 1987. Odour: a mathematical approach to perception and nuisance. In *Environmental Annoyance Characterization, Measurement and Control,* edited by H.S. Koelega. Elsevier Science Publishers, B.V., pp. 75–94.

Hangartner, M., 1990. *Zusammenhang zwischen Geruchsstoffkonzentration, Intensität und hedonischer Geruchswirkung.* Schriftenreihe der VDI-Komm. Reinhaltung der Luft, Bd. 12 (in press).

Katska, J., 1982. *Erfassung und Bewertung von Gerüchen und Ihrer Belästigungswirkung. VDI-Berichte* **416,** 15–28.

Köster, E. P., Punter, P. H., America, A., Schaeffer, J., Maiwald, K. D., 1984. Population panels in odour control—The development of a direct method for judging annoyance caused by odours. In *Proceedings of the International Symposium "Characterization and Control of Odoriferous Pollutants in Process Industries."* SBF, Louvain-la-Neuve.

Köster, E. P., Punter, P. H., Janse, M., Schaeffer, J., Roos, J. C., 1987. Aanvaardbaarheidsgrenzen voor geur, fase 1 en fase 2. Publikatiereeks Lucht Np 70. Ministerie van Volkshuisvesting, Ruimtelijke Ordening en Milieubeheer, 's Gravenhage.

Köster, E. P., Punter, P. H., Maiwald, K. D., Blaauwbroek, J., Schaeffer, J., 1985. Direct scaling of odour annoyance by population panels. In *Odorants VDI Berichte 561*, VDI Verlag Duesseldorf.

Maiwald, K. D., Punter, P. H., Blaauwbroek, J., Schaeffer, J., 1989. Stankbeleving in de woonomgeving (2de fase) Publikatiereeks Lucht Nx 85. Ministerie an Volkshuisvesting, Ruimtelijke Ordening en Milieubeheer, s'Gravenhage.

Perrin, M. L., Jezequel, M., Delpeuch, J. B., Nadal, R., 1989. Étude de la gêne provoquée par des odeurs d'origine industrielle. In L.J. Brasser and W.C. Mulder, eds.: Man and his ecosystem. *Proceedings of the 8th World Clean Air Congress* 1, 111–116.

Punter, P. H., 1987. Setting up and use of population panels for measuring and monitoring annoynce. In *Environmental Annoyance: Characterization, Measurement and Control*, edited by H.S. Koelega. Elsevier Science Publishers, B.V., pp. 15–116.

Punter, P. H., Blaauwbroek, J., 1989. Measurement of odour annoyance: comparison of two different models. In L.J. Braser and W.C. Mulder, eds.: Man and his ecosystem. *Proceedings of the 8th World Clean Air Congress* 1, 123–128.

Punter, P. H., Maiwald, K. D., Blaauwbroek, J., Schaeffer, J., Köster, E. P., 1986. Stankbeleving in de woonomgeving, le Fase. Publikatiereeks Lucht 50, Ministerie van Volkshuisvesting, Ruimtelijke Ordening en Milieubeheer, 's Gravenhage.

Schaal, B., 1988. Olfaction in infants and children: developmental and functional perspectives. *Chem. Senses* 13, 145–190.

Stein, M., Ottenberg, P., Roulet, H., 1958. A study of the development of olfactory preferences, *Arch. Neurol. Psychiat.* 80, 264–266.

Steiner, J. E., 1979. Oral and facial innate motor responses to gustatory and some olfactory stimuli. In *Preference behaviour and chemoreception*, edited by J.H.A. Kroeze. I.R.L., London.

Zoeteman, B.C.J., 1978. Sensory assessment and composition of drinking water. Doctoral Thesis, Utrecht University.

Physical-Chemical Properties of Odorous Products—Solubilization

G. Martin

4.1 Introduction

Olfactory perception as well as most deodorization processes requires the transfer of the stimulus into an aqueous solution and then toward the sensory cells; this transfer requires the product to have certain thermodynamic and chemical properties, such as solubility and adsorbability. Thus, to the author it seemed that odorous effects could be legitimately studied by applying the laws of thermodynamics, but that this would necessitate complementary research on the nature of mucus, receptor chemical structure, and so on. Indeed, one of the objectives of this chapter is to create common areas of interest between physiologists and chemists by establishing an exchange of knowledge.

In deodorization techniques (scrubbing, oxidizing or nonoxidizing, and biological purification) or chemical analysis (reactive bubbling, for example), the transfer of the pollutant to the water comes before the action of the reagents. On the other hand, the emission of odorous compounds occurs in most cases from the aqueous solutions.

It appears that vapor pressure and odorous substance solubility are implicated in various perception phenomena and process engineering. Therefore, this rather theoretical chapter is concerned with elements of pure substances and solution thermodynamics.

4.2 Olfactory Physical Chemistry

Odorous molecules that are going to stimulate the olfactory mucosa must have the following requirements.

1. *Initially be in a gaseous form.* Normally gaseous molecules may be odorous (for example: O_3, H_2S, NH_3 . . .) but liquids or solids that have "sufficient" vapor pressure in the air may be odorous as well. This observation can be summarized by the necessary presence of a minimum number $n_i = C_i V$ per volume unit (V) in the air or at partial pressure P_i (C_i is the concentration). It is known that P_i is dependent on compounds, and it is not surprising that odorous products generally have molecular masses between 17 and 300 g mol^{-1}. (This remark should be used with caution because odorous substances with molecular masses superior to 300 are not exceptional.) Furthermore, P_i varies with the temperature, as will be seen in Section 4.3.

2. *Be transported by inspiration.* Odorous molecules (M) will come into contact with the olfactory mucosa if they are transferred by normal or accelerated (sniffing = forced inspiration) inspiration or by the retronasal passage during tasting. In the human species, normal inspiration occurs at a velocity of 1 m/s. Figure 4.1 illustrates these principles. The odorous flux $N_i = C_i v + J_i$ is the sum total of convection and diffusion. It can be limited to convection (when the concentration of odorous substances is weak, the diffusion pressure gradient is certainly less important). The molecules enter the olfactory slit.

3. *Be "soluble" in the mucus.* The mucus is an aqueous solution that contains other types of solutes (organics, minerals, microorganisms, . . .). The stimuli must cross the gas-mucus interface before coming into con-

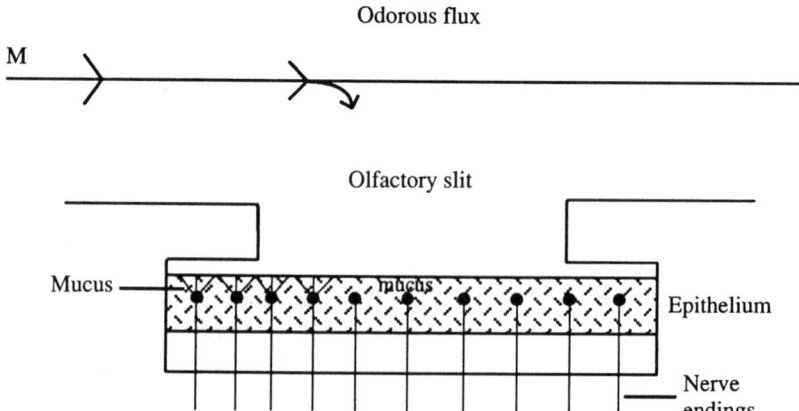

Figure 4.1. Schematics of olfactory stimulation.

tact with the nerve endings. The mucus that covers the olfactory epithelium, produced by Bowman's glands, includes a thin aqueous layer. Therefore, an odorous molecule must be sufficiently hydrosoluble with partition coefficient K_i between air and water playing an important role (cf. Section 4.3).

4. *Be adsorbable at the olfactory cilia surface.* The nerve endings branch out into numerous olfactory cilia which make up a dense feltlike layer due to their quantity and surface area. The cilia are animated by a beating motion that increases their probability of coming into contact with odorous molecules. The olfactory cilia membrane consists of a double lipidic layer including proteins that are used in metabolic and ionic exchanges. These 6–10-nm diameter transmembrane particles play a determining role in chemical reception. The odorous molecule must have an affinity for the receptor proteins. It is thought that it will link with the membrane by reversible physisorption (the energy ranges from 20 to 40 kJ mol^{-1}). Van der Waals bonds are nonspecific and their easy breaking yields back the initial entities.

Another way to describe this interaction of the molecule dissolved in the mucus with the protein is that it must have a double affinity—hydroaffinity and lipoaffinity—or have a hydrophilic-lipophilic balance (HLB). (This is a useful concept in tensio-active theory.)

The odorous molecule–receptor interaction brings about a transmembrane ionic current that can be electrically detected. The recognition and fixation of the odorous molecule on the receptor site implies certain structural appropriateness and also certain physical properties of the molecule (analogous to an enzyme–substrate couple).

1. Its size, cross section, shape, and volume (spherical, flat, . . .) are implied.
2. Its electrical structure is also implied. Most odorous molecules have either

- A nucleophilic nature ($\overline{N}H_3$, R $\overline{N}H_2$, $H_2\overline{S}$, R $\overline{O}H$),
- An unsaturated nature (benzene, alcenes, alcines) (molecule hydrogenization reduces or makes the odorous character disappear), or
- Both characters (for example, thiophene, furan). Aqueous solubility and protein adsorbability can also be interpreted in terms of polarizability (α), permanent bipolar moment (μ), or pK_a.

These few remarks should not be taken as general rules since the example of the tetrahydrothiophene used in the odorization of city gas shows an olfactory threshold 30 times lower than thiophene.

It should be noted that Laffort (1969) established a model relating the odorous character of a molecule to its volume, its polarizability, its Henry's constant, and its pK. Odor can be related a priori to the molecular structure, or rather to the occurence of groups in the molecule giving it the required odorous properties. These osmophoric groups are NH_2, SH, $> = <$, ... It was found that the odorous character requires specific molecular spatial configurations. There must be a stereochemical appropriateness between the molecule and the receptor.

It may be speculative but interesting to predict the possibility that certain substances, in becoming too easily attached to the receptors, may render the molecule–receptor couple stable and perhaps permanent, thus reducing the subject's olfactory sense.

This brief introduction has made it possible to identify several physical and chemical properties that determine odor: vapor pressure, solubility, pK, polarizibility, shape, and stereochemical appropriateness.

4.3 Thermodynamic Potential

When an open system receives heat δq and matter (dn_i) its entropic variation is dS with $\delta S_e = \delta q/T + \Sigma\ S_i\ dn_i$ and δS_i, such that $dS = \delta S_e + \delta S_i$. The irreversible contribution δS_i includes terms of energy dispersion into heat $\delta R/T$, of transfer $\delta q_t/T$ (if T is homogeneous), and of chemical reactions $(\Sigma\mu_i dn_i/T)$. An immobile system receiving heat and work increases its internal energy $dU = \delta q + \delta w = T\ dS + \delta w - TdS_i$ for a closed system and then $dG = -S\ dT + V\ dp - T\delta S_i$.

Let $T\ \delta S_i = -\delta\ \emptyset$, then $dG = -S\ dT + V\ dp + \delta\ \emptyset$. In natural evolution $\delta\ S_i \geq 0$, $\delta\ \emptyset$ is always negative, and $\emptyset$ is only a function of state under particular conditions such as constant T and p; $\emptyset$ is the thermodynamic potential and G is described as the thermodynamic potential at constant T and P.

For a fixed, closed, and nonisolated system which only receives homogeneous pressure work $(\delta w = -p\ dV)$ (T is the same at all points) evolving without dissipating friction and within which chemical species may change (n_i), $dG = -S\ dT + V\ dP - \Sigma\mu_i\ dn_i$ and

$$\mu i\ =\ \left(\frac{\delta G}{\delta n_i}\right)_{T,P,n_j}$$

is the chemical potential of the species i. At constant P and T, one obviously gets $dG_{(TP_i)} = \Sigma_i\mu_i dn_i$. Since G is an additive function of state, one gets $G = \Sigma_i\mu_i n_i$. Hence, at constants T and P, one gets the Gibbs-Duhem relationship: $\Sigma_i n_i d\mu_i = 0$.

4.4 Vapor Pressure of Pure Compounds

Given an immobile pure compound characterized only by the state variables p, V, and T, a reversible heat δq is added and the solution of $\delta W = -p\ dV$ is provided; its internal energy U varies by $dU = \delta Q + \delta W = T\,dS - p\,dV$ and its free enthalpy varies by $dG = -S\,dT + V\,dp$. In the case of an ideal gas at constant temperature, this formula leads to $dG = RT\,(dp/p)$ and is integrated as $G = G^0 + RT\,\ln(P/P_0)$. If the gas is nonideal (real), its fugacity f can be inserted such that $dG = RT\,d\ln f$; hence $G = G^0 + RT\,\ln f/f_0$.

When a pure compound (e.g., water) is in two physical states in equilibrium (e.g., liquid and vapor) it can be shown that the free enthalpy G has the same value in both phases $G_l = G_v$. This equality takes place under precise well-defined conditions of temperature T and pressure P. In other words, the liquid/vapor equilibrium of a pure compound occurs at defined P and at an imposed T (e.g., water boils at 100 °C under $p = 1$ atm); if the temperature varies to dT starting from T, the equilibrium is seen again at $p + dp$ such as that

$$-S_l dT + V_l dp = -S_v dT + V_v dp$$

S_l, S_v, V_l, and V_v are massic entropies and massic volumes of liquid and vapor, respectively. This equation is written $(S_v - S_l)dT = (V_v - V_l)dp$; $S_v - S_l$ is the entropy of vaporization corresponding to the phase change of a predetermined mass of compound (e.g., 1 mole):

$$\frac{\Delta H_v}{T}\,dT = \Delta V_v\,dP \text{ or } \Delta V_v \left(\frac{dp}{dt}\right)_{vap} = \frac{\Delta H_v}{T}$$

$\Delta V = V_v - V_l = (Z_v - Z_l)RT/P$ is calculated by compression factors Z $\Delta Z_{vl} = Z_v - Z_l$. If it is assumed that $V_l \ll V_v$ (which would be true of the compound under far from critical conditions) and that the vapor is similar to an ideal gas, then, for 1 mole:

$$(V_v = \frac{RT}{P}) \text{ which finally yields } \frac{d\ln P^0}{d(1/T)} = \frac{\Delta H_v}{R}$$

N.B.: P is the pressure of the liquid/vapor equilibrium at temperature T and is subsequently distinguished from pressure applied to a system by writing it as P^0.

Vaporization enthalpy depends on the compound being studied and varies with the temperature, but if, for a first approximation, it is assumed to be constant, the result is

$$\ln P^0 = -\frac{\Delta H}{RT} + A \equiv A + \frac{B}{T}$$

This Dupre's equation represents an approximation. A more precise approximation is given by the empirical Antoine's relation:

$$\ln P^0 = A + \frac{B}{T + C}$$

C is an empirical constant, usually assumed to equal 43 K, but when a more precise value is required, one uses:

$$C = \begin{cases} + 0.3 - 0.034T_b \text{ if } T_b < 125 \text{ K} \\ + 18 - 0.19T_b \text{ if } 400 \text{ K} > T_b > 125 \text{ K} \\ 60 - 0.294T_b \text{ if } T_b > 400 \text{ K} \end{cases}$$

T_b is the normal boiling point temperature (under 1 atm) of a pure compound. B is calculated as

$$B = - \frac{1}{\Delta Z_{l,v}} \frac{(Tb + C)^2 \Delta H_v}{RT_b^2}$$

ΔZ_{lv} is the difference between liquid and vapor compressibility coefficients at a normal boiling point temperature T_b ($\Delta Z_v l \# 0.95$) and ΔH_v is the normal vaporization enthalpy.

A, B, and C values of various gases can be found in *l'Encyclopédie des gaz* (1976). Table 4.1 gives the normal pressure P^0 as a function of T, for water. These relationships make it possible to understand that the vapor pressure at a similar temperature is higher for a compound with a weaker vaporization enthalpy. Furthermore, since ΔH_v is proportional to $(T - T_c)^n (n \# 0.38)$, the farther a compound is from its critical state, the higher the vapor pressure.

Table 4.1. Order of Magnitude of Henry's Constant K_i for Several Gases (Pure Water Solvent) [K_i in (Molar Fraction)$^{-1}$ atm.]

Compounds i	K_i at Various Temperatures (t °C)	$K_i = f(T)$
NH$_3$	0.25 (0 °C); 0.73 (20 °C); 1.85 (40 °C)	exp (14.303–4280.5/T)
H$_2$S	268 (0 °C); 483 (20 °C); 745 (40 °C)	exp (13.094–2035.2/T)
CH$_3$SH	112 (15 °C); 207 (25 °C)	
COS	920 (0 °C); 2,190 (20 °C)	
SO$_2$	18 (0 °C); 36 (0 °C); 66 (40 °C)	exp (13.079–2781.2/T)
CCl$_4$	392 (0 °C); 1,270 (20 °C); 3,532 (40 °C)	exp (23.164–4693.2/t)
CHCl=CCl$_2$	201 (0 °C); 540 (20 °C); 1,279 (40 °C)	exp (19.779–3951.6/T)
C$_6$H$_6$	79 (0 °C); 228 (20 °C); 579 (40 °C)	exp (19.986–4864.5/T)
CH$_4$	5,520 (0 °C); 10,200 (20 °C)	
O$_3$	1,940 (0 °C); 3,760 (20 °C); 12,00 (40 °C)	exp (18.536–2920.2/T)

Vapor pressure of pure water as a function of temperature (Antoine's formula)

$$\log P^0 \text{ H}_2\text{O} = 7.966,81 - \frac{1,668.21}{t + 228}$$

4.5 Solubility

Mucus is a rather complex aqueous solvent containing dissolved species. This solvent and preexisting solutes are related by low-energy Van der Waals bonds and by hydrogen bonds that are more specific and have a higher energy (-30 kJ mol^{-1}). Dissolved ionic species are responsible for the ionic strength of the mucus:

$$1 = 1/2 \Sigma\, C_i\, Z_i^2.$$

The study of the solubilization of odorous molecules into the mucus can be approached using thermodynamic concepts.

4.5.1 General Properties of Solutions

4.5.1.1 Gaseous Solutions or Gaseous Mixtures

For the substance i in a solution and by analogy with the calculation of G at constant T for a nonideal (or real), pure gas, one can write

$$d\mu_i = RT\, d(\ln f_i)$$

which is integrated as

$$\mu_i = \mu_i^0 + RT \ln(f_i/f_i^0)$$

where μ_i^0 is the chemical potential of the pure substance i and f_i^0 is its fugacity, at normal pressure (1 bar).
One can choose to write

$$\mu_i = \mu_i^+ + RT \ln \frac{f_i}{f_i^+}$$

by introducing the fugacity f_i^+ of the substance i under pressure P and μ_i^+, the potential of pure i, under pressure P.

The activities $a_i^{0v} = f_i^v/f_i^{0v}$ and $a_i^{+v} = f_i^v/f_i^{+v}$ lead to the formula of vapor-phase chemical potential:

$$\mu_i^v = \mu_i^{0v} + RT \ln a_i^{0v} = \mu_i^{+v} + RT \ln a_i^{+v}$$

A particularly interesting case is when the gaseous phase in equilibrium with the liquid is comparable to an ideal gas. In this phase, $f_i^0 = 1$ bar, and $f_i^+ = P$; $f_i = p_i$ (partial pressure of i, in the gaseous phase). Thus

$$a_i^{0v} = p_i/1 \;,\; a_i^{+v} = P_i/P = y_i$$

by introducing y_i, the molar fraction of i in an ideal gaseous phase. Generally, the activity coefficient $\gamma_i^{+v} = a_i^{+v}/y_i$ is used to compare the actual activity to that of the ideal phase.
Also, one gets $f_i^{+v}/P = \varphi\, i^+$.

4.5.1.2 Condensed Solutions (Liquid or Solid)

The chemical potential of the component (i) of a liquid can be written

$$\mu_i^L = \mu_i^{xL} + RT \ln a_i^{xL}$$

by analogy with what has already been written for gases. As previously shown, several references can be used to characterize chemical potential and activities. If x_i stands for the molar fraction in the liquid phase, the Gibbs-Duhem relationship is written $\Sigma x_i d\mu_i^L = 0$.

4.5.1.3 Liquid-Gas Partition Equilibrium

Let a solute i be present in two homogenous phases in equilibrium, with constants T and P. The equilibrium implicates the equality of the chemical potentials μ_i^L and μ_i^v; if it didn't, the biphasic system would evolve by mole transfer from one phase to the other [e.g., $dG = (\mu_i^v - \mu_i^L)dn_i$]. Thus, $\mu_i^L = \mu_i^v$ with $d\mu_i^v = RT\, d(\ln f_i^v)$. Hence, $d\mu_i^L = RT\, d(\ln f_i^v)$.

The Gibbs-Duhem relationship implies that $\Sigma x_i d\mu_i^v = 0$ or $\Sigma x_i d(\ln f_i^v) = 0$. For example, in the case of a binary system $x_1 d(\ln f_1^v) + x_2 d\, (\ln f_2^v) = 0$. If f_1^v is proportional to x_1 then the same is true of f_2^v with respect to x_2.

In the particular case of an *ideal gas solution* $f_1^v = p_1 = p_i^0 x_1$ and $f_2^v = P_2$, resulting in $p_2 = k_2 x_2$. When the solution (1) follows Raoult's law, the product (2) follows Henry's law. The constant k_2 is known as Henry's constant. This property is equally found if the solution (1) follows Raoult's law and (2) follows Henry's law (see Figure 4.2). If the aqueous solution is ideal, $k_1 = P_1^0$ and $k_2 = P_2^0$; if not, $k_2 = P_2^0$.

Three types of situations can be distinguished (see Figure 4.3). Let the dilute solution of (2) in the solvent (1). In the ranges where Raoult's and Henry's laws are applicable, the solution is an "ideal dilute solution." (see

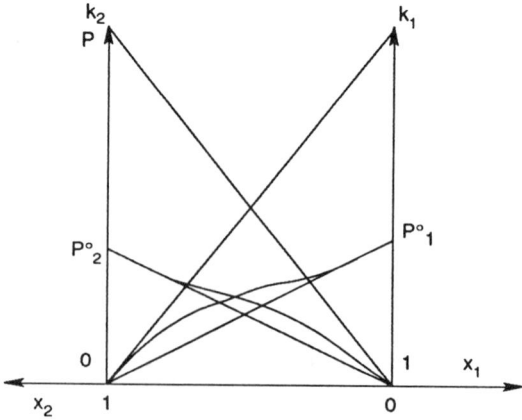

Figure 4.2.   Henry's and Raoult's laws in a nonideal solution.

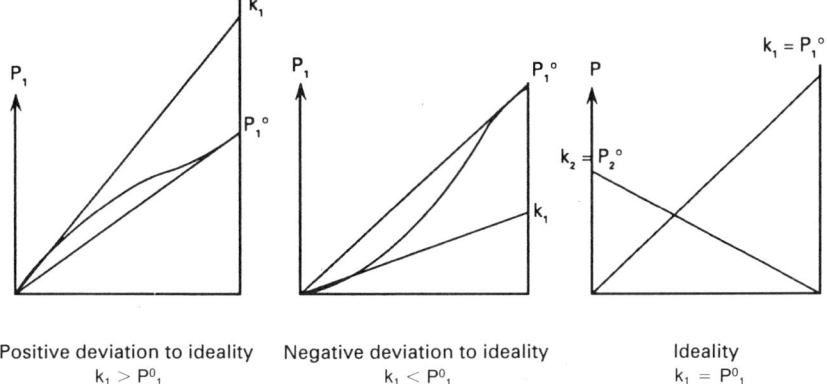

Positive deviation to ideality | Negative deviation to ideality | Ideality
$k_1 > P^0{}_1$ | $k_1 < P^0{}_1$ | $k_1 = P^0{}_1$

Figure 4.3.   Henry's and Raoult's laws in a nonideal solution.

Figure 4.4). In the case of solutes susceptible of giving rise to hydrogen bonding (i.e., amines, . . .), there will be a negative deviation from ideality.

4.5.1.4 Activities

In the case of a vapor/liquid equilibrium

$$\mu_i^v = \mu_i^L$$
$$\mu_i^v = \mu_i^{+v} + RT \ln a_i^{+v}$$

The reference state is as follows: pure compound, at pressure P, $\mu_i^L = \mu_i^{+L} + RT \ln a_i^{+L}$, and temperature T, where

$$a_i^{+v} = \frac{f_i^v}{f_i^{+v}} \; ; \; a_i^{+L} = \frac{f_i^L}{f_i^{+L}}$$
$$f_i^v = f_i^L \, \# P_i$$

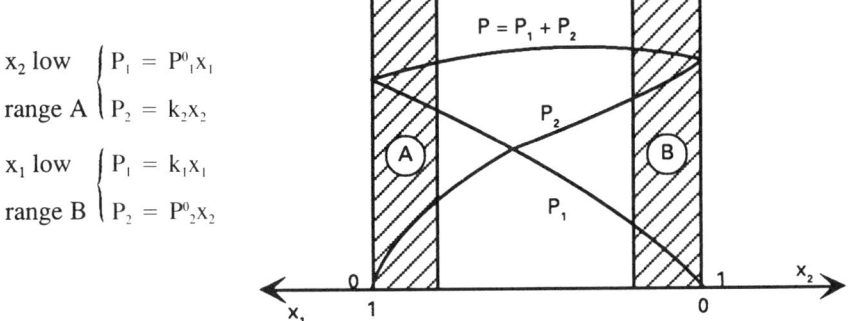

x_2 low $\quad \begin{cases} P_1 = P^0{}_1 x_1 \\ P_2 = k_2 x_2 \end{cases}$
range A

x_1 low $\quad \begin{cases} P_1 = k_1 x_1 \\ P_2 = P^0{}_2 x_2 \end{cases}$
range B

$P = P_1 + P_2$

Figure 4.4.   Total vapor pressure in a nonideal solution.

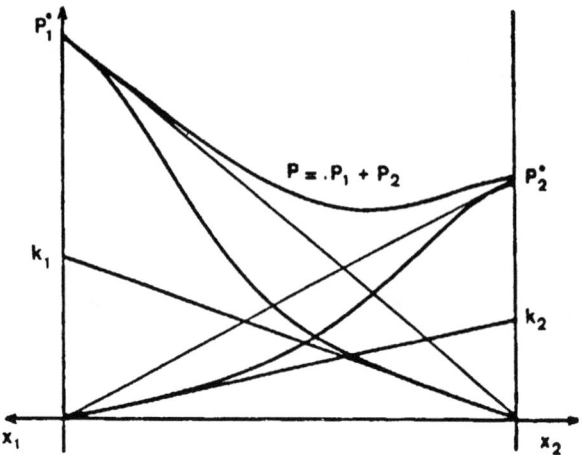

Figure 4.5.   Total vapor pressure in a nonideal solution.

The partition constant is defined as $K_{ai} = a_i^{+v}/a_i^{+L}$.

One gets $\mu_i^{+V} - \mu_i^{+L} = \Delta G_{VI} = - RT \ln K_{a_i} = + RT \ln \dfrac{f_i^{+V}}{f_i^{+L}}$ ΔG_{VI} designates the free enthalpy of vaporization of i at (T, P), it is observed as $f_i^{+V} \# P; f_i^{+L}$.

For f_i^L, one calculates $V_i^L \, dp = RT \, d(\ln f_i^{+L})$; hence

$$f_i^{+L} = f_i^{+L_0} \exp \left[1/RT \int_{P^0}^{P} V_i^L \, dp \right] \text{ for } P > P^0.$$

V_i^L designates the liquid molar volume which can be assumed as constant and leads to the following deduction:

$$RT \ln \frac{f_i^{+L}}{f_i^{+L_0}} = V_i^L (P - P^0{}_i)$$

$P^0{}_i$ being the saturating vapor pressure of i at temperature T (cf. Section 4.4). f_i^{+0} is then close to $P^0{}_i$; also

$$f_i^{+L} \# P^0{}_i \exp \left[\frac{V_i^L (P - P^0{}_i)}{RT} \right]$$

In the present case the total P is the atmosphere, which leads to $f_i^{+L} = P^0{}_i$; that is to say $K_{a_i} = P/P_i$. From this one concludes

$$a_i^{+v} \# \frac{f_i}{f_i^{+v}} \# \frac{P_i}{P} = y_i$$

$$a_i^{+L} \# \frac{f_i}{f_i^{+L}} \# \frac{P_i}{P^0{}_i}$$

The choice of a_i^{+L} concerns the reference (diluted or pure solution)

$$a_i = \gamma_i x_i$$

hence

$$P_i = P^0{}_i a_i = P^0{}_i \gamma_i x_i$$

refers to a pure solution, while

$$P_i = k_i a'_i = k_i \gamma_i' x_i$$

refers to a diluted solution:

$$
\begin{aligned}
\mu_i &= \mu_i^{+v} + RT (\ln y_i) = \mu_i^{0v} + RT (\ln P_i) \\
&= \mu_i^{0v} + RT \ln (P^0{}_i a_i) = \mu_i^{0V} + RT \ln (k_i a'_i) \\
&= \mu_i^{0v} + RT (\ln P^0{}_i) + RT (\ln a_i) = \mu_i^{0v} + RT (\ln k_i) + RT (\ln a'_i)
\end{aligned}
$$

It is noted that

$$\mu_i = \mu_i^{0L} + RT (\ln a_i) = \mu'^{0L} + RT (\ln a'_i)$$

with $\mu^{0L} = \mu_i^{0v} + RT \ln P^0{}_i$ and $\mu'^{0L} + RT \ln k_i$. With $a_i = \gamma_i x_i$ referring to a pure solution,

$$\gamma_i \rightarrow 1 \text{ if } x_i \rightarrow 1$$

and $a'_i = \gamma'_i x_i$ referring to an infinitely diluted solution,

$$\gamma'_i \rightarrow 1 \text{ if } x_i \rightarrow 0$$

$P_i = P^0{}_i a_i$ or $P_i = k_i a'_i$.

Note: If the term $V_i^L / RT (P - P^0{}_i)$ cannot be assumed to be small, that is to say, at high pressure, $(\ln \gamma_i)$ or $(\ln \gamma'_i)$ must be corrected by this value (cf. Section 4.4).

Summary

The reference ideal gas is $\mu^0{}_i^{(v)}$, with a solution of $\mu_i^{(v)} = \mu^0{}_i^{(v)} + RT (\ln P_i)$. *The ideal solution is formed with a pure solvent* (in this case, water): The reference ideal solution is

$$p_i \, p_i^0 \, x_i \text{ (Figure 4.6a)}$$
$$\mu_i = \mu_i^0 + RT (\ln x_i)$$

The real solution is

$$p_i = p^0{}_i a_i$$
$$\mu_i = \mu_i^0 + RT (\ln a_i)$$

The activity coefficient is $\gamma_i = a_i / x_i$

$$\text{if } x_i \rightarrow 1 \qquad \gamma_i \rightarrow 1 \text{ (Raoult's law)}$$

The diluted ideal solution is as follows. For the solute, there are the following relationships:

- The reference ideal solution

$$p_i = k_i x_i \text{ (Figure 4.6b)}$$
$$\mu_i = \mu'^0_i + RT \, (\ln a_i')$$

- The real solution

$$p_i = k_i a_i'$$
$$\mu_i = \mu'_i + RT \, (\ln a_i)$$

- The activity coefficient $\gamma'_i = a'_i / x_i$

if $x_i \to 0$, $\gamma_i' \to 1$ (Henry's law)

Table 4.1 presents a few values for Henry's k_i.

4.5.1.5 *Mixing and Excess Thermodynamic Functions*

The free enthalpy of mixing will correspond to the difference of the free enthalpy of a solution, as compared to its components at the pressure P and temperature T:

$$\Delta G_M = \Sigma \, n_i \, \mu_i - \Sigma \, n_i \, \mu_i^0$$

hence

$$\Delta G_M = \Sigma_i \, n_i \, (\mu_i^0 + RT \ln a_i) - \Sigma n_i \mu_i^0$$
$$= \Sigma_i \, n_i \, RT \ln a_i$$

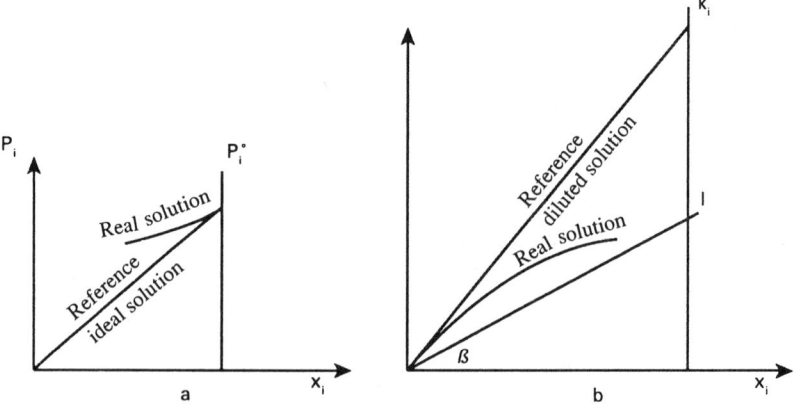

Figure 4.6a. and b.   Definition of localities.

The molar free enthalpy of excess compares the free enthalpy of a solute in a nonideal solution with the free enthalpy of the same solute in an ideal solution. For 1 mole of mixture

$$\Delta G_M = \Sigma \, x_i \, RT \ln a_i$$
$$\Delta G_M = \Sigma \, x_i \, RT \ln x_i + \Delta G_E$$

For the solution, the molar free enthalpy of excess is

$$\Delta \, G_E = \Sigma_i RT x_i \ln \gamma_i$$

The free enthalpy of a real mixture is the sum of the mixing free enthalpy of the ideal mixture and the molar free enthalpy of excess. The assessment of the excess quantities relies on models.

4.5.1.6 Study of Specific Binary Solutions

For a binary mixture, the molar free enthalpy of excess

$$RT \, (x_1 \ln \gamma_i + x_2 \ln \gamma_2) = \Delta G_E$$

can take a symmetrical form for simple solutions. Then

$$\ln \gamma_1 = A \, T \, x_2^2 \text{ and } \ln \gamma_2 = A \, T \, x_1^2$$

This leads to $\Delta G_E = A \, T \, x_1 \, x_2$, which, in turn, leads to

$$\Delta S_E = - \frac{\partial}{\partial T} (\Delta G_E) = \left(\frac{\partial A}{\partial T} \right) x_1 x_2$$

A simple solution can show a complementary property. This is the case of a regular solution, which is defined by the absence of the molar entropy of excess when the mixture is made ($\Delta S_E = 0$), which leads to $A = K/T$; in other words,

$$\Delta G_E = K \, x_1 \, x_2 \text{ and } \Delta S_E = 0.$$

The molar free enthalpy of simple regular solutions is thus written

$$\Delta G_M = RT \, (x_1 \ln x_1 + x_2 \ln x_2) + K x_1 x_2$$

The value of x being smaller than 1, the ideal contribution is negative. If the excess parameter is positive, demixtion can occur. These two contributions are presented in Figure 4.7. The mixing free enthalpy divided by RT depends on the value of K. This is illustrated by Figure 4.8.

The transfer in solution is accompanied by the occurrence of interactions 1↔2, substituting former interactions 1↔1 and 2↔2 in the pure compounds. In regular solutions, these new interactions are less favorable ($K > 0$) than the previous interactions. If the balance is too unfavorable ($K > 2RT$), then miscibility is only partial. (The dotted line in Figure 4.8 has no physical meaning.) This binary solution model can be interpreted by using Van der Waals molecular interactions.

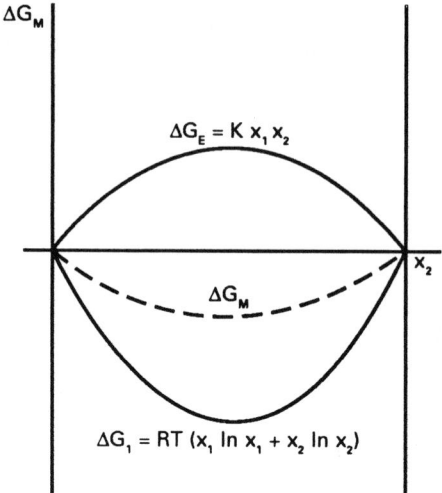

Figure 4.7.   Free enthalpy.

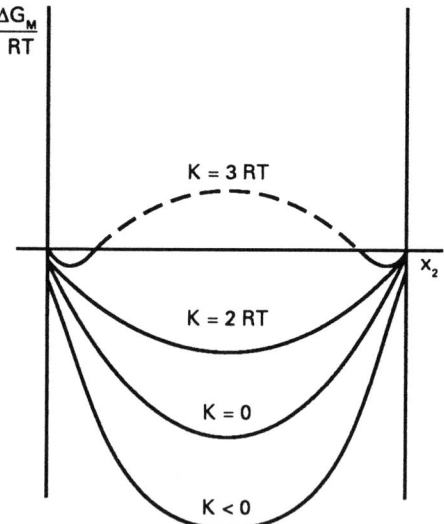

Figure 4.8.   Free enthalpy.

Considering the case of a binary solution composed of two types of 1 and 2 spherical molecules of the same radius, let Z stand for the number of molecules surrounding the spherical molecule ($Z \approx 8$ to 10). If E_{11}, E_{12}, and E_{22} represent Van der Waals' energies, there are surrounding molecules ($1 \leftrightarrow 1$, $1 \leftrightarrow 2$, and $2 \leftrightarrow 2$) (for nonpolarized molecules):

$$E_{11} \neq \alpha_1^2/r^6, \; E_{12} \neq \alpha_1 \, \alpha_2/r^6, \; E_{22} \neq \alpha_2^2/r^6$$

Thus,

$$E_{12} = \sqrt{E_{11} \times E_{22}}$$

This expression is approximatively transposable to polarized molecules.

For a binary mixture containing N_1 and N_2 molecules, the total number of interactions (n) is $n = (Z/2)(N_1 + N_2)$. The molar free energy of excess is calculated by (cf., e.g., Chabanel, 1986)

$$\Delta G_E = K x_1 x_2$$

with $K = 1/2 \, Z \, N(2E_{12} - E_{11} - E_{22})$ (N is Avogadro's number).

In regular solutions, ΔG_E is proportional to $x_1 x_2$ and K is proportional to $2E_{12} - E_{11} - E_{22}$. These results lead to the following precisions:

- If $K = 0$, $\Delta G_E = 0$, the solution is ideal and interactions between species $1 \leftrightarrow 2$ are equal to those between identical molecules $1 \leftrightarrow 1$, $2 \leftrightarrow 2$.
- If $K > 0$, $\Delta G_E > 0$, then ΔG_M is less negative than for an ideal solution. The mixture can occur if K is not too positive.
- If $K < 0$, ΔG_M is more negative than for an ideal solution.

If a nonassociated liquidlike benzene or toluene is introduced to an associated liquidlike water (the mucus), the free enthalpy of excess is highly positive and the H bonds of the water will be destroyed. This deviation from positive ideality can lead to demixture. (See Figure 4.9.)

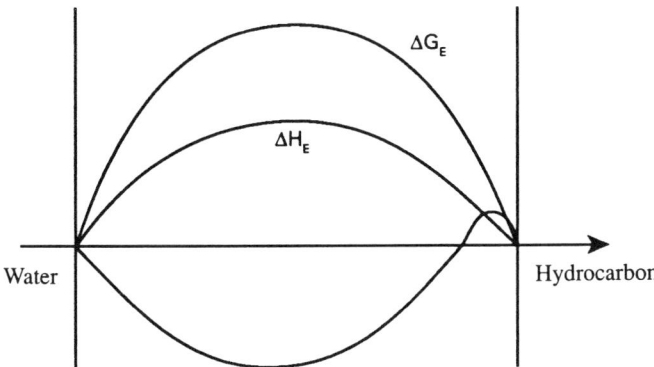

Figure 4.9. Magnitudes of excess.

4.5.2 Solubility and Activities

Odorous molecules in the gaseous phase are generally in very low concentrations and their chemical nature is extremely variable (amines, aldehydes, cetones, acids, and hydrocarbons, for example). In solution in the mucus, they are susceptible to ionization and Henry's law, useful in the study of diluted solutions, and must be explored with caution.

4.5.2.1 Case of a Regular Solution

This section concerns solutions where the entropy of excess is nonexistent and the free enthalpy is symmetrical in comparison to $x_i = 1/2$. In this case, the activity coefficients can be calculated as follows.

$$\ln \gamma_i = \frac{V_i}{RT} (\delta_1 - \delta_2)^2 \tag{1}$$

According to Hildebrand, V_i is calculable (for example, see Ried and Sherwood, 1966); it is often simplified by taking the molar volume of the liquid under experimental conditions and, in practice, the value at 25 °C. γ_1 is the coefficient of the compound 1 in the binary mixture;

$$\delta_i = \frac{\Delta U_{vi}}{V^L_i} \; 1/2$$

is the solubility parameter (Hildebrand's) of component i, ΔU_{vi} is the internal molar energy from i to vaporization ($\Delta U_{vi} = \Delta H_{vi} - RT_b$).

Formula 1 corresponds to the formation of a solution from solute molecules of the same size. For those that differ considerably in size, the formula is written

$$\ln \gamma_1 = \frac{V_1^L}{RT} (\delta_i - \delta_2)^2 + \ln \frac{V_1^L}{V_2^L} + 1 - \frac{V_1^L}{V_2^L} \tag{2}$$

Equations (1) and (2) concern solutions that are made regular; hydrocarbons or other nonpolarized substances in mixtures may justify the use of these relations. In the case of applying them to polarized and polarizable odorous products to be dissolved in the mucus, these relations are not thought to be general enough.

4.5.2.2 Case of an Associated Solvent Without Electrolytes

In the case of associated solvents such as water, Hayduck and Laudie (1973) have presented a correlation giving solubility as a function of the association by hydrogen bonding. If the temperature rises toward critical value, the dissolved molar fraction tends toward a value independant of dissolved species.

4.5.2.3 Incidence of Solute Ionization

Components in the gaseous phase are, within odorous conditions (P# 1 bar, T# 300 K), nonionized. On the other hand, many odorous compounds will be ionized by aqueous solutions. Ionization equilibrium is characterized by

$$S + H_2O \leftrightarrow A^+ + B^-$$

If S is a weak acid AH, this equilibrium is characterized by acidity constant K_a (Table 4.2); the relative dielectric constant of pure water. $\varepsilon_r \# 80$ plays a determining role in ionizing AH species in the mucus.

Here are some more precisions concerning Henry's law for ionizable substances: in the previous part, the case of an aqueous solution without ionization has been illustrated. However, a large number of odor-causing substances such as NH_3, H_2S, CH_3CO_2H, . . . , will become ionized. Assuming that the ionic dissociation leads to the formation of two ions of charges $+e$ and $-e$, respectively, we have

$$AH \xrightarrow[\text{gaseous}]{\text{water}} A^-_{aq} + H^+_{aq}$$

Table 4.2. Actidity Constant of Several Pollutants

Compound	pK_i	pK_2	t °C	Ref.
Ammoniac	9.233	–	20°	1
Methylamine	10.657	–	25°	1
Dimethylamine	10.732	–	25°	1
Trimethylamine	9.810	–	25°	1
Ethylamine	10.807	–	20°	1
Diethylamine	10.489	–	40°	1
Triethylamine	11.01	–	18°	1
Butylamine	10.77	–	20°	1
Di-isobutylamine	10.91	–	21°	1
Pyridine	5.25	–	25°	1
Butyric acid	4.81	–	20°	1
Isobutyric acid	4.84	–	18°	1
Valeric acid	4.82	–	18°	1
Isovaleric acid	4.77	–	25°	1
Hydrogen sulfide	7.04	11.96	18°	1
Hydrogen sulfide		12 to 15		3
Methyl mercaptan	9.701	–	20°	1
Methyl mercaptan	10.3	–	25°	2
Ethyl mercaptan	10.6	–	25°	2
n-propyl mercaptan	10.65	–	25°	2
iso-propyl mercaptan	10.86	–	25°	2
n-butyl mercaptan	10.65	–	25°	2

1. *Handbook of Chemistry and Physics*, 1981.
2. Crampton, 1974.
3. Millero, 1986.

If the diluted solution is ideal at constant T and P, the free enthalpy of this reaction is

$$\Delta G = \mu(AH) - \mu(H^+{}_{aq}) - \mu \ (A^-{}_{aq})$$
$$= \mu^0_{AH} \ (g) + RT \ln p_i$$
$$- [\mu^0_{(H^+{}_{aq})} + RT \ln \gamma_{H^+} \ X_{H^+} \ - \ \mu^0(A^-{}_{aq})$$
$$+ RT \ln \gamma_{A^-} \ x_{A^-}]$$

Hence

$$\Delta G = \Delta G^0 + RT \ln [p_i/(a_{H^+} \ a_{A^-})]$$

At equilibrium $p_i = K \ (a_{H^+} \ a_{A^-})$, with $K = \exp \ (-\Delta G^0/RT)$.

If the activity coefficients are assumed to be close to 1 and the ionization of AH is complete (e.g., HCl) $x = x_{AH} + x_{A^-} \# x_{A^-}$, in the same way, $x \# x_{H^+}$. Thus, the molar fractions x_{H^+} and x_{A^-} being equal, $p_i = Kx^2$. Therefore, Henry's law for ionizable substances is very different from its expression in the case of nonionizable substances ($p_i = Kx$). One can introduce the average activity of the electrolyte, $a_\pm{}^2 = a_+ \ a_-$, and the average activity constant, $\gamma_\pm{}^2 = \gamma_+ \ \gamma_-$. Therefore, $p_i = K \ (\gamma_\pm)x^2$ with $x = x_+$ or x_-.

The activity coefficient $\gamma_\pm$ is calculated according to Huckel's model:

$$\ln \gamma_\pm = -0.505 \ | \ Z^+ \ Z^+| \ \sqrt{l}$$

where l is the ionic strength of the solution: $l = 1/2 \ \Sigma C_i Z_i^2$. For monovalent ions, $\ln \gamma_\pm = 0.505\sqrt{l}$. In the case of weak ionization (e.g., NH_3, H_2S) Henry's law is still written as $p_i = \gamma_i k_i x_i$. Taking the example of H_2S,

$$H_2S + H_2O \rightleftharpoons HS^- + H_3O^+ \ (K_1)$$
$$HS^- + H_2O \rightleftharpoons S^{2-} + H_3O^+ \ (K_2)$$

H_2S should be equated to

$$\frac{S_{total}}{1 + K_2H^+ + K_1K_2/H^{+2}}$$

4.5.2.4 Solutions Containing Electrolytes

Van Krevelen and Hoftlizer (1948) related the solubility in a complex solution (x) to the solubility in a pure solvent (x^0):

$$\ln \ (x/x^0) = h \ l$$

h is a function of the contributions due to the dissolved gas and the electrolytes.

4.6. Conclusion

Thermodynamics can bring valuable information to the body of knowledge concerning olfaction, but its use requires data that is only partially known at present.

Bibliography

Chabanel, M., 1986. *Thermodynamique Chimique*. Ellipses, Paris, pp. 177–180.

Crampton, M. R., 1974. *The Chemistry of the Thial Groups*, Part 1, edited by S. Patai. Wiley and Sons, London, pp. 379–415.

Encyclopédie des Gaz - Air Liquide, 1976. Elsevier Scientific Publishing Company, Amsterdam.

Handbook of Chemistry and Physics, 61st ed., 1981, edited by R. C. Weast et al. CRC Press, Boca Raton, Florida.

Hayduck, W., Laudie, H., 1973. *Al Che* **19**, 1233.

Hildebrand, J. H., Praunitz, J. M., Scott, R. L., 1970. *Regular and Related Solutions*. Van Nostrand, Princeton.

Laffort, P., 1969. A linear relationship between olfactory effectiveness and identified molecular characteristics, extended to fifty substances. In *Olfaction and Taste III*, edited by C. Pfaffman. Rockefeller University Press, New York, pp. 150–157.

Le Magnen, J., 1976. Olfaction. In *Physiologie II. Système Nerveux-Muscle*, edited by Ch. Kayser. Flammarion, Paris, pp. 933–984.

Martin, G., 1989. Cours de thermodynamique. ENSC Rennes (in press).

Millero, F. J., 1986. The thermodynamics and kinetics of the hydrogen sulfide system in natural waters. *Marine Chemistry* **18**, 121–147.

Ried, R. C., Sherwood, T. K., 1966. *The Properties of Gases and Liquids*. McGraw Hill, New York, pp. 346–349.

Van Krevelen, D. W., Hoftlizer, P. J., 1948. Kinetics of gas liquid reactions. I. General theory. *Recueils Trav. Chim. Pays-Bas* **67**, 563.

Uziel, A., 1983. Physiologie de l'Olfaction. In *Physiologie Neurosensorielle en Oto-Rhino-Laryngologie*, edited by Y. Guerrier and A. Uziel. Masson, Paris, pp. 2–29.

Olfactory Communication and Facets of Odorousness

P. Laffort

5.1 General Points

5.1.1 Olfaction Among the Other Chemical Sensitivities

External chemical receptivity is that set of functions that allow a living being to communicate with its environment by capturing chemical signals. Three classical modalities can be distinguished (Le Magnen, 1976):

1. *Olfaction* or the sense of smell, characterized by a very acute ability to differenciate the most subtle molecular structures and to detect them at very low concentrations. It could be said that this is an analytical sense.
2. *Gustation* or the sense of taste, characterized by overall perception of four fundamental tastes: sweet, salty, acidic, and bitter. It could be said that this is an integrative sense.
3. *Common chemical sensitivity* or trigeminal sensitivity (conveyed by the trigeminal nerve), characterized by perception of irritating, stinging, and suffocating stimuli.

In higher vertebrates, each of these modalities correspond to well-separated and localized anatomical structures (nasal fossae, tongue, etc.). In invertebrates, localization of different receptors is much less clear: insect and crustacean antennae are polysensorial and olfactory and gustatory receptors are also found on their legs.

There is no clear barrier between each of the families of stimuli; some substances can stimulate several of the three sensorial modalities. For example, for humans, acetic acid is odorous, irritating, and has an acidic taste. In marine animals, where the aquatic environment serves as transport, the overlap between modalities is, of course, even larger.

Olfaction, taken in the broad sense, can be divided into two main categories that have sometimes been considered as two distinct sensorial modalities:

1. *Specialist olfactory sensitivity,* which is found in insects and a number of marine invertebrates; all the information is present as soon as it is received by the neuroreceptors, which are highly specific. The intermediate steps of the nervous system are thus only used to transmit the information, not to process it. This type of olfaction, very efficient in its narrow sensitivity spectrum, is very close to classic pharmacological phenomena.

2. *Generalist olfactory sensitivity,* which occurs in vertebrates as well as in certain insects by way of neuroreceptors that are not very specific. The information acquires its significance and specificity by parallel processing of a large number of signals by the brain which plays the role of a computer. Contrary to the specialist olfactory system, the generalist olfactory system makes it possible to perceive all surrounding chemical substances, somewhat like vision, which makes it possible to perceive all the material objects in the field of vision. On the other hand, its performance in detecting very low concentrations is poorer than that of specialist sensitivity.

In fact, the boundary between the two types of olfactory sensitivity is not always as clear as it has sometimes been thought. Be that as it may, only generalist olfaction, as found in humans, will be discussed in this chapter.

5.1.2 Some Vocabulary

In the strict sense of the term, *coding* is called "The transformation of a message according to a collection of conventions or a dictionary of equivalents between two languages" (Robert, 1973) or "system of letters, symbols and rules for their association to transmit messages secretly or briefly" (Collins Thesaurus, 1991). Applied to sensorial physiology, this concept concerns the transformation of information outside a living being into information perceived by that being, in other words, translation of a physical "language" into a sensorial "language." Due to a shift in the meaning of words, the term sensorial coding and thus olfactory coding is often used to mean *coding mechanisms.* This is rather like confusing hardware and software in computer science when these terms refer to distinctly different things: the same computer language can be applied to different types of machine and the same computer can use different languages.

In order to banish any ambiguity, the coding mechanisms (the hardware) were discussed in Chapter 1 and coding itself (the software) is discussed in this chapter and Chapter 6. This chapter will be concerned with the description of the olfactory language itself and the following chapter with current attempts to connect this language with pertinent molecular properties.

5.1.3 Olfactory Communication: A Language

Keeping this pattern in mind will help remove another ambiguity. Linguists distinguish the content of an information that they call the "signified" from its support that they call the "signifier." Confusion between these two aspects of language, especially in popular publications, can frequently be noted, even though they correspond to totally different things.

For example, even though all the sensorial modalities ("the five senses," as said in the past) are brought into play in individual recognition, the role of the sense of smell dominates in the majority of animal species, with the exception of diurnal birds and marine mammals, who mainly use hearing. The privileged role of vision belongs exclusively to several primates, including the human species (Leroy, 1987, pp. 150–158, Bruyer, 1988). The photomontage in Figure 5.1 shows more or less the vision most animals would have of a group of people posing in front of the camera. Clothing details (or sometimes even facial details such as glasses, hair color, etc. which are not shown here) would certainly help identification, but it would not be estab-

Figure 5.1. Photomontage illustrating the difficulty of individual recognition in the animal world by vision alone. (Photo INRA - Bures-sur-Yvette, 1982.)

lished except on the basis of tone of voice and, above all, the odor unique to each person.

5.1.3.1 Information Carried by the Olfactory Message (the Signified)

It is relatively easy to know the content of olfactory information when one is dealing with humans: it is sufficient to ask them. It is quite different in animal experimentation where one is most often reduced to observing spontaneous behavior, in other words, motor activity. By simplifying matters to the extreme, behavior is either attraction or repulsion. For example, in unicellular organisms (protozoa), this would be the only way of understanding their chemical sensitivity, called chemotaxy for this reason.

The elementary attraction and repulsion behavior of protozoa is linked to individual survival: the search for food and flight from danger. They are found all along the animal scale in more complex forms known as alimentary and alarm behavior. There are also those related to the survival of the species: reproductive and social behavior. All complex behavior observed in higher vertebrates in their relationships within or outside of their own species (display, mating, nesting, motherhood, migration, play, tracking, exploration, territory marking, hunting, camouflage, agression, group life, etc.) can be classified in one of four main categories. This type of classification is not useless, as far as each animal species having its own "habits and customs"; scientific literature having to do with behavior related to olfactory perception is extremely abundant (Leroy, 1987). This systematization makes it possible to find some sort of unity in the midst of such profusion.

In the human species, with the addition of information given by spoken and written languages, the sense of smell has ceased to have a vital significance; it is no more than a source of pleasure or displeasure in the midst of the four main behavior groups:

1. *Alimentary behavior.* It is probably in connection with food that the human sense of smell is most implicated, together with gustation and other sensitivities of the oral-pharyngeal cavity (thermal, mechanical, and common chemical). The overall set of properties of food and drink correspond to these sensorial perceptions in what should be called their organoleptic properties and is often improperly called their "taste" (Le Magnen, 1965; Laffort and Hoehn, 1987). The time, know-how, and energy devoted to food preparation, that is, to the bringing out of their organoleptic qualities, are considerable and an indication of the cultural, affective, and social importance that food consumption represents for the human species.

2. *Alarm behavior.* There are still a few remaining examples that concern humans and the role of odors as an alarm system. First of all, there is the odor of burning, on a stove to alert the cook as well as in a town or a

forest to indicate a fire. One can equally note the alert or warning concerning spoiled food, usually meat or fish. Finally, there is the example of city gas distribution companies who odorize the gas they distribute with a malodorous substance (generally thiophane) so as to alert the consumer in case of a leak.

3. *Reproductive behavior.* Different studies, partially evoked in Chapter 2 have been carried out concerning the relationship between olfaction and sexuality. For example, the influence of the level of female sex hormones on olfactory sensitivity and preference-aversion responses with regard to vaginal or underarm secretions (Doty, 1976). In a more general way, it could be said that body odors play a role in intimate relations like other sensory stimuli, but a relatively minor role. Perfume can be considered as much a seduction accesory as clothing, but fantasies about "love potion" perfumes and/or odors that reappear periodically in newspapers should be firmly rejected.

4. *Social behavior.* Generally, in this category one can place parental behavior while the young still need their mother and group behavior when individuals are living together. For the human species, apart from the role as object identifier played in everyday life, the sense of smell is implicated in social life relatively little. One could point out the fact that children are calmed by the odor of their mother, which they will recognize from among others until the age of three (Montagner, 1978; Schaal, 1988). There is also the role of smell as a clinical diagnostic aid, for example, breath, urine, or fecal odors (Winter, 1978; Doty, 1981; O'Neill et al., 1988). One could equally underline the part played by olfaction in the characterization of groups of humans belonging to different cultures, notably because of culinary odors that can impregnate habitations and clothing. For a long time, not without reason but wrongly in a univocal manner, sickness and stench have been associated, to the point where development of public hygiene in the 19th century was carried out very much like a battle for city deodorization (Corbin, 1986). Curiously, after more than 100 years of Pasteurian culture, confusion between odorous properties on one hand and the beneficial or harmful effects on the other hand still remains strong in the mind of the public: on one hand, one willingly speaks of "aromatherapy" and on the other atmospheric pollution is more likely to be judged on its bad odor than its toxicity. An extension of this attitude is manifested by the development of intensive use of personal deodorants, which is unjustified from a hygienic point of view. This could be spoken of, as by Corbin (1986), as a social phenomenon that is less innocent than it appears at first (the rich "smell good" and the poor "smell bad").

It has been pointed out at the beginning of section 5.1.3.1 that, in animal experimentation, the content of olfactory information is generally understood by behavior observation. However, one should remember that this is not a one-to-one relationship: a given odor may be perceived and have a significance (alimentary, for example) without necessarily triggering behavior. This behavior would depend on several factors and notably in this case the individual's satiation. Later in this chapter, ways of overcoming this difficulty by using behavior conditioning techniques will be discussed.

In several *exceptional cases,* electrophysiological recording (in other words, nervous sytem electrical activity) not only gives information on the sensation but also on the significance of the sensation: by means of retroactive loops in the olfactory system, the two types of information are mixed together. One of the most spectacular known examples is the electrical response in a salmon's olfactory bulb when it smells the odor of its spawning place (Hara, 1970).

5.1.3.2 Olfactory Signifiers or Facets of Odorousness

When a sensorial communication is established, it is the result of a stimulus acting on a receptor. Depending on one's point of view, one speaks of stimulating properties or sensorial properties, but one cannot exist without the other. For example, the taste and the texture of a piece of food correspond to the gustative and mechanical sensitivities of the oral cavity as the sound and the odor do during auditory and olfactory perception. This is how the French term *odorité* has been defined as the property of a substance or a mixture of substances to have an odor, that is, to be an olfactory stimulus (Le Magnen, 1962). A suggested translation for this neologism is "odorousness" ("odority"). Therefore, its different aspects correspond to the support of the olfactory information. Generally, four of them are outlined:

1. *Qualitative aspects.* The concept of "olfactory pattern" has been developed in Chapter 1: the olfactory neuroreceptors work as transducers, each sending a signal to the central nervous system. The spatial distribution of these signals determine a kind of "pattern," whose definition is progressively refined by the olfactory bulb, then by the olfactory centers. This olfactory pattern qualifies or discriminates the stimuli: identical patterns correspond to undiscriminated odors, whereas very different patterns qualify perfectly discriminated odors.
2. *Intensitive aspects.* The olfactory system also provides intensitive information resulting from the sum of each signals according to a process by the central nervous system, which will be discussed later. The partition of the information into qualitative and intensitive aspects, convenient for the experimenter, might as well rely on the physiological reality of bulb neurone specialization into one tract or the other, at least in mammals. A recent study of the rat olfactory bulb can be interperted as follows (Döving, 1987): about 45% of the mitral cells concerned with a given odor would be of the qualitative kind, whereas 34% would be of the intensitive kind and 18% of the adaptive kind, that is, reporting information

variation (for this last item, see Chapter 2 for discussions of adaptative phenomenon); only 3% of the investigated cells were of an undefined kind.

3. *Hedonic character* or affective tonality. It is named so, in order to differentiate this qualitative character from the first one. Here "quality" has its other meaning, found, for instance, in *Agence pour la Qualité de l'Air* (Agency for Air Quality). Applied to olfaction, this character corresponds to a more or less pleasant odor, as developed in Chapter 3. Somehow, the hedonic or preference-aversion character is part of the information content and not its support, thereby relevent to section 5.1.3.1 as well. In fact, it has already been evoked in connection with protozoa chemotactic behavior. In reality, any classification necessarily involves an arbitrary character and no borderline is exempt from challenge in whichever domain: in section 5.1.3.1, motherhood was classified into the social behavior, but could have been considered as an extension of the reproductive behavior as well. The odorous alert of spoiled food was included in the alarm category, but it is also relevant to alimentary behavior. This is also the case of the hedonic character of odors: it indeed brings some information related to the subject's and group's recent and past history (see Chapter 3), but for the most part it is also linked to the qualitative character proper for a given species, as shown by several correlations obtained between these two characters (Dravnieks et al., 1984). It is consistent with significant correlations found in the human species between preference-aversion responses and EEG recordings (Brandl et al., 1980) and the fact that, as observed since antiquity, the hedonic character is more strongly associated with the chemical senses than to the physical ones (e.g., for Plato, there were only two main classes of odors: stinky and pleasant). Beyond the inevitable debates on the respective roles of the acquired and innate preference-aversion reactions with respect to odors (Steiner, 1979), it should be remembered that for a species, an individual, and a given moment, "hedonic space" is a perfectly determined part of "qualitative space" at a given intensitive level. The general meaning of "odor space" will be discussed later in the chapter.

4. *The problem of mixtures.* The intensitive and qualitative aspects of the olfactory message, which will be developed in the next section, are related either to a pure substance or to a mixture with a precise composition. In fact, in nature there are only mixtures; therefore, the problem of comparing odorous properties of a mixture to that of its components arises; the comparison itself can then be seen from a qualitative or intensitive point of view. Few authors have studied mixtures from a qualitative angle at the present time, and their work does not provide clear information at the present time; therefore, this aspect will be passed over for the time being. On the other hand, significant progress has been made in recent years on the intensitive angle of mixtures, which will be discussed in Chapter 7.

5.2 Intensitive Aspects of Olfactory Coding

5.2.1 Evaluation Methods

Several methods allowing the evaluation of olfactory perception from an intensitive point of view have been described in Chapter 1 for animal experimentation and in Chapter 2 for human psychophysical experimentation. They will be evoked here briefly, and possibly completed.

5.2.1.1 Psychophysical

The different components of intensitive coding have been evaluated by questioning human subjects by applying experimental procedures and statistically processing the responses so that interferences from psychological factors are neutralized. The data obtained in this way can be considered purely sensorial. These different components are those that define the totality of the stimulus-response curve (threshold and supraliminal perception) on one hand and, on the other hand, the olfactory system's aptitude to perceive stimulation differences, either in time (adaptation phenomenon) or at the moment (differential threshold). Data banks of standardized average intensitivities have also been suggested.

5.2.1.2 Animal Electrophysiology

Generally speaking, recordings of electrophysiological responses, that is, nervous system electrical activity, allow reliable evaluation of the intensitive supraliminal sensation; on the other hand, they are not well adapted for threshold value evaluation. Contrary to single unit recording techniques used in a qualitative coding perspective, global recording techniques use large electrodes (macroelectrodes), which make it possible to collect responses from a large number of cells. There are essentially two types of response.

1. *Slow potentials.* In the most simple cases, one observes a monophasic negative polarity wave with an amplitude of several millivolts lasting several seconds. The maximum amplitude of this wave is taken as a significant parameter. The electroolfactogram (EOG) described in Chapter 1 and the electroantennogram (EAG), its equivalent in invertebrates, are related to this data group. These two types of response reflect a sum of potentials called receptor cell "generators," but through mechanisms that are not totally identical. It could be said that slow potential variations are olfactory information coding by *amplitude modulation.* The first EOG recordings are credited to Ottoson (1956) and the first EAG recordings by Schneider (1957).
2. *Action potentials.* These are rapid phenomena (10–50 hertz) that occur at the trigger level (axon insertion point on the cell) under the effect of

dendritic generator potentials. They result from the transformation of an amplitude modulation into a frequency modulation, which is more reliable for information preservation during transmission along the axons. The olfactory nerve would lend itself rather well to multiunit discharge recording if it were not scattered into many nerve bundles in most species. Two notable exceptions have allowed recordings in the entire nerve, which indicate the overall sensation rather well for the animal: The *Gopherus polypheus* tortoise (Tucker, 1963) and the *Rana catesbeiana* giant frog (Mozell et al., 1984; Kurtz and Mozell, 1985). Some multiunit recordings have also been obtained in the olfactory bulb, but the response is less representative of overall sensation, which depends more on electrode location.

5.2.1.3 Evoked Cortical Potentials and EOG in Humans

Plattig and Kobal (1977) were the first to succeed in collecting olfactory evoked cortical potentials (OEP) in humans by appropriate processing of electroencephalogram (EEG) recordings. In the case of chemical sensitivity, this technique, currently used for other sensory modes, notably vision and audition, comes up against a time uncertainty at the beginning of stimulation of about 1/10th of a second whereas it is about a millisecond for auditive and visual stimuli. During this lapse of time uncertainty, numerous incidents occur in the neocortex, notably of somatic origin (e.g., blinking in some subjects). This difficulty can be overcome by rigorous subject selection, implementing numerous measures in such a way so as to extract the olfactory information from "background noise," and by placing the stimulator tip in the immediate proximity of the olfactory mucosa.

OEP recording has made a certain number of studies possible, notably the previously mentioned observation of hedonic response (Brandl et al., 1980), interactions between pain and olfactory sensitivity (Schwarze and Kobal, 1984), and finally confirmation of the central character, not peripheral of adaptation (Kobal, 1980) by comparison with the human EOG (which is also an exploit). The most recent publications on this topic are those by Kobal et al. (1992) and Prah and Benignus (1992). Nevertheless, these techniques are too much a matter of experimental prowess to be used clinically or in the systematic exploration of odorousness (odority).

5.2.1.4 Conditioned Behavior

In animal experimentation, associative conditioning completes the contribution of electrophysiological techniques in sensorial evaluation perception. The information that it provides is totally integrated by the central nervous system, whereas electrophysiological experimentation gives information on the peripheral level and intermediate integration (it has already been seen how difficult it is to record evoked cortical potentials in humans). Conditioned behavior makes it possible to establish olfactory thresholds that are not accessible by electrical response. They can also be used in the qualitative domain; but on the contrary, it is badly suited to intensitive supraliminal evaluation.

In the broad sense of the term, conditioning means any relatively durable modification to the behavior of individuals resulting from their sensorial experience (Delacour, 1981). There are associative and nonassociative forms of conditioning, but only the former have any interest for sensorial investigation. Associative conditioned behavior is the response to a stimulus specific to another when they have both been perceived in association. For example, in Pavlov's classic experiments, the dog salivates when it hears a sound if that sound has been associated many times with food presentation. The extremely numerous experimental procedures can be divided into three main types:

1. *Classical conditioning* was formalized by Pavlov (1928). An unconditional stimulus *US* (food) provokes a response *R* (salivation), another stimulus, originally neutral (the sound of a metronome), is associated with food presentation and ends up provoking the response *R*; thus it becomes a conditional stimulus *CS*. All procedures in which the response is that habitually provoked by *US,* even if it is a motor response instead of a reflex response, can be included in this group. An example of application to sensorial investigation can be cited [see Schwartz's (1955) determination of olfactory thresholds in honeybees for fifteen floral substances]. A determined odor is associated with the presence of sugar in an "artificial flower" placed among others that have no odor or sugar; as soon as training is finished (which is very fast in honeybees), the odor is diluted progressively until the point at which the flowers are visited by chance is reached (see Chapter 2 for the definition of threshold concentration).

2. *Instrumental or operating conditioning.* According to this procedure, recommended by Skinner (1938) for the first time, the animal must make a specific response *R* to obtain a "reward" (positive reinforcement) or avoid a "punishment" (negative reinforcement). For example, it must press a treadle associated with a visual stimulus if it wants to obtain food or move to another compartment in a cage if it wants to avoid an electric shock. This type of procedure differs from the previous one by its artificial character which links reinforcement (food or avoidance) with motor activity. Orientation in a labyrinth (often in the form of T or Y), frequently used in spontaneous behavior triggered by olfaction, can be equally related to classical conditioning or instrumental conditioning, according to an adopted experimental procedure. Moulton and Marshall (1976) applied instrumental conditioning to olfactory threshold measurement in dogs by using an olfactometric chamber like the one shown in

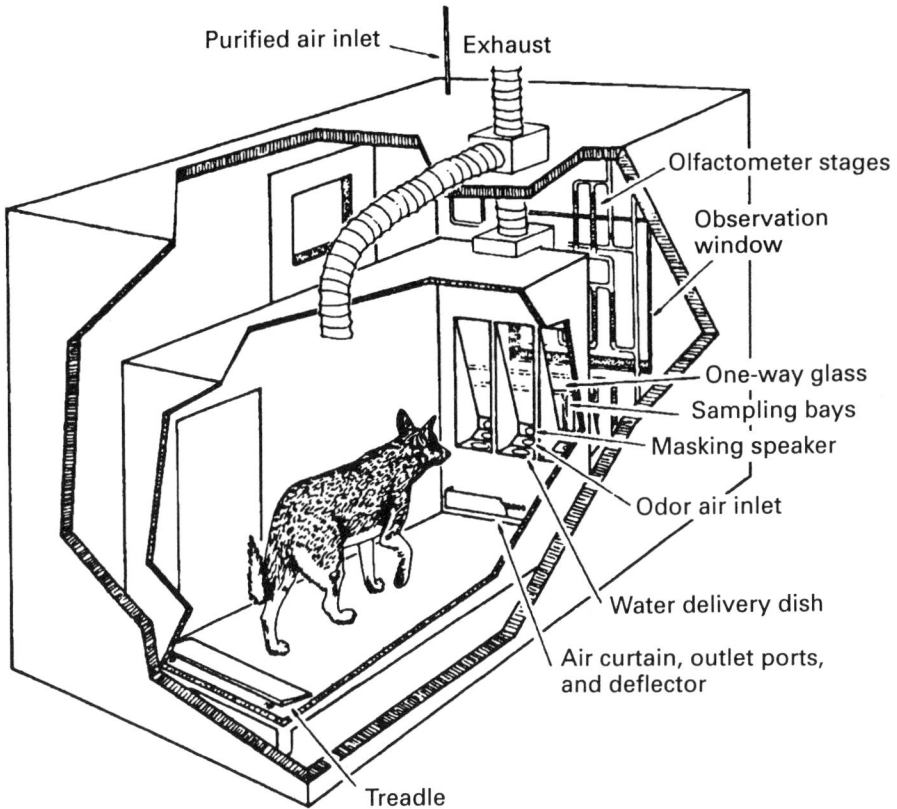

Figure 5.2. Simplified diagram of an experimental chamber allowing testing of olfactory acuteness in dogs by instrumental conditioning (Moulton and Marshall, 1976). The system could possibly be used with human subjects: in this case, the dishes of water for thirsty dogs would be replaced by coins!

Figure 5.2. The animal is previously made thirsty; it then learns to press on a treadle situated at the rear of the chamber to open three bays at the front, only one of which is odorous. The interruption of an infrared beam by the dog's muzzle will allow access to a dish of water only when it makes a "correct choice." In the present case, the olfactory threshold corresponds to an average of two correct choices out of three for a sufficient number of tests (corrected percentage is equal to 50%: see Chapter 2).

3. *Ingestive aversion acquired by nosogenic treatment.* This type of training is the combination of other more general types:

  • on one hand, nosogenic conditioning, that is, that which induces a temporary disturbance in the organism; for example, avoidance of an electric shock as in instrumental conditioning, and

• on the other hand, ingestive conditioning, some of which uses aversion to bitter (but not necessarily toxic) tastes, which is genetically determined in many species. Le Magnen (1950) evaluated odorous discrimination to 11 substance pairs in rats by using classical association conditioning between bitter or sweet tastes in sweetened drinking water flavored with the aromas being studied.

Acquired ingestive aversion by associated nosogenic treatment, often called conditioned taste aversion, was investigated for the first time by Nachman (1963) and was the subject of many works (see reviews in Riley and Baril, 1976 and Royet, 1983). The principle is as follows: after ingestion of an inoffensive substance *A* (sweetened water, for example) a treatment *T* provoking gastrointestinal disturbances is administered (generally a parenteral lithium chloride injection): then a stable aversion with respect to *A* develops. This type of training is very efficient with rats in order to assess the gustatory sensitivity, especially susceptibility to sweetness and, to a lesser extent, olfactory sensitivity. On the contrary, visual stimuli are the most involved in triggering of ingestive aversion for diurnal birds and apes. A feature of this type of training is the time interval between the two stimuli *US* and *CS* that can reach several hours, whereas in the two other types of training there must be time contiguity.

As has been seen in the previous examples, the differences between the three suggested types of associative conditioning do not always seem obvious as far as their nature is concerned: it would be sufficient to short circuit the treadle in the Moulton and Marshall experimental procedure to transform the testing chamber for dogs into an olfactometer with classical choice relevant to type I training. The acquired ingestive aversion by associated nosogenic treatment is sometimes considered a specific type of training, as it does not belong to the conditioning category. The arguments in favor of this point of view partially rely on the fact that aversions are established and resistant to extinction after only one exposition to the situation. In fact, these characterstics are found in experimentation on honeybees, particularly in that based on the proboscis reflex which will be exposed under qualitative coding aspects. In reality, these classifications have been established by vertebrate specialists whose first preoccupation was a better understanding of learning and memorization mechanisms. Invertebrate specialists, who were the first to experimentally provoke conditioned behavior with Von Frisch in 1919, generally speak of "associative conditioning" without further precision. From a strictly sensorial point of view, such as the one presented here, one could simply distinguish two associative conditioning groups with, of course, intermediate situations. On one hand, one would have rapid and durable conditioning (such as ingestive conditioning associated with LiCl in rats and the proboscis reflex in honeybees), for which the name *genetically "prepared" training* has been suggested from time to time (Delacour, 1981), which are the most efficient for sensorial studies. On the other hand, there are types of conditioning that take longer to implement and are more fragile. They are sometimes at the origin of highly complex training, made possible by sophisticated operating procedures [e.g., going so far as to teach chimpanzees a language with 160 signs (Gardner and Gardner, 1969; Premack, 1971)]; however, they are less efficient for sensorial studies.

5.2.2 Stimulus-Response Relationship Models

There is about a century separating the first psychophysical sensorial perception model (Fechner, 1860) from the first *physiological* model applied to the chemical senses (physiochemical, as it were) (Beidler, 1954). This interval, which corresponds to that of technical mastery, largely explains the low amount of overlapping between the two approaches tending to bring together stimulus intensity and sensorial response amplitude, whether in general or in the more restricted area of olfaction. The two traditions will be successively explored here, as well as how they complement each other.

5.2.2.1 Psychophysical Approach

If an abstraction is made of more complex models with a less general value, it could be said that there has been a lengthy competition between two laws describing sensorial perception development as a function of stimulus intensity: Fechner's law and Stevens' law. In olfaction, stimulus intensity is most often represented by an "all things otherwise being equal" concentration; stimulus variables other than concentration will be more precisely described later. Let C stand for concentration, R for sensorial response intensity, and log for the logarithm decimal function; the two "laws" are respectively expressed by the following equations:

$$R - R_0 = n (\log C - \log C_0) \quad \text{(Fechner)} \qquad (1)$$
$$\log R - \log R_0 = n (\log C - \log C_0) \quad \text{(Stevens)} \qquad (2)$$

In other words, with Fechner's law one would have a linear function of the sensorial response as a function of the concentration logarithm, whereas with Stevens' law, it would be necessary to use a log-log coordinate system to obtain a straight line. The first function is said to be logarithmic and the second a power function, since one can also write

$$R/R_0 = (C/C_0)^n \qquad (3)$$

The logarithmic function relies on two postulates:

1. The correspondance that would occur between differential threshold fraction $\Delta C/C$; in other words, the relative precision of a stimulus estimate (see Chapter 2) and sensorial perception increase ΔR.

2. Bouger-Weber's "law" of constancy of the $\Delta C/C$ differential fraction. This gives

$$\Delta R = a \, \Delta C/C \to dR = a \, dC/C \to R = a \log C + b$$

In fact, the first postulate is not always verified experimentally and it has been shown since then that the differential fraction $\Delta C/C$ varies with the concentration. Therefore, the logarithmic function cannot be theoretically justified.

From a purely experimental point of view, one of the supraliminal olfactory perception evaluation methods described in Chapter 2, category scaling, has been used to justify a posteriori Fechner's logarithmic function. It should be remembered that according to this method, which was applied to olfaction by Allison and Katz (1919) for the first time, the subjects are asked to classify odor intensities in one of these classes: odorless, very weak, weak, mean, strong, and very strong (sometimes a sixth class is added: extremely strong). By attributing a value such as 0, 1, 2, 3, 4, and 5, (6) to each of these classes, one notices a linearity between these numbers and the concentration logarithm, but this is begging the question (moral conflict) in the totally arbitrary correspondance between semantic descriptors and numerical values that are supposed to represent a perception intensity.

On the other hand, the direct evaluation method (advocated by Stevens beginning in 1936) of sensation magnitude by the subjects ended up justifying the power function (suggested by Plateau in 1872). The first direct evaluations in olfaction date from the end of the 1950s (Jones, 1958a and 1958b) and have been developed since then. Consequently, the validity of this last function [Equations (2) and (3)] is considered as valid, under the condition that very strong stimulation, generally situated in the painful sensitivity zone, is excluded from experimentation and where an olfactory sensitivity plateau can be observed (Katz and Talbert, 1930).

It should be noted that exponent n in equation (3) is nearly always smaller than 1 in olfaction and varies with the substances used (Patte et al., 1975). It follows that the concept of concentration threshold multiples, sometimes called "odor units," is in itself devoid of interest.

5.2.2.2 Electrophysiological Approach

Two tendencies can be observed in attempts to model the electric response as a function of the concentration of an olfactory stimulus. First of all, there is the tendency to include all the elementary parameters of the phenomenon, such as dendrite length and diameter, membrane conductance, each potential generator's maximum amplitude, etc. These "theoretical" models offer some interest to help mechanism comprehension, but the number of biological parameters and the complexity of the suggested equations are such that their validation on experimental bases is made problematical. They will not be discussed here. The second tendency is much more pragmatic, even if theoretical justifications are not always absent. Models belonging to this tendancy do not include more than three parameters each and therefore can be easily compared with experimental data. Electrophysiological responses as a function of concentration are represented by sigmoidal curves in logarithmic-linear coordinates (log-lin) and in a hyperbolic arc, one of whose branches is slanted with a variable slope and the other being horizontal when both coordinates are logarithmic. Figure 5.3 presents these two graphical representations as well as the trace of the different parameters corresponding to the three main models described so far.

The first model suggested is to Beidler's credit (1954). It is based on the "mass action" law. The equilibrium reaction between stimulus and receptor is symbolized in the following manner, with K as the equilibrium constant, R_M as the maximum response, and a as the proportional constant (sometimes the constant $K' = 1/K$ is used instead of this last constant):

$$\text{Stimulus} + \text{Receptor} \underset{}{\overset{K}{\rightleftharpoons}} \text{Stimulus-Receptor}$$
$$C - aR \quad a(R_M - R) \qquad\qquad aR$$

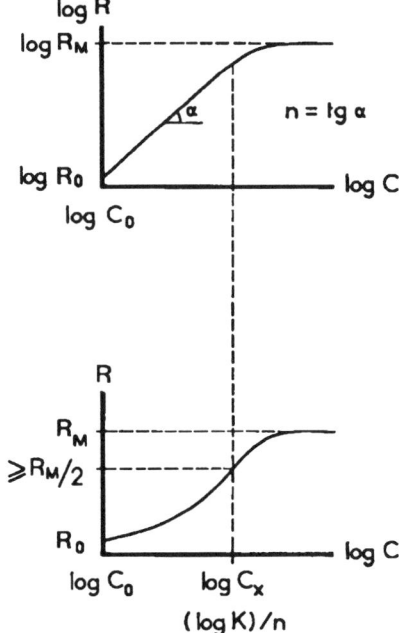

Figure 5.3. Diagram of log-lin and log-log coordinates of olfactory response as a function of concentration. The parameters are those involved in the different model descriptions (Patte et al., 1989).

from which comes

$$K = \frac{aR}{a \, (R_M - R) \, (C - aR)} = \frac{R}{(R_M - R) \, (C - aR)} \tag{4}$$

In assuming the infinite availability of the stimulus, that is, the nonexhaustion of C through its linked part, aR becomes negligible with respect to C. Equation (4) then becomes

$$K = \frac{R}{(R_M - R) \, C} \tag{5}$$

from which

$$R = \frac{R_M \, K \, C}{1 + K \, C} = \frac{R_M \, C/K'}{1 + C/K'} \quad \text{(Beidler)} \tag{6}$$

This relation has a form that is identical to Langmuir's adsorption isotherm.

For low concentrations, when the product KC is negligible with respect to 1, one would have

$$R = K \, C \, R_M \tag{7}$$

that is,

$$\log R = \log C + \log K \, R_M \tag{8}$$

In other terms, the regression coefficient of $\log R$ as a function of $\log C$ must be equal to 1. This is what is often observed in gustation, where this model has been applied with as much success on isolated fibers as on the entire nerve or the psychophysical response. This is not the case in olfaction, where slopes n smaller than 1 are generally observed, not only in the psychophysical response as has been seen above, but in electrophsysiological responses as well.

To overcome this difficulty, Tateda (1967) suggested enlarging the equilibrium reaction as follows:

$$n \text{ Stimulus } + \text{ Receptor } \underset{}{\overset{K}{\rightleftarrows}} \text{ Stimulus}_n\text{–Receptor}$$

which, by a similar calculation to the one previously described, gives the equation

$$R = \frac{R_M \, K \, C^n}{1 + K \, C^n} = \frac{R_M \, C^n K'}{1 + C^n K'} \quad \text{(Hill)} \tag{9}$$

which can also be written

$$R = \frac{(C/C_x)^n R_M}{1 + (C/C_x)^n} \tag{10}$$

in which the point at C_x, $R_M/2$ coordinates corresponds to the inflexion point of the sigmoidal line in log-lin (semi-log) coordinates.

Equation (9) is similar to that empirically suggested by Hill (1913); that is, to take the hemoglobin and oxygen combination into account. It also takes experimental reality in olfaction into account rather well. For example, when KC^n is small with respect to 1 (the case of low concentrations), one has

$$\log R = n \log C + \log(K R_M) \tag{11}$$

The oblique branch of the log-log trace in Figure 5.3 can have a variable slope (in conformity with the experiment), which would not be possible with Equation (8). The main inconvenience of Hill's model is that it imposes an inflexion point on the sigmoidal line (see Figure 5.3) for a response that is equal to half of the maximum response. This is not always the case in experimentation. The main advantage of this model is that it easily lends itself to iterative programs that make it possible to adjust the n, R_M, and K parameters in such a way that the curves are better fitted to the experimental data. Last, from a theoretical point of view, the chemical assumption on which this model relies is strongly questionable: neuroreceptors do not have the same sensitivity spectrum. The theoretical model remaining to be developed is certainly of the statistical kind, by considering the neuroreceptors as a population of individuals.

Independantly of any theoretical base, Laffort (1966, 1968, 1977) suggested a model, slightly different from the former, that allows for inflexion points corresponding to responses higher or equal to $R_M/2$, more consistent with experimental data (see Figure 5.3). When plateau values R_M are known with precision, this model performs better than Hill's; but this is not the case when R_M is not known with precision, as difficulties arise in applying iterative programs to adjust the parameter values to the experimental data. The equation corresponding to this model is as follows

$$R = \frac{R_0}{[C_0/C + (R_0/R_M)^{1/n}]^n} \qquad \text{(Laffort)} \tag{12}$$

in which C_0 and R_0 are, respectively, the concentration value and the corresponding response for any anchoring point of the straight part of the trace in log-log coordinates. This anchoring point or reference can be the threshold value.

One should notice that Equation (6) (Beidler) is a special case of Equations (9) and (11), where the exponent value is set to 1. In this case (and only in this case) the following correspondance between the parameters occurs:

$$1/K = K' = C_x = C_0 R_M/R_0 \tag{13}$$

In the more general case where $n - 1$, Laffort's model's parameters cannot be related to the dissociation constant K (or K'). However, C_x is equal to $K'^{1/n}$ (and not to K', as it has sometimes been written).

As a conclusion to this subsection devoted to the modeling of the stimulus-response relationship, it can be stated that Equations (10) and (12) are quite close to the experimental reality in olfaction, as much for electrophysiological data as for the psychophysical data. Stevens' model [Equation (3)], which is used in psychophysics, is in fact a particular case of both. Fechner's model [Equation (1)], even though still in favor in some industrial circles, should be abandoned. Last, it should be remembered that no theoretical model has been validated to date.

5.2.3 Variables Influencing Stimulus Effectiveness

The possible influences of the physiological state of subjects on their olfactory responses have been reviewed in Chapter 2. More specifically, the role of masking odors worn by the subject, age, state of satiation, awareness, and sexual hormonal impregnation have been studied. These items will not be reviewed here. The variables influencing the stimulus and susceptible to modify the olfactory response remain to be studied. These two groups of factors should be kept in mind when designing an experimental procedure for intensitive measurements of odorousness. The gaseous phase concentrations will be considered as known and solely involved in the following development. Mass transfer from the liquid phase to the gaseous phase has been discussed in Chapter 4 and the special case of olfactory perception through air, of liquid and solid aerosols, will not be reviewed.

5.2.3.1 Dynamic Variables

Like Mozell et al. (1984) and Kurtz and Mozell (1985), one can consider that olfactory responses are related to the following:

- *Three independent variables*—the number of odorous molecules in the sniffed sample N , time of sniffing T, and the volume of the sniffed sample V
- *Three derived variables*—the concentration $C = N/V$, the flowrate $F = N/T$, and the delivery rate $D = N/T$

According to the electrical response from the frog's whole nerve obtained by these authors after olfactory stimulation with octane, it can be deduced that the effective variable X is expressed by the following equation:

$$X = NT^{0.6} V^{-0.6} = N/F^{0.6} \tag{14}$$

Work by Mozell and collaborators are the most recent on this question. The first study on the role of the inhaled volume is credited to Berthelot (1901). This author showed significant differences in the olfactory perception for 1 and 40 mL of gas at the same concentration, which can be easily understood because of the dilution of small volumes in the nasal cavity. The olfactometric principle developed by Elsberg and Levy (1935) originates from this observation. This principle was somehow successful in the clinical tests: in order to study deviations of the olfactory sensitivity of patients with respect to "normality," perception thresholds were expressed in terms of mL for a given constant concentration.

The important role of the volumetric flowrate (or inhaling rate) F is commonly observed: in order to better perceive an odor, mammals use sniffing, that is, high velocities of the respiratory stream during very short durations. This sniffing has the effect of transforming the air flow from laminar to turbulent, thereby enhancing the transfer of the odorous molecules to the olfactory mucosa (Zwaardemaker, 1925).

To these purely anatomical reasons for the influence of the inhaled volume and volumetric flowrate, a phenomenon of relative reaction rate between stimulus and receptor at the olfactory mucosa level is added, as suggested for the first time by Le Magnen (1947), and confirmed by later psychophysical works (Mullins, 1955; Stuiver, 1958; Schneider et al., 1966; Rehn, 1978), and by electrophysiological results (Tucker, 1963a and 1963b). A survey of known results has been submitted (Laffort, 1966b), according to which the efficient variable would not be the same at low levels and high levels. At very high inhaling velocities ($F-$ natural sniffing rate), the significative variable would be the concentration C , whereas at intermediate velocities it would be the delivery rate D and at very short stimulation durations ($T < 0.2$ seconds) it would be the total number of molecules N. This scheme seems to be challenged by recent findings by Mozell et al. As a matter of fact, these authors only studied the intermediate zone as described above ($T > 0.2$ seconds; $F < 150$ mL/mn), within which they refined the preceding observations by varying the three independent variables N, T, and V instead of two in the preceding works. It would be interesting to confirm the general validity of these results by the use of other odorous substances and other measuring techniques (especially psychophysical techniques). The first and third zones could also be explored through the use of the same principle of variation of the three independent variables.

In the absence of more thorough knowledge on this subject, it is necessary to precisely know, or better yet to keep constant, all variables except the concentration in all intensitive olfactory measurements. This is especially true for the odorous flowrate.

Last, it should be noted that the role of the stimulation duration has been developed here only for values smaller than a few seconds. Beyond this value, adaptation and recuperation phenomena occur, not very sensitive at the peripheral level but important at the central level, which have been already developed in Chapter 2.

5.2.3.2 Temperature

Temperature effects on olfactory sensations are multiple (Grundvig et al., 1967; Laffort, 1969). First of all, temperature acts on odorous source volatility in solid or liquid states. For example, repercussions on atmospheric odorous pollution that are spread over the same area will differ in a hot or cold climatic context, but this goes back to Chapter 4. The necessity of thermoregulation of olfactory stimulators when they include solid or liquid odorous sources should be equally underlined, since it is often neglected by experimenters.

Another temperature influence, the opposite of the preceding one, is proportion modification of odorous substances that are dissolved in the mucus covering the neuroreceptor dendrites. The same Van't Hoff's law implicated in volatility rules this dissolution:

$$\log K \, {}^{\text{mucus}}_{\text{air}} = \frac{\Delta H}{2.3 \, RT} + Cte \tag{15}$$

in which K is the partition coefficient for diluted solutions (or Henry's coefficient), ΔH is the vaporization enthalpy, R is the perfect gas constant, and T is the absolute temperature.

According to this equation, a lower temperature should lead to olfactory response increase. In fact, this is what is noted in experimentation on cold blooded animals: for example, the tortoise (Tucker, 1963a).

In warm blooded animals and notably in humans, several effects are combined whose result varies according to the case. For example, on one hand, there is the nonconstant temperature of inhaled air and, on the other hand, there is peripheral vasodilation, which, within certain limits, causes concomitant hypersensitivity. It should be noted that this peripheral vasodilation can be added consecutively to previous time spent in a hot atmosphere or at physical exercise, from which come the recommendations for all human subjects to observe a rest period in a thermally regulated atmosphere before participating in olfactory experimentation. In practice, a temperature of 25–30 °C can be considered optimal for olfactory perception; this is the temperature used for "hot dish" consumption, slightly higher than the "thermal comfort" temperature of air (20 °C).

In conclusion, one should remember that all living environments adapt poorly to extreme temperatures and the tendancies described above are only observed within relatively narrow limits: below 4 °C tissues are anesthetized and above 45 °C they present cellular lesions accompanied by a sensation of pain.

5.2.3.3 Humidity

Insects possess specific humidity receptors in their anntenae. Therefore, it is best to maintain a constant relative humidity during evaluation of their olfactory sensitivity by means of electrophysiological techniques; this type of work is generally carried out in a dry atmosphere.

In terrestrial vertebrates, no humidity influence on olfactory sensitivity has been found (Tucker, 1963a), provided that olfactory mucosa dessication is not provoked by overextended dry air stimulation. Naturally, one should take notice of the well-known phenomenon of earthy odors after the first rainfall of a thunderstorm or the moment when the dew falls after a sunny day. In these instances, one has an "extraction" of odorous molecules by drops of water, which is once again a return to Chapter 4.

5.2.3.4 Atmospheric Pressure

It has been shown (Laffort and Gortan, 1987) that in a hyperbaric atmosphere such as that experienced by deep sea divers, certain odorless gases such as methane, totally odorless at normal pressure, "become" strongly odorous. This phenomenon can be very simply explained by an increase in the number of molecules per volume unit that reach the olfactory neuroreceptors. The demonstration of this limited phenomenon, which is interesting from a theoretical point of view, is nevertheless without repercussions in habitual olfactory experimentation. Variations in atmospheric pressure under normal conditions, even with a change in altitude, never exceed several tenths of an atmosphere, whereas in a hyperbaric caisson they exceed many dozens of atmospheres.

5.2.4 Relationships Between the Stimulus-Response Curve Parameters

5.2.4.1 Perceived Intensity at Threshold

Going back to the power law used in psychophysics [Equation (3)],

$$R/R_0 = (C/C_0)^n,$$

in which the coordinate point C_0, R_0 corresponds to a reference point which can be chosen as the threshold value. In this case, one of the problems is to know if the perceived intensity R_0 is the same for all odorous substances. Although threshold establishment methods and direct evaluation of perceived intensities ("magnitude estimation") are totally different in nature, the following two part reasoning makes it possible to answer the question.

1. The rectilinear lines in log-log coordinates obtained by the direct evaluation method are comparable to category lin-log coordinate lines (which amounts to considering the category scale as logarithmic). Contrary to the direct evaluation scale, the category scale crosses the threshold value, since it begins at odorlessness.
2. Concentrations corresponding to the interpolated category 0.5 (zero being "odorless" and 1 the "very weak" category) are values comparable to those obtained by classic threshold establishment methods (50% positive responses) (Patte et al., 1975).

Therefore, it seems legitimate, although this concept is not classic in the least, to consider the perceived intensity at threshold R_0 as constant for all substances and consequently express psychophysical perceived intensities in multiples of R_0. By posing $R_0 = 1$, Equation (3) thus becomes

$$R = (C/C_0)^n \tag{16}$$

5.2.4.2 Relationships Between Thresholds and Power Law Exponents

A very significant correlation ($P < 0.001$) has been established between olfactory thresholds and power law exponents obtained through psychophysical experimentation in humans (Laffort et al., 1974). In other words, one observes a tendency to converge for log-log traces: the lower the thresholds are the lower the slopes are. This phenomenon, which was confirmed on several data sets from the literature, has not been explained by physiological mechanisms to date. The theoretical model for stimulus-response relationships that still remains to be designed, as was discussed earlier, should take these physiological mechanisms into account.

Furthermore, the set of stimulus-response curves obtained with honeybees for about 60 substances through electroanntenographic response allowed the establishment of the two following equations with Pearson's correlation coefficients close to 1 (0.99 and 0.97, respectively) (Patte et al., 1984 and 1989):

$$\log C_x = \log \text{SVP} + \frac{\log R_M}{n} - 2.33 \tag{17}$$

$$\log C_0 = \log \text{SVP} + \frac{\log R_0}{n} - 2.33 \tag{18}$$

In these equations SVP is the saturated vapor pressure at room temperature and C_0 corresponds to a response R_0 of 0.1 mv. Equation (17) suggests that, in Hill's model, two parameters are sufficient to describe the experimental curves: the maximum response R_M and the power law exponent n. Equation (18) confirms results observed in 1974 for psychophysical responses, provided that the threshold concentrations in the air are replaced by partial pressures, that is, the concentrations in an ideal solution according to Raoult's law (see Chapter 4 for the definition of an ideal solution). Figure 5.4 shows the comparison between the two types of data (electroanntenographic response in honeybees and psychophysical response) and complements it by a calculated threshold concentration in a hypothetical polar phase. The properties of this phase have been calculated á priori for the studied substances according to their solubility parameters as described in Chapter 6.

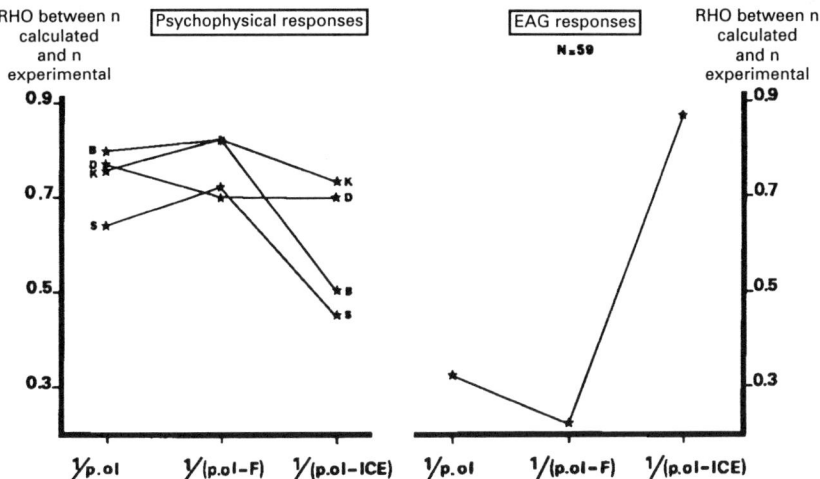

Figure 5.4. Spearman's correlations (RHO) between the experimental power law exponents and those derived from olfactory threshold values or p.ol (olfactory power). These last values are expressed in concentrations in the air weighed down by a "filter effect" of the mucus F and by the saturating vapor pressure or ICE (index of cohesive energy). One can observe completely reversed tendancies between EAG responses in honeybees and psychophysical responses after the work of several authors designated here by the letters B, D, K, and S, the number of substances N being between 22 and 82 (Laffort and Patte, 1987).

Reversed tendencies are observed for the two types of data. The best correlations are obtained for concentrations in the polar phase in psychophysical measurements and for the ideal phase in EAG measurements. The psychophysical result is classical and is not surprising: in order to reach neuroreceptors, odorous molecules must be dissolved into the aqueous mucus (Laffort, 1963). The results obtained with honeybees are somehow unexpected and seem to be explained only by "dry" transfer into "pores-tubules" as had been shown by several anatomical studies (Keil, 1982). Naturally, a minimum amount of water is necessary at the dendritic cilia surface so that depolarization of the cell membranes can occur and therefore the ionic exchange takes place. However, this water layer could be thin enough so that the membrane system behaves like an ideal solution.

An objection that could be made with respect to this work relies on the comparison between completely integrated data by the central nervous system (psychophysical data) and pheripheral data (EAG). Validation of the results by similar peripheral measurements on different animal species is necessary prior to accepting them as definitive. Their possible confirmation would shed light on the involved mechanisms and would be especially interesting if it were possible to deduce the exponent values in a simple way directly from the threshold values.

5.3 Qualitative Aspects of the Olfactory Code

5.3.1 General Points

As we have just seen, the intensitive character of the olfactory response can rather easily be approached and the number of works on this topic is relatively large. The situation is the reverse for the qualitative aspects, as much for their concepts as for the number of publications.

The methods of characterization of the olfactory quality can be sorted into two main families: the profile methods (absolute type) and the direct comparison methods (relative type). They will be successively reviewed, along with the methods to go from one to the other and how to go from either to what can be called an odor space. To accomplish this, Figure 5.5 will be useful for all the following steps.

From left to right, a rectangular matrix of vertical profile, a similarity triangular matrix, and an odor space are represented. A more or less high proximity of two points in the odor space corresponds to two more or less

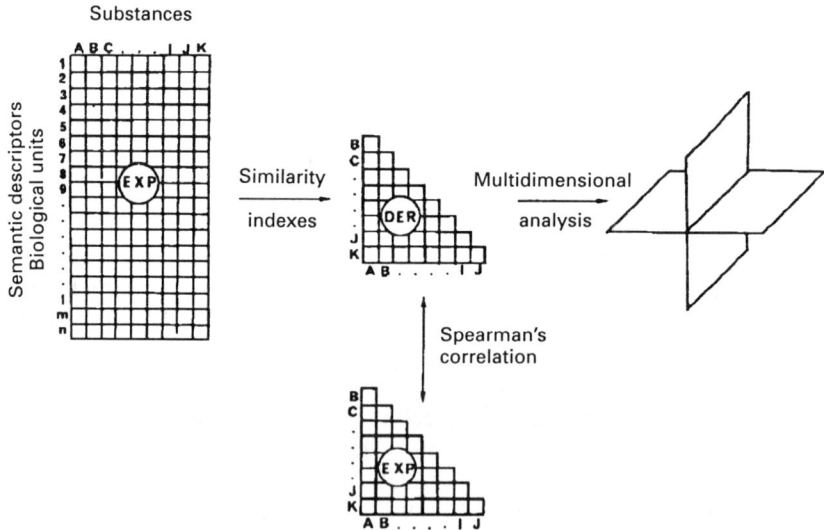

Figure 5.5. Chart of two families of methods making it possible to evaluate the odorous quality (taken in the sense of discriminablity): the rectangular and triangular data tables. The triangular tables can be directly obtained or derived from the rectangular tables. Finally, both experimental designs can be processed by multidimensional analysis (EXP and DER represent experimental and derived data tables respectively) (Figure presented by Patte and Rouault, 1989)

similar profiles in the rectangular matrix and a more or less high similarity value in the triangular matrix. Last, the triangular matrices can be obtained either directly or be deduced from rectangular matrices.

5.3.2 Profile Methods (Rectangular Data Matrixes)

The concept of "olfactory pattern" has already been mentioned; it should be recalled that this is how the set of neuroreceptors stimulated by a given compound (pure substance or mixture thereof) is referred to. This pattern, blurred at the beginning, is progressively refined by central nervous system processing, especially by contrast enhancement and apparently by anatomic grouping of the active neurons as well (see Chapter 1). Each one of the columns in the rectangular matrix of Figure 5.5 represents a specific olfactory pattern characteristic of substances A, B, K. Rectangular matrices can be established either by animal or psychophysical experimentation.

5.3.2.1 Single Unit Electrophysiology

This type of experimentation, initiated by Döving (1965, 1966), uses the single unit electrical response of the frog's olfactory mucosa or olfactory bulb (electrophysiological methods design: Gesteland et al., 1963 for the mucosa method and Döving, 1964 for the bulb method). Developed by Holley, Duchamp, Revial and Sicard at Lyon, these experimental procedures and their application are described in detail in Chapter 1 and will not be discussed here further. One should only keep in mind that the multidimensional analysis of the resulting numerical data has made it possible for these authors to point out odorous substance grouping (arylic, camphorous, etc.) for the animal species in question.

5.3.2.2 Other Mapping Techniques

Other techniques have been explored for the purpose of somehow "materializing" the olfactory pattern produced by a determined substance at the olfactory bulb level. These techniques are mainly selective degeneration, metabolic marking, and voltage dependant fluorescent dye. The second two methods have been described in Chapter 1; selective degeneration of bulbar cells by prolonged exposure of rats to odorous stimulation has been suggested by Döving and Pinching (1973) and Pinching and Döving (1974) as a way of materializing the bulbar olfactory pattern. Without getting into details, it should be specified that olfactory bulb "maps" representing areas concerned by a given odor have indeed been obtained (Jourdan, 1982, 1985; Royet et al., 1987) but at the present time, none of these three techniques have been proven to be sufficiently accurate for the generation of reproducible data rectangular matrices. Thus, as far as mapping techiques are concerned, we are witnessing hopeful developments especially with respect to the use of fluorescence.

5.3.2.3 Psychophysics

Two procedures for odor profile establishment by psychophysical interrogation can be distingushed: one by comparison with a limited number of standard substances (4–30) and another by notation of a more extensive number of descriptive semantics (44–146). In each case, the comparison is carried out on the basis of a small number of grades (6–9: the category method).

Crocker and Henderson (1927) were the first to suggest an odor profile method. These authors thought they could characterize all odorous substances with the help of four descriptors: "fragrant," "acid," "burnt," and "caprylic or goatlike," each modulated on a scale from 0 to 8 and respectively represented by four standard substances. For example, according to these authors, the odor of ethanol was characterized by the number 5414 and that of rose by 6423.

Schutz (1964), Wright and Michels (1964), and Amoore and Venström (1967) have also suggested limited series of standard substances (7–9) justified by prior experimentation that will not be discussed here. By comparing a substance or an unknown compound to each of these standards, these authors thought they had a sufficiently reliable tool for comparing the odor of beef meat before and after radiation (Shutz), for example, or determining a large number of pure substance profiles: 50 for Wright and 107 for Amoore and Venström. Still using this method, but extended to 30 standard substances or essential oils, Boelens and Haring (1980) determined odor profiles for 319 compounds used in the perfume or food industries.

Procedures undertaken by Harper et al. (1968) and Harper (1977), then Dravnieks et al. (1978) and Dravnieks (1982), differ from those previously mentioned by the abandonment of reference substances and by an important increase in profile descriptors: 44 for Harper and 146 for Dravnieks. These numbers result from a compromise between the totality of descriptors in use according to a survey (more than 800) and their implementation frequency. Harper's method, extended and systemized by Dravnieks, was adopted by the American Normalization Comission ASTM E-1804; it includes the list of 146 semantic descriptors, the experimental procedure, and statistical data processing method for raw results. Applying this method, Dravnieks (1985) has published average profiles for 160 odorous compounds.

This quick review does not cover all the aspects of work done in this field, the major part (carried out in industry) not having been either published or available. It seems that the two described methods are equally implemented, each having its advantages and disadvantages: heavier manipulation of standard substances on one hand and more subjectivity in descriptor choice on the other. It is clear that the ideal method would be a superimposition of these two methods, with each standard substance corresponding to a very precise descriptor. For that matter, an intermediate method can be considered: descriptor "objectivization" for standard substances as a first step, followed by a semantic profile starting from an "objectivized" memory.

5.3.3 Similarity Index Methods (Triangular Data Matrixes)

Determination of odorous similarities for a set of studied compounds taken by pairs is certainly the most objective way to investigate the odorous quality taken in its "discriminative" sense. However, its use runs up against the lack of experimental flexibility, since the number of experiments N increases as the square of the number of substances X [more precisely, $N = X(X - 1)/2$]. This explains why the published triangular matrices generally have about 20 odorous substances ($N = 190$). This type of data can be established by psychophysical experimentation. In animal experimentation, electrophysiological response does not lend itself to direct triangular matrix obtention; only associative conditioning is efficient. Finally, indirect methods can also be considered.

5.3.3.1 *Associative Conditioning*

Except for work carried out by Le Magnen (1950) with rats, which has already been referred to, publications on this subject are mainly concerned with proboscis reflex use in honeybees. When the antennae or the extremities of the legs (tarsus) of certain insects come into contact with a sweet solution, a tongue (proboscis) extension reflex occurs. If a whiff of an odorous substance on the antennae is associated with this sweet contact stimulation, the association of the two is generally very fast; proboscis extension occurs when the conditioning odor is presented again, even in the absence of a sweet reward (reinforcement). The presentation of odors perceived as more or less close to each other induces a number of confusions that can be assimilated to an index of qualitative similarity. Few results have been published at this time; Vareschi (1971) and Laffort et al. (1988) can be cited.

5.3.3.2 *Psychophysics (Category Scaling)*

Published data is slightly more extensive. Yoshida (1964), Döving and Lange (1967), Woskow (1968), Berglund et al. (1973), Dravnieks (1974), and Dravnieks et al. (1978) can be cited. The number of substances per matrix ranges from 10 to 25. The experimental procedure is identical to the one used for the establishment of rectangular matrices with standard substances.

5.3.3.3 *Indirect Determination*

Two indirect methods have been prescribed to reflect the more or less important overlap of olfactory patterns: the index of binary mixture interaction and the cross adaptation index.

The procedure allowing the obtention of odorous interaction indexes for substance mixtures, that is, synergy and inhibition indexes, is very recent and will be studied in Chapter 7. The first results suggest that for neuroreceptors (peripheral pattern) there is no interaction. In other words, indexes must be obtained at higher integration levels (for instance, through the use of psychophysics) in order to reflect qualitative nuances. This is consistent with the fact that, at the peripheral level, the olfactory patterns are not very contrasted: taken as pairs, they are very close to being totally similar.

The crossadaptation studied in Chapter 2 has been used as well in an attempt to characterize odorous similarites. It has been observed that after extended exposition to an odorous sustance A, the sensitivity loss for A is higher than for a different substance B, in other words autoadaptation is higher than crossadaptation. Therefore, one could hope that the extent of crossadaption could be related to the extent of qualitative similarity. This problem is quite similar to the problem of synergy indexes, as the succession of different odorous sensations can be considered as a "time mixture." As in the case of mixtures, the quasi-absence of peripheral adaptation is noted. This fact limits the use of this approach to very integrated levels. In practice, no symmetry is observed ($A \rightarrow B \neq B \rightarrow A$). As shown in Chapter 2, Moncrieff (1957) proposed a qualitative similarity index that combines four types of sensitivity loss ($A \rightarrow A, B \rightarrow B, A \rightarrow B, B \rightarrow A$). The consistency between values obtained in this way and directly established similarity indexes has not been demonstrated. Thus, it is probably necessary to reevaluate the modeling of the phenomenon.

5.3.4 Relationships Between the Profile Method and Similarity Indexes

This problem is somewhat analogous to that posed by robot portrait construction. As a first step, the witness(es) are asked to describe a subject (usually a suspect) by means of a semantic descriptor profile: eye and hair color, facial shape, characteristic details, etc. Then a portrait is constructed with the help of these elements and the witness(es) are asked if the portrait resembles the person in question; a final result is generally obtained by successive approximations. In a more general way, two aspects are in question: validation of the number and quality of chosen descriptors and the *coding* used by the brain to go from an absolute description (a profile) to a relative description (a resemblance). Returning to the rectangular and triangular matrices described above, which are objective tools making it possible to materialize these concepts, the question of the relation(s) that link them remains, that is, the mathematical model(s) that make it possible to go from one to the other.

The necessity of validation has not escaped notice by certain authors who recommended the profile method (Schutz, 1964; Dravnieks et al., 1978) but to do this, they arbitrarily used a principal component multidimentional analysis (PCA); this is equivalent to relying on the Euclidean distance in normalized profiles to transform rectangular matrices into "sensorial space." On the other hand, some authors use the X^2 distance which is the basis of factorial correspondance analysis (FCA). Is one of these methods more justified than the other? Are there other more appropriate methods? Such are the questions posed by certain authors (Patte and Rouault, 1989; Rouault and Laffort, 1993).

The means used as an attempt to respond to these questions is as follows, in accordance with that represented in Figure 5.5: as a first step, the triangular matrices have been derived from the profile matrix established by Dravnieks (1985) by the means of a number of proximity, distance, or similarity mathematical models. As a second step, the derived matrices were compared to triangular matrices directly obtained by different authors by means of Spearman's correlation coefficient. It should be remembered that Spearman's correlation coefficient *RHO* is expressed by the equation

$$RHO = 1 - \frac{6 \, \Sigma di^2}{(N^2 - 1) \, N} \tag{19}$$

in which *di* designates the difference in rank of the elements in each of the series considered and *N* is the number of cases.

From this work, a first sorting has made it possible to select five mathematical models among thirty tested: a distance *RD* proposed by Rouault (unpublished), an index *TI* proposed by Tanimoto (1959), Pearson's correlation coefficient *r*, the distance "city block" *CB* (which is used to describe the length of a residential street between right angle cross streets in American cities), and finally the distance X^2. These five models are expressed by the following equations:

$$RD = \frac{\Sigma | X/\overline{X} - Y/\overline{Y}|}{2N} \quad \text{(Rouault)} \tag{20}$$

$$TI = 1 - \frac{\Sigma \min (X, Y)}{\Sigma \max (X, Y)} \quad \text{(Tanimoto)} \tag{21}$$

$$\frac{1 - r}{2} = \frac{1}{2} \left[1 - \frac{\Sigma (X - \overline{X})(Y - \overline{Y})}{\sqrt{\Sigma (X - \overline{X})^2 (Y - \overline{Y})^2}} \right] \quad \text{(Pearson)} \tag{22}$$

$$CB = \frac{2}{N} [\Sigma | X - Y|] \quad \text{("City block")} \tag{23}$$

$$d = \frac{1}{N} \sqrt{\Sigma \frac{T}{S} \left(\frac{X}{\overline{X}} - \frac{Y}{\overline{Y}} \right)^2} \quad \text{(distance of } \chi^2) \tag{24}$$

in which X and Y are elements of the two columns of the same line, $\overline{X}$ and $\overline{Y}$ are average values of the elements in the two considered columns, N is the number of lines, S is the sum of the elements of the same line, and T is the total sum of the table.

The correlations represented in Figure 5.6 have the interesting character-istic of being relatively high (0.7 and above) for experimental data which have also been proven to be mutually very significantly correlated (Wright and Michels, 1964; Dravnieks et al., 1978; Woskow, 1968) and, frankly, quite poor for those not correlated with other authors' findings (Boelens and Haring, 1980). Berglund et al. (1973) and Yoshida's (1964) data correspond to intermediate situations. Therefore, the totality of the two types of results presents a high degree of consistency. It should be noted that, contrary to the other authors at the origin of triangular as well as rectangular matrices,

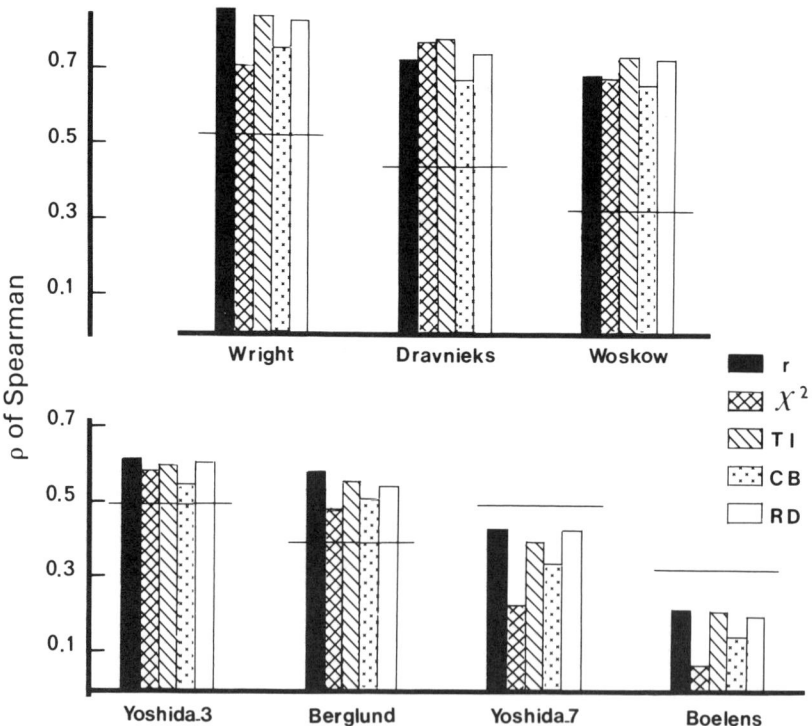

Figure 5.6. Spearman's rank correlations between odor similarities directly estab-lished by seven authors and derived odor similarities (by the use of five different mathematical models) from the atlas of odor profiles established by Dravnieks (1985). In each of seven cases the horizontal straight line corresponds to a probability that the results are due to chance <0.001 (figure presented by Patte and Rouault, 1989).

Boelens and Haring used perfume industry professionals as subjects, which may explain the nonsuperimposition of results, not that this necessarily implicates a value judgement with respect to their work.

The second remark with regard to Figure 5.6 is that out of the five models tested, there are two that never finish "in front" for each of the seven directly established matrices: the distances "city block" and X^2. On the other hand, on the basis of this data, it is not possible to come to a conclusion between the other three models. The fact that they are somewhat equally placed allows one to presume that a new, more efficient model is still to be found.

It should be equally noted that the two types of matrices cannot be obtained by animal experimentation at the same integration level (pheripheral and central); comparisons are not possible except for psychophysical data. On the other hand, a problem that has only been partially elucidated though it could be more so, is that of comparison (according to the same principle) between rectangular matrices obtained by single unit electrophysiological exploration at two different integration levels: neuroreceptors and bulb (Duchamp, 1982).

5.4 Conclusion

This chapter could be concluded by the introduction to Chapter 6, knowing that the establishment of correlations between odorous and molecular properties can only rely on quantitative biology (olfactology), that is, reproducible numerical data banks. To establish these olfactory data banks, it is first necessary to define what one means by "odorousness," which is what has been attempted in the preceding pages. It is because this has been ignored that some "theories of olfaction" and correlations depend on too fragile a base.

Whether in this structure-activity perspective or simply for the purpose of phenomenon description, one can regret the too frequent confusion between qualitative, intensitive, hedonic properties, and, even, volatility. In order to keep the principal preoccupations of this work in mind, it is certain that a concept such as "olfactory nuisance" must be clarified into its olfactory intensitive component, its olfactory hedonic component and its trigeminal component.

The description of present knowledge about olfactory coding makes it possible to justify certain prescribed procedures in the olfactometric norms (such as, for instance, the French norm AFNOR NF X 43–101). In addition, it should allow a more thorough investigation of certain research directions among which one should certainly include animal experimentation in a psychophysical framework.

Bibliography

AENOR, 1986. Qualité de l'air. Méthode de mesurage de l'odeur d'un effluent gazeux. Détermination du facteur de dilution au seuil de perception. Norme X 43-101, *AFNOR*, Tour Europe Cedex 7, 92080 Paris la Défense, France, p. 13.

Allison, V. C., Katz, S. H., 1919. An investigation of stenches and odors for industrial purposes. *J. Industr. Eng. Chem.* **11**, 336–338.

Amoore, J. E., Venström, D., 1967. Correlations between stereochemical assessments and organoleptic analysis of odorous compounds, In *Olfaction and Taste II*, edited by T. Hayashi. Pergamon Press, Oxford, pp. 3–17.

Beidler, L. M., 1954. A theory of taste stimulation, *J. Gen. Physiol.* **38**, 133–139.

Berglund, B., Berglund, U., Engen, T., Ekman, G., 1973. Multidimensional analysis of twenty-one odors. *Scand. J. Psychol.* **14**, 131–137.

Berthelot, M., 1901. Observations sur les procédés propres à déterminer les limites de la sensibilité olfactive. *Ann. Chim. Phys.* 7, 460–464.

Boelens, H., Haring, H. G., 1980. *Molecular Structure and Olfactive Quality.* Naarden International, P.O. Box 2, 1400 CA Bussum, The Netherlands.

Brandi, U., Kobal, G., Plattig, K. H., 1980. EEG-Correlates of olfactory annoyance in man. In *Olfaction and Taste VII*, edited by H. Van der Starre. Noordwijkerhout, The Netherlands, pp. 401–404.

Bruyer, R., 1988. La reconnaissance des visages. *La Recherche* **200**, 774–783.

Collins Concise Dictionary and Thesaurus (1991). HarperCollins Publishers.

Corbin, A., 1986. *The Foul and the Fragrant: Odour and the French Social Imagination.* Harvard University Press, Cambridge, MA (French edition 1982).

Crocker, E. C., Henderson, L. F., 1927. Analysis and classification of odors, an effort to develop a workable method. *Am. Perfumer Essent. Oil Rev.* **22**, 325–327, 356.

Delacour, J., 1981. *Conditionnement et Biologie—Une introduction à la Neurobiologie de l'apprentissage.* Masson, Paris.

Doty, R. L., 1976. (Ed.) *Mammalian Olfaction, Reproductive Processes and Behavior.* Academic, New York.

Doty, R. L., 1981, Olfactory communication in humans, *Chem. Senses* **6**, 351–376.

Döving, K. B., 1964. Studies of the relation between the frog's electroolfactogram (EOG) and single unit activity in the olfactory bulb. *Acta Physiol. Scand.* **60**, 150–163.

Döving, K. B., 1965. Studies on the responses of bulbar neurons of frog to different odour stimuli. *Rev. Laryng.*, October Suppl., pp. 845–854.

Döving, K. B., 1966. An electrophysiological study of odour similarities of homologous substances, *J. Physiol.* London **186**, 97–109.

Döving, K. B., Lange, A. L., 1967. Comparative studies of sensory relatedness of odours. *Scand. J. Psychol.* **8**, 47–51.

Döving, K. B., Pinching, A. J., 1973. Selective degeneration of neurons in the olfactory bulb following prolonged odour exposure. *Brain Res.* **52**, 115–129.

Döving, K. B., 1987. Response properties of neurones in the rat olfactory bulb to various parameters of odour stimulation. *Acta. Physiol. Scand.* **130**, 285–298.

Dravnieks, A., 1974. A building-block model for the characterization of odorant molecules and their odors. *Ann. N.Y. Acad. Sci.* **237**, 144–163.

Dravnieks, A., 1982. Odor quality: semantically generated multidimensional profiles are stable. *Science* **218**, 799–801.

Dravnieks, A., 1985. *Atlas of Odor Character Profiles.* ASTM data series, DS 61, Philadelphia, PA, 354 pp.

Dravnieks, A., Bock, F. C., Powers, J. J., Tibbetts, M., Ford, M., 1978. Comparison of odors directly and through profiling. *Chem. Senses Flavour* **3**, 191–225.

Dravnieks, A., Masurat, T., Lamm, R. A., 1984. Hedonics of odors and odor descriptors, *APCA J.* **34**, 752–755.

Duchamp, A., 1982. Electrophysiological responses of olfactory bulb neurons to odour stimuli in the frog. A comparison with receptor cells. *Chem. Senses* **7**, 191–210.

Elsberg, C. A., Levy, L., 1935. The sense of smell. A new and simple method of quantitative olfactometry. *Bull. Neurol. Inst. N.Y.* **4**, 5–19.

Fechner, G. T., 1860. *Elements der Psychophysik.* Breitkopf and Härtel, Leipzig, 907 pp.

Frisch, K. Von, 1919. Uber den Geruchssim der Biene und seine blütenbiologischen Bedeutung. *Zool. Jahrb.* **37**, 1–238.

Gardner, R. A., Gardner, B.T., 1969. Teaching sign language to a Chimpanzee. *Science* **165**, 664–672.

Gesteland, R. C., Lettving, J. Y., Pitts, V. H., Rojas, A., 1963. Odors specificities of the frog's olfactory receptor. In *Olfaction and Taste I,* edited by Y. Zotterman. Pergamon Press, Oxford, pp. 19–34.

Grundvig, J. L., Dustman, R. E., Beck, E. C., 1967. The relationship of olfactory receptor stimulation to stimulus environmental temperature. *Exp. Neurol.* **18**, 416–428.

Hara, T. J., 1970. An electrophysiological basis for olfactory discrimination in homing salmon: a review. *J. Fish. Res. Bd. Canada* **27**, 565–586.

Harper, R., Bate-Smith, E. C., Land, D. G., Griffiths, N. M., 1968. Glossary of odor stimuli and their qualities. *Pref. Essent. Oil Rec.* **59**, 22–37.

Harper, R., 1977. Flavour characterization and its development. In *Olfaction and Taste VI,* edited by J. LeMagnen, London, Information Retrieval Ltd., pp. 393–400.

Hill, A. V., 1913. The combinations of haemoglobin with oxygen and with carbon monooxide. *Biochem. J.* **7**, 471–480.

Jones, F. N., 1958a. Scales of subjective intensity for odors of diverse chemical nature. *Am. J. Psychol.* **71**, 305–310.

Jones, F. N., 1958b. Subjective scales of intensity for the three odors. *Am. J. Psychol.* **71**, 423–425.

Jourdan, F., 1982. Spatial dimension in olfactory coding: a representation of the 2-Deoxyglucose patterns of glomerular labelling in the olfactory bulb, *Brain Res.* **240**, 341–344.

Jourdan, F., 1985. L'image des odeurs. *Le Courrier du CNRS* **66-67-68**, p. 76.

Katz, S. H., Talbert, E. J., 1930. Intensities of odors and irritating effects of warning agents for inflammable, poisonous gases. *U.S. Dept. Comm. Bureau Mines, Technical Paper* **480**, 1–37.

Keil, T. A., 1982. Contacts of pore tubules and sensory dendrites in antennal chemosensilla of a silkmoth: Demonstration of a possible pathway for olfactory molecules. *Tissue Cell* **14**, 451–462.

Kobal, G., 1980. Adaptive properties of the electro-olfactogram (FOG) in Man. In *Olfaction and Taste VII*, edited by H. Van der Starre Noordwijkerhout, The Netherlands, p. 209

Kobal, G., Hummel,T., Van Toller, S., 1992. Differences in human chemosensory evoked potentials to olfactory and somatosensory chemical stimuli presented to left and right nostrils. *Chem. Senses* **17**, 233–244.

Kurtz, D. B., Mozell, M. M., 1985. Olfactory stimulation variables. Which model predicts the olfactory nerve response? *J. Gen. Physiol.* **86**, 329–352.

Laffort, P., 1963. Mise en évidence de relations linéaires entre l'activité odorante des molécules et certaines de leurs caractéristiques physico-chimiques. *C.R. Acad. Sci. Paris* **256**, 5618–5621.

Laffort, P., 1966a. Recherche d'une loi de l'intensité odorante supraliminaire, conforme aux diverses données expérimentales. *J. Physiol. Paris* **86**, p. 551.

Laffort, P., 1966b. Les facteurs de l'activité quantitative des molécules odorantes. Thèse Paris-Orsay, 69 pp.

Laffort, P., 1968. Some new data on the physicochemical determinants of the relative effectiveness of odorants. In *Theories of Odors and Odor Measurement*, Proceedings Nato Summer School, Istanbul 1966, edited by N. Tanyolaç. Maidenhead (U.K.) Technivision, pp. 247–270.

Laffort, P., 1969. Température et qualités organoleptiques. *Ann. Nutr. Alim.* **23**, 63–77.

Laffort, P., 1977. Some aspects of molecular recognition by chemoreceptors. In *Olfaction and Taste VI*, edited by J. LeMagnen. IRL Press, London, pp. 17–25.

Laffort, P., Gortan, C., 1987. Olfactory properties of some gases in hyperbaric atmosphere. *Chem. Senses* **12**, 139–142.

Laffort, P., Hoehn, R. C., 1987. *Physiological Mechanisms Involved in Olfaction, Taste, and Other Oral Sensitivities*, edited by J. Mallevialle and I. H. Suffet, Lyonnaise des Eaux and American Water Works Association, 6666 West Quincy Ave., Denver (Colorado), pp. 15–33.

Laffort, P., Marfaing, P., Devos, M., 1988. Olfactory qualitative confusion in Honey-bee. ECRO 8'88 Congress, July 18–22, Warwick, U.K. *Chem. Senses* **14**, 216.

Laffort, P., Patte, F., 1987. Derivation of power law exponents from olfactory thresholds for pure substances. *Ann. N.Y. Acad. Sci.* **510**, 436–439.

Laffort, P., Patte, F., Etcheto, M., 1974. Olfactory coding on the basis of physicochemical properties. *Ann. N.Y. Acad. Sci.* **237**, 193–208.

Le Magnen J., 1947. Étude sur l'excitation olfactive. *C.R. Acad. Sci. Paris* **225**, 1378–1380.

Le Magnen, J., 1950. La discrimination olfactive par le rat blanc de substances à odeurs homologues. *C. R. Soc. Biol.* **144**, 1319–1320.

Le Magnen, J. (Ed.), 1962. Vocabulaire technique des caractères organoleptiques et de la dégustation des produits alimentaires. *Cahiers techniques du CNERNA*, CNRS, 15 Quai Anatole France, Paris, France, 86 pp.

Le Magnen, J., 1965. Les bases sensorielles de l'analyse des qualités organoleptiques. *Ann. Nutrit. Alim.* **19**, A11–A54.

Le Magnen, J., 1976. Olfaction. In *Physiologie II. Système Nerveux-Muscle,* edited by C. Kayser, Flammarion, Paris, pp. 933–984.

Leroy, Y., 1987. *L'Univers Odorant de L'Animal.* Sociète nouvelle des èditions Boubée, Paris.

Moncrieff, R.W., 1957. Olfactory adaptation and odor intensity. *Am. J. Psychol.* **70**, 1–20.

Montagner, H., 1978. *L'Enfant et la Communication,* edited by Pernoud Stock, Paris.

Moulton, D. G., Marshall, D. A., 1976. The performance of dogs in detecting α-ionone in the vapor phase. *J. Comp. Physiol.* **110**, 287–306.

Mozell, M. M., Sheehe, P. R., Swieck, S. W., Jr., Kurtz, D. B., Hornung, D. E., 1984. A parametric study of the stimulation variables affecting the magnitude of the olfactory nerve response. *J. Gen. Physiol.* **83**, 233–267.

Mullins, L.J., 1955. Olfaction. *Ann. N.Y. Acad. Sci.* **27**, 247–276.

Nachman, M., 1963. Learned aversion to the taste of lithium chloride and generalization to other salts. *J. Comp. Physiol. Psychol.* **56**, 343–349.

O'Neill, H. J., Gordon, S. M., O'Neill, M. H., Gibbons, R. D., Szidon, J. P. (1988). A computerized classification technique for screening for the presence of breath biomarkers in lung cancer. *Clin. Chem.* **33**, 1613–1618.

Ottoson, D., 1956. Analysis of the electrical activity of the olfactory epithelium. *Acta Physiol. Scand.* **35**, 1–83.

Patte, F., Etcheto, M., Laffort, P., 1975. Selected and standardized values of suprathreshold odor intensities for 110 substances. *Chem. Senses Flavor* **1**, 283–305.

Patte, F., Etcheto, M., Marfaing, P., 1984. Étude comparative des réponses électroantennographiques de l'Abeille à 59 substances. *J. Physiol. Paris* **79**, 67A.

Patte, F., Etcheto, M., Marfaing, P., Laffort, P., 1989. EAG stimulus-response curves for 59 odorants in the Honey-bee. *J. Insect. Physiol.* (to be published.)

Patte, F., Rouault, J., 1989. Is olfactory space of the "Euclidean type"? ECRO 8'88 Congress, July 18–22, 1988, Warwick, U.K. *Chem. Senses* **14**, 191–192.

Pavlov, I. P., 1928. *Lectures on Conditioned Reflexes,* Translated by W. H. Gantt. Intern. Publ., New York, p. 414.

Pinching, A. J., Döving, K. B., 1974. Selective degeneration in the rat olfactory bulb following exposure to different odours, *Brain Res.* **82**, 195–204.

Plateau, J., 1872. Sur la mesure des sensations physiques et sur la loi qui relie l'intensité de ces sensations à l'intensité de la cause excitante. *Bull. Acad. Sci. Belg.* **33**, 376–388.

Plattig, K. H., Kobal, G., 1977. Olfactory evoked brainpotentials and electro-olfactogram in man. In *Olfaction and Taste VI,* edited by J. Le Magnen. Information Retrieval Ltd., London, p. 203.

Prah, J. D., Benignus, V. A. (1992). Olfactory evoked responses to odorous stimuli of different intensities. *Chem. Senses* **17**, 417–425.

Premack, D., 1971. Language in Chimpanzees. *Science* **172**, 808–822.

Rehn, T., 1978. Perceived odor intensity as a function of air flow through the nose. *Sensory Processes* **2**, 198–205.

Riley, A. L., Baril, L. L., 1976. Conditioning taste aversion: a bibliography. *Anim. Learn. Behav.* **4**, 15–135.

Robert, P. (Ed.), 1973. *Dictionnaire Alphabétique et Analogique de la Langue Française.* Sociéte du Nouveau Littré, Paris, 1972 p.

Rouault, J., Laffort, P., 1993. Le cerveau olfactif utiliserait-il une distance mathématique?, in Les Sciences Cognitives, 4th conference of AIDRI, Lyon, June 11–12, 1992 (in press).

Royet, J. P., 1983. Les aspects comportementaux de l'aversion conditionnée et de la néophobie. *Ann. Biol.* **22**, 113–167.

Royet, J. P., Sicard, G., Souchier, C., Jourdan, F., 1987. Specificity of tial patterns of glomerular activation in the mouse olfactory bulb: computer-assisted image analysis of 2-Deoxyglucose autoradiograms. *Brain Research* **417**, 1–11.

Schaal, B., 1988. Olfaction in infants and children: developmental and functional perspectives. *Chem. Senses* **13**, 145–190.

Schneider, D., 1957. Elektrophysiologische Untersuchungen von Chemo-und Mechanorezeptoren der Antenne des Seidenspinners. *Bombyx mori L.-Z. Vergl. Physiol.* **40**, 8–41.

Schneider, R. A., Schmidt, C. E., Costiloe, J. P., 1966. Relation of odor flow rate and duration to stimulus intensity needed for perception. *J. Appl. Physiol.* **21**, 10–14.

Schutz, H. G., 1964. A matching-standards method for characterizing odor qualities. *Ann. N.Y. Acad. Sci.* **116**, 517–526.

Schwarz, R., 1955. Uber die Riechschärfe der Honigbiene. *Z. Vergl. Physiol.* **37**, 180–210.

Schwarze, C., Kobal, G., 1984. *Olfactory and Somatosensory Evoked Potentials to Different Odorants and Carbon Dioxide.* VIth Congress of ECRO, Lyon, 4–7 September, 1984, p. 103.

Skinner, B. F., 1938. *The Behavior of Organisms.* Appleton-Century-Crofts, New York.

Stevens, S. S., 1936. A scale for the measurement of a psychological magnitude: loudness. *Psychol. Rev.* **43**, 405–416.

Stuiver, M., 1958. Biophysics of the sense of smell. Thesis Gröningen, The Netherlands, 99 pp.

Tanimoto, T. T., 1959. *An Elementary Mathematical Theory of Classification and Prediction.* IBM Program IBCLF.

Tateda, H., 1967. Sugar receptor and α-amino acid in the rat. In *Olfaction and Taste II,* edited by T. Hayashi. Pergamon Press, Oxford, pp. 383–397.

Tucker, D., 1963a. Physical variables in the olfactory stimulation process. *J. Gen. Physiol.* **46**, 453–489.

Tucker, D., 1963b. Olfactory, vomeronasal and trigeminal receptor responses to odorants. In *Olfaction and Taste I,* edited by Y. Zotterman, Pergamon Press, Oxford, pp. 45–69.

Vareschi, E., 1971. Duftunterscheidung bei der Honigbiene: Einzelzell-Ableitungen und Verhaltensreaktionen. *Z. Vergl. Physiol.* **75**, 143–173.

Winter, R., 1978. *Le Livre des Odeurs,* translated from English by M.F. Palomera. Ed. du Seuil, Paris, 173 pp.

Woskow, M. H., 1968. Multidimensional scaling of odors. In *Theories of Odor and Odor Measurements*. Proceedings of the Nato Summer School, Istanbul 1966, edited by N. Tanyolaç. Maidenhead (U.K.) Technivision, pp. 147–188.

Wright, R. H., Michels, K. M., 1964. Evaluation of far infrared relations to odor by a standards similarity method. *Ann. N.Y. Acad. Sci.* **116**, 535–551.

Yoshida, M., 1964. Studies in Psychometric classification of odors (3). *Jpn. J. Psychol.* **35**, 1–17.

Zwaardemaker, H., 1925. *L'Odorat.* Doin, Paris, 305 pp.

CHAPTER

6

Relationships Between Molecular Structure and Olfactory Activity

P. Laffort

6.1 General Points

6.1.1 Strategies for Decoding Olfaction

Returning to the analogy of olfactory communication with a language that was adopted in Chapter 5, one can take an example of written decoding: the Rosetta Stone. The beginning of this chapter is similar to the situation Jean-François Champollion found himself in when he undertook the task of deciphering ancient Egyptian hieroglyphics. He had the use of a text of this type and its Greek translation, that is, its significance (a sacerdotal decree in honor of Ptolemy V). The equivalent is represented by Chapters 1, 2, and 5, data banks for each of the different facets of olfaction and knowledge of the possible content of olfactory signals as well as mechanisms used by living beings to capture and process the molecular messages of odorous stimuli. What remains to be found are the syntax rules of olfactory information, as discussed by Champollion (1824) for written hieroglyphics.

The strategy that seems to impose itself is the establishment of quantitative structure-activity relationships (QSAR). In fact, as will be seen later, the numerous works related to "olfactory decoding" do not all rely on such clear dispositions. First of all, there is the fact that, for certain authors, the "quantitative" part of QSAR is avoided, the studies often remaining descriptive or claiming to be explicative before having proven themselves. There is also the concept of "molecular structure," which does not have exactly the same significance for organic chemists, physical chemists or topological mathematicians. There is equally biological activity, which is sometimes defined in a rather blurred way. In short, one can speak of one possible strategy in the broad sense of the term, but the plural is necessary as soon as an attempt is made to specify the different tracks explored, since they sometimes give the impression of going in different directions. At this point it seems useful to be reminded of the main ones, because the problem has not been resolved at the present time and it is not sure which direction will turn out to be the most fruitful. On the other hand, it can be maintained that the solution will come from a pluridisciplinary approach. Hermann Ludwig von Helmoltz, mid-19th century physiologist and physicist, is certainly among those who have caused the considerable advance made in the connected areas of vision and audition on olfaction. As our knowledge stands at the present time, cooperation should be increased between different research teams who are commonly working in the areas of odors and the sense of smell. This book is a non-negligible contribution to this perspective and perhaps the beginning of greater experimental cooperation.

6.1.2 Olfactory Coding Among Other Biological Coding

Without abusing analogies because they have limits, a last comparison between olfactory communication and language will make it possible to better understand the contradictory core between different "theories of olfaction" described below. Whatever the civilization, the history of human writing shows that it began by being pictographic, that is, each signified object corresponded to a drawing. In evolving, these signs, becoming more and more stylized, were used to designate ideas, then sounds, syllables and finally the alphabetic letters with which we are familiar. Nevertheless, pictograms and ideograms continue to be used, even in the occidental world, such as highway code signs or as "logos" for a number of companies or organizations (the Red Cross, for example).

Genetic coding, in DNA, is obviously alphabetical, since it results from a succession of three-letter (three of the nitrogenous bases A, T, C, G) words (codons). On the other hand, antigen-antibody recognition, based on shape complementarity in the physical sense of the term, is ideographical.

For their relational needs, humans have also invented numbers; it could be said that numerical coding is accomplished by taking only one variable (only one letter) and modulating it intensitively, either in space or in time. For example, such is the case of the image of a "black and white" object on the retina. If the object is colored, the retina image will be made as a function of two criteria (two letters): luminous intensity and wavelength, each numerically modulated. There are human languages of this type: for example, those of the Extreme Orient, where syllables are modulated in four tonal pitches, each indicated by a difference of accent in the alphabetic transcription, when it exists.

Thus, the question is "which type of coding is olfactory coding is related to?" Odorous molecules are too small, at least in gaseous phase olfaction, for the coding to be the alphabetical type. According to some theories, it could be the ideographical type, that is, based on recognition by a relatively high number (about 100) of high specific receptor proteins. On the contrary, olfactory coding could be alphanumerical, that is, based on barely specific affinities with the receptor proteins, bringing a limited number of molecular characteristics (about 10 or less) into play, each one being modulated in intensity. This last example is, as will be seen, the most likely, but it cannot be excluded since the two types may sometimes be jointly implicated.

6.1.3 Difficulties of Bibliographic Structuring

In an attempt to inventory structure-activity relationships in human chemoreception, Beets published a work in 1978 including no less than 492 bibliographical references. Even if one considers that only half of the references implicate olfaction (the other half have to do with taste), this still respresents a large bibliography which, more than ten years later, still remains to be completed. This abundance and the necessity to avoid excluding any approach does not make a structured synthesis easy. Therefore, by necessity, this chapter is presented as much in an enumerative form as a logical connection.

6.2 "Artificial Noses"

The first time the concept and the term were expressed seems to have been in 1923 with Jean Henri Fabre who, in his *Souvenirs Entomologiques,* hoped for the occurence of what he called an "odor radiograph." In fact, keeping in mind current technology in detection and physicochemical analysis, the conception of an "artificial nose" that mimics nature perfectly would not à priori present any problem if olfactory coding rules were known. On the other hand, wanting to conceive this type of equipment without previously having deciphered the code amounts to begging the question. This has not prevented a certain profusion of devices with names that are sometimes quite suggestive (osmopile, electro-odocell, odotron, médor, odorograph, "charm" analyzer, sentor, etc.), which cover different realities.

6.2.1 Ultrarapid and Sensitive Detectors

Independently of its particular coding characteristics, that is, its specificity, two other properties characterize the olfactory system that are only found in part in physicochemical analysis devices: very acute sensitivity (for some substances) and highly rapid response. At the present time, these two properties are antinomic in a bionic perspective, that is, an imitation of living systems. In fact, the lower sensitivity of currently existing physicochemical detectors can only be compensated by previous atmosphere concentrations, which necessarily leads to response delay. On the other hand, the search for a certain specificity such as that coming from retention indexes on chromotographic columns also necessarily leads to a response delay that, much like the previous example, can reach one to several hours. The solution to this last problem seems to reside in the use of a large number of captors placed in parallel and connected to a neuromimetic processing network. This seems to be the objective (probably far from being reached) of a group at Warwick in Great Britain (Shurmer, 1987).

In regard to this ambitious approach, a whole series of organic vapor detector cells must be mentioned, relatively sensitive and with a very low response time, which were suggested in the 1950s and 1960s, like many "artificial noses." These cells having a proper specificity, their use as "objective odor analyzers" can only be accomplished by coupling with chromotographic columns and possible mass spectrographs. It was also at about this period that detectors specific to gas chromotography were being fine tuned (catharometry, flame ionization, electron capture, etc.). Most of the cells proposed as artificial noses, whether in the liquid phase (Berton, 1959; Wilkens and Hartman, 1964) or in the solid phase (Dravnieks, 1965; Tanyolaç, 1965) used modifications of surface electric properties consecutive to organic molecule adsorption. Changing the surface tension of liquids has also been used (Tanyolaç and Eaton, 1950), as well as modification of the thermal conductivity of thermoresistors covered with a liquid film (Moncrieff, 1961; Friedman et al., 1964).

It is certain that the operation of many of these cells presented a spectacular side by itself: placing a Q-tip impregnated with an organic substance near the detector provoked an immediate and reproducible response, more or less characteristic of the substance under consideration and its concentration. Simply bringing a hand close to the detector also provoked a response, though not that of a mineral object. Curiously, between 1965 and 1987, it seems that the ultrarapid and sensitive artificial nose thematic was not the object of publications. Probably an ambitious project such as that of Shurmer (1987) had to use palettes of different types of captors, inspired by those suggested in the 1950s and 1960s to succeed. Nevertheless, one can already predict certain difficulties to be overcome, for example, the fact that liquids covering thermoresistors can be evaporated and that their replacement by polymers or "grafted" silica structures would lead to a lesser discriminatory ability, etc. Close cooperation between technologists and "olfactologists" would very likely contribute to the chances of success for this type of project.

To finish, one should take note of the presentation at the Salon Milipol-Paris 89 (equipment for the police and the army at Bourget, November 7–

10, 1989) of two pieces of equipment constructed by the American firm Thermedics, one of which detects heroin or cocaine ("Sentor") and the other plastic type explosives ("Egis") at the extraordinarily low concentration of 10^{-14} molar fraction in less than 30 seconds. These performances and the announced specificity are indeed remarkable and if they correspond to reality, should turn out to be of highly practical interest. Of course, the firm remains very discreet concerning the implementation of technology. A more recent multielement odor detector based on conducting polymer sensors that has been used as a truffle-hunting device, among other applications, should also be mentioned (Persaud et al., 1991; Talou, 1992).

6.2.2 Physicochemical Analyzer and Biological Detector Coupling

There was at least one scientific instrument company (Delsi Instruments) that commercialized a gas chromotograph outlet accessory called a "heated splitter for olfaction." This device makes it possible for a human subject to characterize the olfactory properties (in intensity and quality) of each of the gaseous effluents as soon as they are recorded by the usual physicochemical detector. In this way, one can obtain olfactively compensated chromotograms, which Dravnieks and O'Donnell (1971) have suggested calling "odorograms."

Before this process became ordinary, it was tested in a rather "handmade" way with a number of variants. The first description is due to Fuller et al. (1964) who recommended a simple derivation by way of dry heat and applied it to the perfume industry. This method was resumed in the food industry (Bayonove and Cordonnier, 1970) and is still used as is at the present time (Etievant et al., 1983). Dravnieks and O'Donnell (1971) suggested mixing the effluent with humidified air at room temperature to avoid irritating the subject's nasal mucosa. In fact, dilution makes it possible to deliver gaseous effluents to the subjects at a comfortable temperature while avoiding condensation at the same time. Another advantage of this process is the increase in odorous flux velocity which constitutes a favorable olfactory condition in spite of concentration decrease (see Chapter 5). Acree et al. (1984) kept the principle of wet dilution while standardizing a method that they called "charm analysis." "Charm" values are chromatographic peaks weighed by their olfactory threshold, which are in turn determined by the successive dilution method applied to the mixture to be analyzed. A microcomputer graphically provides the weighed chromatograms. This procedure is still in use as routine analysis by these authors (e.g., Marin et al., 1988). At this point, the theoretical difficulty associated with the weighing of a concentration by an olfactory threshold, while ignoring the importance of the exponent in the power law for supraliminal values should be noted (see Chapter 5).

The use of an animal (nonhuman) detector at the outlet of a chromato-graph had been implemented for the first time in 1964 by Ottoson and von Sydow. These authors weighed the chromatographic peaks by the amplitude of the associated frog EOG. They suggested calling chromatograms weighed in this way "olfactograms." As a matter of fact, this process never passed the research laboratory stage and was never used industrially. The reason is probably the relatively poor knowledge of the olfactory sensitivity spectrum for the considered animal with respect to human sensitivity rather than the high technical nature of the method. This preliminary step could be the ob-ject of a specific research effort.

6.2.3 "Médor" Type Technical Solutions

Different analytical devices with suggestive names ("Odotron," "Médor," [*Médor* is equivalent to the common dog's name "Rover" in the United States; it is also a contraction of *MEasure of oDOR* (Translator's note)] etc.) were patented during the 1970s whose purpose was to identify and measure the concentration of compounds in the mixture being tested followed by odo-rous efficiency weighing of each of them by referring to a memory (where olfactory thresholds and power law exponents of a certain number of sub-stances are stored) and an "odor addition" depending on this or that "model" (see Chapter 7). This type of apparatus, which is by definition in-capable of "learning," can be suitable for specific industrial environments provided that analysis is limited to a restricted number of compounds. This situation can be encountered in the gas industry, for example (Archis and Charron, 1975).

6.2.4 First Development of Physical Olfactometry

Etcheto (1975) and Laffort (1976) suggested the name "odorograph" for an experimental device allowing the determination of several molecular prop-erties for pure substances that the authors believed to be implicated in ol-factory receptor adsorbability. The apparatus calculated a "predicted human olfactory threshold" from the physicochemical properties thus established. The ins and outs that made it possible to design and operate this apparatus will be exposed and discussed later.

6.3 Principal Theories of Olfaction

Until about 1972, there were a certain number of hypotheses concerning both olfactory coding and its mechanism under the somewhat immodest name of "theories of olfaction" (this is still controversial at the present time, but the expression is no longer in use). Even though these theories lend themselves to being strongly questioned, it is useful to review them briefly, not only in a historical perspective but because some of them have made it possible to establish interesting correlations. In other words, the validity of shown experimental results remains, even if the hypotheses that led to their achievement were partially or totally false.

6.3.1 Radiative Theories

When Fabre (1923) wished for an "odor radiograph," he implicitly admitted that olfactory receptors are sensitive to radiations. Among these theories, it is advisable to distinguish those in which contact between molecules and receptors is necessary from those in which electromagnetic radiation acts at a distance.

6.3.1.1 Radiation at a Distance Theory

This is the oldest theory, since for Aristotle (Treatise of the Soul, about 335 B.C.) only a theory of this type made it possible to explain how vultures could detect a carcass at a considerable distance when no loss of matter could be observed. In the technical context at his disposal, his acute knowledge of the divisible limit of the matter could only come to this conclusion. This theory was conveyed to the Middle Ages by way of Thomas Aquinas' (about 1270) translations and commentaries and taken up again in 1949 by Miles and Beck. These last authors claimed to have observed that honeybees placed in front of two hermetically sealed recipients containing honey showed a significant preference for the one equipped with a window that allowed the far infrared spectrum (between 700 and 1,250 cm^{-1}) to pass. In fact, this experience has never been reproduced, either with the honeybee itself (Johnston, 1953), with human subjects (Shkapenko and Gerebtzoff, 1951) or with frog electrical response (Ottoson, 1956). The scientific community has considered the case closed since 1957 (Thompson).

6.3.1.2 Molecule/Receptor Resonance Theory

This theory was formalized and developed above all by Robert Hamilton Wright beginning in 1954. It is known that when organic molecules are exposed to infrared radiation, they absorb certain frequencies that are characteristic to them, the absorbed energy then being transformed into rotation and vibration movements. Wright's theory supposes that the olfactory receptors selectively enter into resonance with the odorous molecules, this phenomenon being at the origin of the nervous influx of the olfactory pattern (all sensorially equipped animals would emit a sufficient amount of infrared radiation to allow this molecular resonance; this would also occur in the absence of exterior light).

This mechanism hypothesis is perhaps inspired by differential detection of sounds in the part of the inner ear called the cochlea, which functions like a series of specific acoustic resonators. Nevertheless, what is known about molecular interactions does not allow this radiative theory to be accepted as is. Its main interest lies in the important significant correlations that it has made possible, which can be explained perfectly by considering the infrared spectrums for what they are as well: a reflection of molecular structure.

Molecular rotation and vibration movements can also be represented by specific lines in Raman diffusion spectra, which are obtained by exposing the substances to be studied with monochromatic light. The first attempt to link molecular odorous properties and the nature of their Raman spectrum was suggested by Dyson (1937) in a rather unconvincing manner. Within the area defined as "osmic," between 1,400 and 3,500 cm^{-1}, the counterexamples were apparently as numerous as the examples and no definite correlation has been established.

Beginning in 1954, Wright's work was much more structured and evolved, becoming more and more refined for more than 20 years. Figure 6.1 illustrates the method as it has been exposed in this author's most recent publications: on one hand, applied to the odorous quality perceived by humans, and on the other hand, to the power of attraction in different species of insects (Burgess and Wright,1974; Wright, 1981). The procedure can be compared to those that would later be called "expert systems" in artificial intelligence. Absorption spectrums in far infrared (between 100 and 500 cm^{-1}) are established for a certain number of compounds, preferably several multiples of 10, which have the same olfactory nuance in common (in this case, musk,

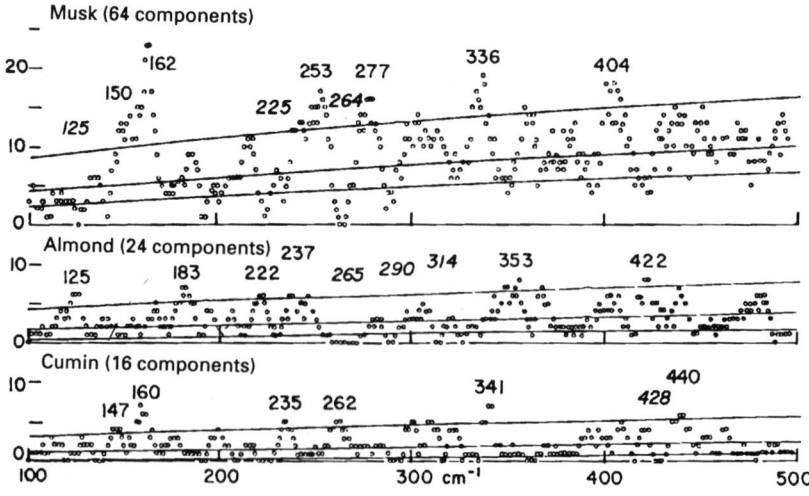

Figure 6.1. Use of the far infrared spectrum in view of olfactory predictability (see explanation in the text) (after Burgess and Wright, 1974).

almond, or cumin). The spectrum is divided into 7 cm^{-1} sections and the number of peaks within each section are counted. The oblique line corresponds to an average number of peaks based on many hundreds of substances chosen at random; it is contained between two other lines that limit the confidence intervals, respectively equal to two standard deviation units toward the top and one toward the bottom. Favorable frequencies (i.e., they are encountered in the spectrums of a large number of substances having the odorous nuance under consideration) and unfavorable frequencies (present in only a few cases) appear on either side of these two areas. When applying a substance to the thus calibrated test, it is sufficient to count the favorable and unfavorable frequencies and "vote" on this basis to estimate the probability that the unknown substance will have the considered nuance. This totally pragmatic procedure leaves the passionate character attached to mechanism theories off to the side; it has allowed fine tuning of nontoxic industrial attracting substances used in the selective biological fight against certain insect species.

In the 100–1,000 cm^{-1} range and by 75 cm^{-1} sections, Raman diffusion spectrums have been used in a slightly different way by an author with nearly the same name as the preceding one: Robert Huey Wright, in collaboration with K. H. Michels (1964) as well as with Schiffman (1974) and Schiffman et al. (1977). These last authors have reconstructed odorous quality similarity matrices (see Figure 5.5) on the basis of Raman frequencies for 19 substances (or 171 distances). They found a correlation coefficient of 0.69 with the experimental matrix, which in this domain can be considered relatively high ($P < 0.001$).

As far as the intensity aspect is concerned, Wright and Burgess (1971) brought forward a very good correlation between the olfactory thresholds of 12 substances and the value of their lowest fundamental frequency (it should be remembered that the total infrared absorption spectrum is made up of fundamental frequencies, harmonic frequencies, and combination bands whose frequencies are equal to the sum of two fundamental frequencies). It is unfortunate that the demonstration only concerned a relatively restricted sample of odorous substances, whereas olfactory thresholds are known for many hundreds of odorants.

Objections that could be made about obtained correlations in the qualitative field concern, first of all, optic isomers, whose spectra are rigorously identical whereas in some cases (but not always) odors differ intensitively as well as qualitatively. As Wright himself observed, there is nothing preventing information coming from spectrum data on the substances to be studied from being completed in other ways of measuring their rotatory power. Another objection concerns hydrogen atom replacement by a deuterium atom which generally results in slight peak shifts without changes in odor quality. In fact, the 7 cm^{-1} bandwidth would correspond to slight peak shifts of this type, which are insignificant from an olfactory point of view.

In summary, limited to its purely experimental aspect, the use of far infrared and Raman spectrums seems to present a certain interest and it is surprising that the procedure has not been taken up by the many authors who have expressed an interest in it.

6.3.2 Thermodynamic Activity Theory

Until the beginning of the 1960s, biological efficiency of a chemical stimulus was often expressed in terms of "thermodynamic activity referred to a pure substance." A ficticious concentration in liquid phase a_i' deduced from the partial pressure P_i (or concentration in vapor phase) and the saturation vapor pressure P_i^0 is expressed by the equation

$$a_i' = P_i/P_i^0 \tag{1}$$

The activity a_i' is not equal to the effective concentration N_i in the liquid phase except for "ideal" solutions, that is, those following Raoult's law. For real solutions, the difference from ideality is given by the activity coefficient γ_i' (equal to 1 in the case of ideal solutions), which is expressed by the equation

$$\gamma_i' = a_i'/N_i \tag{2}$$

These concepts are discussed in more detail in Chapter 4.

Thermodynamic activity was introduced to pharmacodynamics in 1939 by Ferguson. This author noticed that narcotic efficiency thresholds of a large number of substances were grouped in a narrow band of values when expressed in activity, i.e., as a fraction of saturated vapor. He called substances whose thresholds A expressed in this way were between 1 and 1/100 physical or "inert" narcotics and those who stayed outside this range ($A <$ 1/100) chemical or "different" narcotics. The same principle applied to olfaction (Gavaudan et al., 1948; Mullins, 1955; Ottoson, 1958; Higashino et al., 1961) does not allow observation of value grouping, strictly speaking, but of maximum curves for a number of carbon atoms equal to 5 or 6 within a homologous series. As far as olfactory efficiency is concerned, this phenomenon could immediately be interpreted as the occurence of an optimum value for a molecular property related to the size of the molecule when it corresponds to 5 or 6 atoms of carbon.

The overall balance of this approach was established by Laffort (1965), both on a theoretical and experimental point of view. As of today, this approach is totally abandoned as far as the chemical senses are concerned. What can be said about it is that, on one hand, the living environment is very often aqueous and, on the other hand, for lots of chemical stimuli the activity coefficient γ_i' is very often far away from 1 in water such as, for instance, 10^{-6}. Therefore, the a priori assumption that the living environment can be considered as "ideal" as a first approximation and in all circumstances is not theoretically justified. It can possibly be noted a posteriori in very specific cases. This is the case of the honeybee's electroanntenographic response to olfactory stimuli as reported in Chapter 5. This is also possibly the case, at least partially, for the so-called "physical" narcotics.

More generally, an a priori estimate of the concentration in a liquid phase from a known concentration in another phase in equilibrium with the first relies on the use of partition coefficients for infinitely diluted solutions. In practice, their value remains constant for molar fractions less than 1/1,000, which is often the case of biological stimuli. In the special case where one of the two phases is the vapor, the partition coefficient is called Henry's coefficient.

6.3.3 Chromatographic Theory

This mechanism, already suggested in Chapter 1, 1993), is mainly due to Mozell (1964, 1970). This author observed, for the giant frog *Rana catesbieana,* a relatively greater efficiency of hydrosoluble substances on the external branch of the olfactory nerve and a relatively greater efficiency of liposoluble substances on the inner branch of the olfactory nerve, the phenomenon being reversed by inversion of the odorous gaseous flow (retronasal tract). The "hydrosolubility-liposolubility" criteria were determined from retention times in a gas chromatographic column packed with Carbowax 20M. Mozell and Jagodowicz (1973) complemented this electrophysiological experimentation by another quite extraordinary one in which the classic chromatograhic column was replaced by seven in-vivo frogs' mucosa, one after the other. The hydrosolubility-liposolubility criteria or "polarity" criteria thus obtained made it possible to improve the correlation with the anteroposterior sensitivity ratio (The term "polarity," normally used in chromatography, is hardly related to the dipole moment and is defined experimentally.) Later, Laffort and Patte (1987) refined the definition of the chromatographic properties of the frog's mucosa based on these results as well as their own chromatographic work. They found that the mucosa was equivalent to the difference between a polar phase (trimeric acid) and a nonpolar phase (squalane). The acidic character thus defined is especially favorable to the capture of odorous substances such as amines.

The phenomenon implied by the correlation observed by Mozell and coworkers is a time-space distribution of the adsorption of odorants on olfactory neuroreceptors. This distribution evolves according to the classical scheme of chromatography: a barely soluble molecule in the liquid phase remains available for neuroreceptors located downstream of the gaseous flow. It reaches these "downstream" neuroreceptors more quickly than a very soluble molecule that tends to be held by upstream neuroreceptors. This phenomenon must have an influence on the intensive performance of odorants (see Chapter 5). On the contrary, according to Laing (1988), interactions in mixtures do not seem to occur due to a difference in the distribution of hydrophilic and lipophilic substances on the mucosa. This is contrary to a hypothesis previously published by the same author (Laing, 1987).

The question that remains is to know whether or not the observed chromatographic phenomenon is involved in qualitative discrimination of odors. As explained in Chapter 1, the scientific community is rather reserved about the subject. First of all, there is everyone's everyday experience: for instance, a lemon or anis odor once sniffed remains a lemon or anis odor in the mouth, even though Mozell (1973) believed that this so-called sensorial evidence deserved psychophysical experimentation. There have been a certain number of electrophysiological experiments performed on salamanders (Mackay-Sim et al., 1982) in which the chromatographic phenomenon was supressed by local olfactory stimulation close to the electrode. An important topographic specificity of the mucosa was pointed out, which corresponds to a hydrophilic-lipophilic antero-posterior "chemotopy" consistent with that found by Mozell and co-workers by normal nasal stimulation and reversed with respect to that found in the retronasal tract. In other words, everything occurs as if in batrachians, for outward olfaction (as opposed to food palatability), the relative specialization of the neuroreceptors (inherent chemotopy) is such that it optimizes the responses to a passive molecular distribution due to the chromatographic phenomenon (imposed chemotopy). Latest findings by Mozell et al. (1987) are consistent with this scheme.

In conclusion to this set of works, it can be stated that they represent a rather good demonstration of environmental adaptability (probably genetically determined), as well as a thorough knowledge of the physicochemical properties of the frog's olfactory mucosa. In order to extrapolate these findings to human olfaction, it would certainly be very interesting to have some measurements of the chromatographic properties of small mammals' olfactory mucosa. On the other hand, the role played by the chromatographic phenomenon in qualitative discrimination is very unlikely. Even if it were otherwise, this phenomenon could only partially describe discrimination, which excludes the application of the terms "chromatographic theory" or "chromatographic model."

6.3.4 Membrane Penetration Theory

When Hodgkin and Katz began working on giant squid axon in 1949, they established that the intracellular potassium level was higher than the sodium level, which is contrary to what occurs on the outside of the cell. This unequal ion distribution across the membrane is maintained by an energetic process: the "sodium-potassium pump." These authors have also shown that nerve cell excitation consists of a sodium ion intake followed by a potassium ion loss.

Davies (1953) used this mechanism to support an olfaction theory based on odorous molecules' more or less pronounced aptitude to diffuse through the neuroreceptor membrane. The canals thus formed by odorous molecules in the membrane close slowly, allowing ions to pass, triggering the nerve impulse. At the present time it is known that odorous molecules do not penetrate the membrane to form a canal. Their action is external: they are adsorbed on a receptor site which is functionally coupled with an ionic canal (see Chapter 1).

The main interest of this rather old work, now outdated, is that starting from a complicated theoretical formalism which will not be discussed here, it is at the origin of the first predictive model of human olfactory thresholds (Davies and Taylor, 1959). Later, using the same principle Davies (1965) would develop a less convincing qualitative model, due to the fact that he did not apply it to experimental data. These works were regrouped and synthesized by Davies (1971).

The two parameters that Davies and co-workers considered to be dominant were, on one hand, the size of the membrane canal, thought to be in proportion to the section of the molecular model when the molecule is oriented in a hydrophilic-lipophilic direction and, on the other hand, diffusion rate through the membrane, thought to be in proportion to the desorption free enthalpy in the air starting from a water-oil interface.

As far as this theory not recognizing the possible existence of more or less specific receptor proteins, the scientific community is in agreement in considering that it cannot (like the chromatographic theory) take qualitative discrimination into account. On the other hand, in the area of odorous intensity, one can find elements of hydrosolubility, liposolubility, molecular size, and partition coefficients, equally present in more recent predictive models. In this latter model, there is the additional concept of polarizability, which alone can take the higher olfactory efficiency of certain small odorants (notably sulfurous molecules) into account.

6.3.5 Primary Odors Concept

Even though Amoore (1970) was not the only author to have defended the primary odor concept, his name remains associated with this mechanism hypothesis, through what he called the "stereochemical theory of olfaction" and "specific anosmia," two totally different approaches that nevertheless overlap, according to the author. They will be discussed separately.

6.3.5.1 Stereochemical Theory of Olfaction

The hypothesis of a link between molecular shape and their odor was, according to Lucretius (47 B.C.) and already formulated by Epicurus, pungent odors corresponding to sharp shaped molecules (in fact, called "atoms") and sweet and agreeable to round shapes (incidentally, the difference in the olfactory stimulation concept for Epicurus and Aristotle, respectively, "atomist" and "radiative," should be noted).

More recently, the hypothesis according to which odorous quality would depend as much on molecular shape as on functional grouping seems to have been expressed for the first time by Pauling (1946), but it was Amoore (1952) who formalized and then developed the concept over fifteen years by applying it to important experimentation (Amoore et al., 1967). According to this theory, seven types of "receptor sites," located in the neuroreceptor membrane and functioning as "locks," would be accessible to as many "keys"— the seven primary odors. Appropriateness between these "locks" and "keys" is established by complementary shape for five of them and by electrophilic or nuclephilic character for the other two. All odorants belong, more or less, to one or the other categories or to several primary odors, and would thus be susceptible to adsorption in variable proportions on each of the seven types of receptor sites.

As a first step, Amoore built receptacles in molded material adapted to each of the five substances he considered as best representing the five first types of primary odors. The filling rate of these receptacles by molecular models of substances to be tested (estimated by water displacement in the receptacles) was considered in proportion to shape similarity. As a second step, Amoore and co-workers evaluated shape similarities by covering rates of projected shapes of molecular models according to three orthogonal axes. Finally, as a third step, this last method was computerized with the help of a machine called PAPA (Probalistic Automatic Pattern Analyzer). The substances that were considered as best reflecting each of the primary odors changed during these steps.

The objections that can be made to this theory are numerous. The first is that there are many examples of substances that have similar shapes and very different odors because of a difference in the functional group and, on the other hand, some substances have similar odors and totally different molecular shapes. One of the most recently cited examples of this last type of phenomenon is the case of ortho-amino acetophenone, which, according to human sensorial appreciation as well as honeybee behavior, mimics keto-9-decene-trans-oic acid, the main component of royal pheromone, quite well (Southwick, 1989). Of course, the shape and molecular structure of these two substances are totally different.

Without going into the details of the objections made by different authors, the correlations given by Amoore and Venström (1967) between shape similarity (in the most recent version of the method) and odor similarity for 107 odorous substances show both the interest and limitations of such an approach. The correlations are situated between 0.52 and 0.66, dimensions already mentioned in connection with Wright's theory, which will be referred to later as far as other molecular characteristics, that is, they are quite significant, but not at all "predictive."

The main reproach that I would make regarding the different methods of molecular shape similarity evaluation recommended by Amoore and co-workers is that in reality they give, in unknown proportions, a mixture of shape and size similarity and that this last factor is recognized by most authors as having a definite impact on olfactory coding, intensive as well as qualitative. To fully justify these methods, the supplement of information that they provide should have been evaluated with respect to simply taking molecular size into consideration. Finally, functional groups taken into account by this author uniquely for "pungent" and "putrid" forms (and, in part, "minty") are obviously tributary to the molecular shape for their efficiency—the well-known "steric hindrance"—but there are much more precise physicochemical methods for this purpose that will be mentioned later.

As far as the number of the seven primary odors is concerned, Amoore himself had to abandon this idea beginning in 1968, as will be further discussed. The most regrettable part of this moment in olfactory history was its extraordinarily mediatized character during the 1960s and the consequences of false scientific information that continued for a long time afterwards. In France, for example, the very serious International Encyclopedia of Sciences and Techniques (Encyclopédie internationale des sciences et des techniques, Morvan, 1972) considered this theory of seven primary odors as the only valid one. Therefore, in the recent past (Delsaux, 1985), a journal for high school students known as *Phosphore* took up the same idea and a writer even used this "theory" as the subject for a novel. Results as fragile and partial as these should have been accompanied by more modesty in their publicity with respect to the nonspecialist public.

6.3.5.2 Specific "Anosmia"

In the strict sense of the term, anosmia is a total absence of olfactory sensitivity and specific anosmia is a total absence of sensitivity to one odorant or a restrained group of odorants. The first deficiency really exists, but, concerning the second, generally only partial olfactory defects are noticed. It would therefore be preferable to use the term "specific hyposmia."

Guillot (1948a and 1948b) was the first to study specific hyposmias for pure substances and to suggest using these sensorial deficiencies to determine what he called "fundamental odors." His observations were at the origin of one of the most important criticisms of Amoore's first theory, which was based on the hypothesis of characteristic receptor sites for a given odorous quality. There are three known families of substances that have a musky odor: the macrocyclics, the steroids, and the nitrocompounds. Among other examples of dissociation between specific hyposmia and quality, Guillot observed specific sensitivity defects to the first two groups of musks associated with a normal sensitivity to the third group. Perception of musky odors therefore cannot correspond to the same "fundamental" at the receptor level.

However, it was Amoore, beginning in 1968, who experimented with Guillot's suggestions on a grand scale. Figure 6.2 illustrates the method: for certain subjects, olfactory deficiencies are found to be that much more important when the tested substances approach isovaleric acid both by the number of carbon atoms and functional groups. From this, the author deduced that isovaleric acid corresponds to a primary odor. The number of primary odors thus determined by Amoore and co-workers (Amoore, 1977; Pelosi and Viti, 1978; Pelosi and Piranelli, 1981) would be, at the present time, 8 out of a total that they evaluate at 32 (sweat, sperm, rotten fish, malt, urine, musk, menthol, and camphor, respectively represented by isovaleric acid, 1-pyroline, tributylamine, isobutyraldehyde, androstenone, pentadecanolide, L. carvone, and cineole). The objective proposed to the international community would be to finish indentifying the totality of the 32 primary odors followed by establishment of shape similarity for substances to be tested with each one of them, which would somehow form an "olfactory pattern" on a molecular basis.

At the present time, objections that could be made to this approach come more from convergent unfavorable arguments rather than evidence, strictly speaking. These arguments are quite numerous and have been the subject of a synthesis by Bernuzeau (1986); two of them will be discussed here. The first objection comes from "olfactory patterns" obtained from electrophysiological neuroreceptor responses in vertebrates (see Chapter 1). It cannot be excluded that this low cellular specificity is the result of the juxtaposition

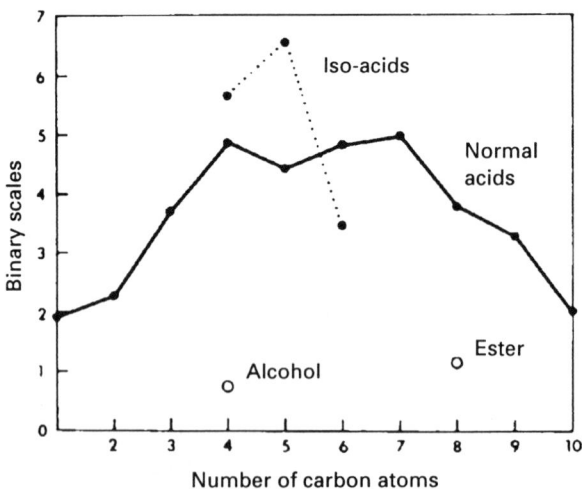

Figure 6.2. Olfactory defects (in terms of threshold concentrations) for a group of subjects initially found to be hyposmic to isobutyric acid, after Amoore (1968). According to the author, the results show that isovaleric acid corresponds to a "primary" or "fundamental" odor.

of highly specific proteins in variable proportions within each receptor cell. However, this remains to be demonstrated and seems unlikely in any case. In fact, at this time and in spite of many attempts, none of these hypothetically highly specific proteins has been identified by concentration and protein purification techniques. This last point will be discussed again later.

The second argument against the concept of primary odors concerns the use of olfactory thresholds and not those of perceived intensities to study olfactory deficiencies. Figure 6.3 illustrates how different values of the power function exponent can lead to considerable differences in thresholds for a nonspecific hyposmia. In fact, the concentration scale being logarithmic, a slope twice as low for substance A than for substance B, for example, would indicate a deficit 100 times larger.

A bank of standardized power function exponents concerning 191 odorants (Devos et al., in preparation) will provide the beginnings of experimental justification to this hypothesis. In fact, among the olfactory sensitivity deficiencies observed by Amoore and co-workers, one of the most spectacular concerns trimethylamine (threshold ratio equal to 830). This deficiency diminishes regularly for tertiary amines along with the lengthening of the carbon chain to be practically indistinguishable from a background starting at C9 (Amoore and Forrester, 1976). For that matter, what does a study of standardized exponents show? A regular value increase for tertiary amines as a function of chain lengthening, which goes from 0.23 for trimethylamine to 0.85 for C12 and 0.54 for C6. In other words, nearly constant olfactory

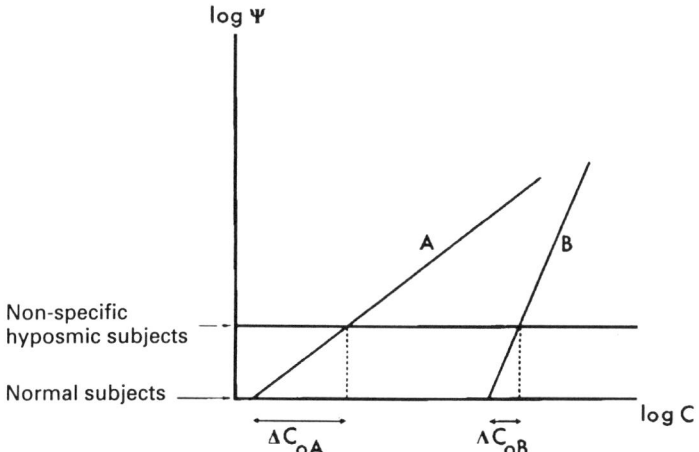

Figure 6.3. Log-log coordinate representation of perceived odorous intensity as a function of concentration, for substances with different Stevens' slopes. The sensorial defect ΔC_0 between nonspecific hyposmic subjects and normal subjects, when it is expressed in threshold concentrations, is noticeably higher for low slope substances than for high slope substances.

defects are noticed, or at least much lower fluctuations, when they are corrected by standardized power function exponent values.

A generalization of this observation would make it possible to bring the concept of primary odors totally in question. For example, the present case would no longer be hyposomia to trimethylamine but the totality of tertiary amines, or perhaps even in a more general way, to the "proton recepetor" character. There would not be "primary substances" but "primary properties." Such is the debate referred to in the introduction to this chapter concerning the "alphanumerical" and "ideogrammatical" coding concepts.

Another experimental argument which would encourage substituting intensity measurements for threshold measurements concerns the variation of olfactory sensitivity in women during the ovarian cycle (see Chapter 2). This phenomenon, demonstrated for the first time by Le Magnen (1952) by threshold measurement, seemed to be specific for one of the studied substances, pentadecanolide, which has a musky odor, variations for other sustances with nonmusky odors being much lower (safrol, gaiacol, amyl salicylate, eucalyptol). In fact, a behavioral response experience with female rats by Pietras and Moulton (1974), not in terms of concentration thresholds but in response percentages to given concentrations (analogous to intensity evaluations), has thoroughly confirmed cyclic sensitivity variations, but has not made it possible to establish a difference between pentadecanolide and other substances with nonmusky odors (cyclopentanol, eugenol, α-ionone).

The restrictions on the primary odor concept having been formulated, olfactory decoding by studying specific hyposmias is nevertheless a very interesting approach that could be improved by the use of perceived intensities instead of olfactory thresholds, as has been seen. However, its implementation remains difficult: some human specific hyposmias are only observed in 2% of the population and animal experimentation depends on mutated strains of specific hyposmias, which are quite rare. Studies that can be cited include those carried out by Wysocki et al. (1977) and Price (1977) on mice and Venard and Pichon (1984) on fruit flies, which remain preliminary for the moment.

6.3.6 Other Theories

The above list of "theories of olfaction" is not exhaustive, strictly speaking, but it does cover nearly all those that constitute an attempt to quantatively decode olfaction. The others have more to do with mechanism hypotheses and, because of this fact, would be more appropriate either in Chapter 1 or in another chapter with a purely historical character, which is not included in this book. However, one of them will be mentioned, under the name "immunochemical theory," even though the reporting authors have never used this expression.

As has been seen in Chapters 3 and 5 the pleasant-unpleasant components of certain odors has been the object of numerous debates on the subject of its innate and/or acquired character, which are not completely closed at the present time. Another aspect of the question is to know how much olfactory sensitivity itself, in its intensive and qualitative components, is genetically transmitted or acquired by an individual.

The first experimental work related to this problem was performed by Le Magnen (1949) who showed that intramuscular injection of odorous substances into human subjects increased their olfactory sensitivity to those substances (amyl salicylate, eucalyptol, camphor, ethylic ether) in a durable way. The explanation hypothesis was that odorous substances could behave like antigens, that is to say, "fabricating" olfactory receptor proteins by imposing their imprint on previously nonspecific proteins.

Forty years later, Wysocki et al. (1989) seem to have confirmed this point of view even though they apparently had no knowledge of Le Magnen's work. These authors succeeded in lowering the olfactory threshold for androstenone for nearly half of the subjects tested, previously hyposomic to this substance, by exposing them to its odor over a period of six weeks for three times 3 min. per day. On the contrary, they did not observe any modification of the amyl acetate threshold. Thus, these very interesting results do not allow the conclusion that olfactory sensitivity results uniquely from an acquisition process. Besides, good common sense is in conflict with this idea since unknown odors are detected at their first presentation.

Another problem concerns cognitive (identification) or affective (emotional) processing of information provided by the odorous molecule. Significance is generally acquired when a previously unencountered odorant becomes a symbol during a learning process. On the other hand, it is known that fear behavior can be triggered in animal species brought up in laboratories for several generations, who therefore have not been in contact with their natural predators, by exposing them to the odor of these predators (Vernet-Maury et al., 1968). This type of behavior necessarily involves a genetically transmitted aptitute to recognize the informative content of the signal. Also Beauchamp's surprising findings demonstrating a tight relationship between the individual odor of a mouse and its histocompatibility major complex (HMC), which is equivalent to human HLA, should be noted. Therefore, it seems cautious to conclude that innate and acquired processes are both involved in olfactory information without being able to specify in which respective proportion they occur.

6.4 Current Pragmatic Approaches

In opposition to ambitious attempts constituted by the theories to suggest global explanations, a certain number of pragmatic approaches to relationships between structure and activity were developed and overwhelmed the theories with time.

6.4.1 Classic Organic Chemists' Approaches

As it has been seen in Chapter 4 and as will be seen again here, for physical chemists what characterizes a substance is the set of its properties, or "potentialities" of the substance when placed in a given environment. They name these potentialities chemical potential, intermolecular forces, energy, entropy, solubility parameters, activity coefficient, etc. It is very different for organic chemists, for whom a molecule's "signature" is its structure in the most basic sense of the word, that is, the space arrangement of its constitutive atoms. Organic chemists who became interested in the structure activity in olfaction did not escape this approach. An example illustrates this statement rather well: in the already cited work on structure-activity relationships in human chemoreception, Beets (1978) included no less than 814 figures and tables among which more than 90% are related to developed chemical formulas. Even though this book is rather "ecumenical," since all approaches are treated, Beets' pedagogy is affected by the fact that he is himself an organic chemist.

Another characteristic of these organic chemists involved in olfaction is that they almost all come from the aroma and perfume industry. Thus, for these researchers, the objective is first to observe whether this or that limited structural modification of a given substance improves or deteriorates its olfactory properties, then to attempt to predict the phenomenon. The olfactory properties are generally assessed according to a protocol from the perfume industry: the substance, after dilution in a solvent, is presented on a strip of filter paper called "mouillette," which is sniffed most often after evaporation of the solvent. This process, very logical from an operational point of view, stops being very sound when attempting to deduce information on olfactory stimulation, as changes in the molecular structure of the odorous substance can influence not only intramolecular forces binding this molecule to its physical support or to the solvent, but mainly its volatility. Odorousness of so-called odorless substances (Stoll, 1965) has been demonstrated this way without difficulty by confining a few drops of the substance in a closed 1 L recipient for 24 h. The ratio of hyposmic subjects to one of two very structurally close substances (two epimeric compounds of sclareol-lactone) was found in direct relationship to the difference in their fusion temperature (Laffort, 1968; Stoll, 1968).

Despite this limitation in principle, the very pragmatic approach used by the organic chemists made the synthesis of numerous substances possible with interesting odorous properties. Only a few general rules came out of this abundant work. Ohloff's "triaxial" rule (1971) can be cited, according to which the necessary and sufficient condition for a compound to present ambergris' six characteristic nuances is to be based on a transdecaline system whose substituants 1, 2, and 4 are axial, one of them including oxygen (see Fig 6.4).

Of course, the success of this rule in this very specific case emphasizes the remaining work to be done: several hundreds of similar rules would be

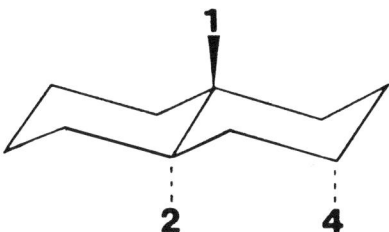

Figure 6.4. Diagram illustrating the (1,2,4) triaxial rule according to Ohloff (1971) (in this case, the molecular axis coincides with the vertical axis), so that an organic compound has the odor of ambergris (see text also).

needed to cover the whole domain of olfactory nuances. Organic chemists' works also point out, in partial opposition with Amoore's, that functional groups and spatial molecular geometry are both of the utmost importance in their interaction with olfactory receptors (Beets,1968).

6.4.2 Biochemical Approaches

Three main routes are being explored by biochemists.

6.4.2.1 Biochemical Mechanisms of Ionic Membrane Canal Openings

A large part of the present research in biochemistry aims at knowledge of the mechanism of information transduction by receptor proteins, which are included in the neuroreceptor membranes (with an intracellular and an extracellular part) that trigger the opening of an ionic canal when an odorous molecule adsorbs on them. According to the most recent surveys (Lancet and Pace, 1987; Chanel, 1987; Shirley and Persaud, 1990), this occurs within an enzymatic "cascade" in which an intracellular protein (second messenger or protein G) plays an important role in the release of cyclic AMP (adenosine monophosphate) and/or IP3 (inositol triphosphate), which in turn triggers the canal opening, the first step in the electrophysiological transduction process. As can be seen, this is dealing with stepwise mechanisms transforming a chemical signal into an electric signal. These mechanisms, at the molecular level, are not directly relevant to this chapter and will not be developed further. Besides, they have already been suggested in Chapter 1.

6.4.2.2 Isolation and Identification of Receptor Proteins

This theme would be of the utmost importance if worked out completely in the establishment of a basis for in vitro olfactory patterns. Unfortunately, all attempts carried out to date have failed, at least as far as the isolation and the identification of highly specific proteins are concerned. The first attempt was performed by Ash (1968) who would have isolated a protein specific of lavander odor by the grinding of rabbit olfactory mucosae followed by fractionation. This result has not been confirmed. Since then, a certain number of authors have claimed the occurence of proteins extracted from olfactory mucosa of several animal species, capable of binding to odorous substances [odorant binding proteins (OBP)]. Several classes of techniques were used: UV spectrum modification, enzymatic activity modification, affinity chromatography, radioactive labeling, manufacturing of poly- and monoclonal antibodies used as electrical response inhibitors, etc. These experiments were performed on dogs, rabbits, rats, frogs, cows, sows, and mice and the isolated proteins were found to display some binding potential to androstenol, anisole, benzaldehyde, camphor, pyrazines, and thiazoles. From all these works, the most recent being Bignetti et al. (1987), Anholt (1987), and Price and Willey (1988), it can be outlined that the specificity is far from what was primarily expected as substances different by their odor and structure, such as, for instance, different ketones, amyl acetate, and anisole, bind to the same protein. Furthermore, the role played by OBP as receptors is far from being demonstrated, even though they are not found in the respiratory mucosa: it is at least debatable. They could be proteins in charge of carrying the odorous substances through the mucus in steps preceding and/or following the excitation. Its role in the latter case would essentially be "house keeping." In conclusion, it can be said that the content of this paragraph bears lots of promising perspectives, but that results are, for now, quite limited.

As expressed by Le Magnen in the preface of the present book, great expectations are placed in the knowledge of gene encoding for the synthesis of the olfactory receptor proteins. According to Buck and Axel (1991), that should be effective for 18 different members of an extremely large muligene family that encodes seven transmembrane domain proteins, whose expression is restricted to the olfactory epithelium. However, as of today, the following step consisting of values of relative binding of these proteins with sets of odorants in order to generate in vitro olfactory patterns, is not yet given. This very interesting study must therefore be considered as preliminary for the moment.

6.4.2.3 Modifications of Mucosa Properties

Starting from the hypothesis that the membrane of each cell bears several receptor proteins each accepting different odorous substances, two types of modifications of olfactory mucosa properties have been used so far in order to experimentally provoke specific hyposmia.

In the first method, an odorous substance is used as a "shield" for receptor proteins on which it adsorbs while a "protein poison" is applied to alter all the others. After rinsing, it is shown that only the substances close to these used as "shields" can trigger an unchanged global electrical response from the mucosa (EOG). This method has been suggested and used for the first time by Getchell and Gesteland (1972) on frogs, with butyl *n*-butyrate as a "shield" and *N*-ethylmaleimide as "poison." The electric response was unchanged for butyl *n*-butyrate itself and methyl *n*-butyrate, but strongly altered for cis dichloro-1,2 ethane, anisole, L limonene, and ethyl acetate. This method raised great hopes when published but, curiously, it has not been used very much since (Menevse et al., 1978; Delaleu and Holley, 1980).

In the second procedure, there is an attempt to selectively prevent the activation of receptor proteins and the responses are compared for selected series of odorous substances before and after deactivation. There is no longer shielding by odorous substances. One must look for partial and different "bans." Reactive odorous substances, lectines, and UV radiations have been used for this purpose (see the review in Lancet, 1984). The use of lectines have been the object of recent important developments: the effect of concanavaline A on rat's EOG has been measured for 129 odorous substances among which 8 have been each presented at different concentrations (Shirley et al., 1987a and 1987b; Polak et al., 1989). Processing of these results only allows for a few general trends: as far as the length of the carbon chain is concerned, a maximum EOG amplitude deviation occurs between C4 and C6 and among families of substances, thiols, carboxylic acids, and hydrocarbons are the most affected.

It looks as if, through one or the other of these methods, the modification of the properties of the mucosa by reagents remains a very promising exploratory technique.

6.4.3 Exclusive Use of Multidimensional Analysis

In Chapter 5, it has been seen that odorous quality could only be numerically characterized by rectangular or triangular matrices, triangular matrices being either derived from rectangular ones or obtained experimentally. The normal procedure for establishing quantitative structure-activity relationships in the qualitative domain is, needless to say, to make two similarity or distance triangular matrices coincide, one being of an olfactory nature and the other of a molecular nature. As a matter of fact, several authors have limited themselves in a first step of determining the olfactory matrix and submitting it to a multidimensional analysis, that is, to recreate a virtual space in which distances are best respected (upper part of Figure 5.5). Purely phenomenological descriptions are thus obtained that sometimes enable the experimenter to find certain analogies. Thus, in experiments on frog neuro-receptors referred to in Chapter 1, a group of substances presenting the occurence of a benzene ring in their formula as being in common was pointed out (Duchamp et al., 1974). However, the other groups pointed out in this publication and the set of related works (original and literature data in Sicard and Holley, 1984) did not make it possible to suggest other common molecular characteristics until Laffort (1977), Chastrette (1981), and Eminet and Chastrette (1983) proceeded to establish proper quantitative structure-activity relationships.

In practice, all investigators cited in Chapter 5 as having established similarities and odor profiles submitted their data to multidimensional analysis. This method has been also used in data processing in the food industry where useful relationships were found, for instance, between certain methods of wine making and organoleptic properties (Etievant and Issanchou, 1987). In the perfume industry, this method allowed for the grouping of olfactory nuances in larger sets: fat notes, sweet, agressive, round, lively, etc. (Doré et al., 1984, Jaubert et al., 1987).

6.4.4 Use of a Limited Number of Molecular Parameters

This section corresponds the most to the general title of this chapter, as the structure-activity relationships are quantitative. The works that refer to this section have a certain parenthood with Hansch's, well known in pharmacology, even though authors of the two fields often ignored each other. Parameters that are used are either empirical, experimental, or theoretical. They have been applied to narrow families of compounds as well as other types of compounds, to although the two aproaches have very different angles of interest. Several data processing tools are used for this purpose. In the intensitive field, the stepwise multilinear regression is the most important tool when a variable that is sought to be predicted, an olfactory threshold, for instance, is compared to molecular parameters supposed to be pertinent. Then the most efficient of them are automatically selected by the program. In the qualitative field, the principles of software packages have been explained in Chapter 5 (Figure 5.5) (see also Schiffmann et al., 1977). Discriminant analyses are also used that make it possible to classify a set of compounds in several categories (generally 2: occurrence or absence of odorous note).

6.4.4.1 Hansch-like Parameters

According to Hansch (1969, 1973), the response of a biological system to a biologically active agent is mainly the function of three properties of this agent: a hydrophobic factor, a steric factor, and an electronic factor. Other factors such as those of hydrogen bonding or polarizability can be added in specific cases.

- Among the hydrophobic factors, the most used is the octanol-water partition coefficient P. Nonpolar phase chromatographic retention indexes, parachor, and surface active potential are used as well. (Parachor and surface active potential are both related to surface tension, but in a different way.)

- Among steric parameters, a variable corresponding to the size of the molecule (molecular volume or Van der Walls' mean radius) or an index corresponding to its nonsphericity (Van der Walls' radii along three orthogonal axes, for instance) are used.
- Among the empirical electronic parameters, the most known are "Hammett's constants," related to the acid dissociation constant pKa.

At least one can calculate the molecular electronic parameters, that is, their capacity to yield or accept an electron, through quantic chemistry calculations. With the exception of the latest, all the useful parameters can easily be obtained from physicochemical constant handbooks.

In practice, only a few authors established quantitative correlations in olfaction, explicitly referring to Hansch's method. Among them, Wolkowski et al. (1977) and Greensberg (1979, 1981) regressed olfactory thresholds between the octanol-water partition coefficient P for restricted families of compounds, most often belonging to homologous series. As a matter of fact, this type of work is hardly useful, as any property related to the molecular size (as is the case for the partition coefficient P) would have given the same result. Boelens (1976) found a significant correlation between the more or less pronounced odorous character of two families of compounds with a musky odor and an almond odor, respectively, and an equation involving the octanol-water partition coefficient P and a steric character parameter obtained from molecular models.

When the historical succession of correlations found in olfaction with respect to physicochemical parameters is examined, it is found that they are consistent with Hansch's general concept, especially with the extended five parameter version (hydrophobic, electronic, steric, hydrogen bond, polarizability.

- The first correlation between olfactory properties and molecular properties was suggested by Davies and Taylor (1959), whose work has already been mentioned concerning the membrane penetration theory. The two molecular properties used in this work were a steric index obtained from molecular models and a hydrophobic index (a partition coefficient between air and oil-water interface).
- In 1963, for thresholds, Laffort took into account a hydrophobic index (air-water partition coefficient), a steric index (molecular volume), and an electronic coefficient (pKa).
- In 1964, Beck established a relationship between olfactory thresholds and a "dynamic surface" concept derived from molecular models, but only for homologous series, which limits the applicability of this concept.
- Laffort (1969) and Laffort and Dravnieks (1973) introduced for the first time a hydrogen bond index and a molecular polarizability index that they called ε as in electronic, in their later publication as this factor is related to the possibility of giving electrons. These new factors were added to the

preceding factors (air-water partition coefficient and molecular volume). They applied these parameters to olfactory thresholds and power law exponent predictibilities.

- In 1981, Chastrette suggested taking into account the molar volume, the volumic polarizability, the octanol-water partition coefficient, a steric coefficient derived from molecular models, and the electron acceptor-donor properties to represent the results found by animal experimentation by Duchamp et al.(1974) involving 20 substances.
- In 1982, Wright proposed an interesting relationship between the power law exponents and the parachor.
- In 1983, Eminet and Chastrette found a very good consistency between the "camphoraceous/noncamphoraceous" character of 98 substances, taking into account only 2 of the 6 parameters defined in 1981 by Chastrette: the steric factor and the molar volume (10% of the substances were "misplaced").

6.4.4.2 Pattern Recognition Techniques

This is a set of computer techniques that make it possible to recognize and to discriminate objects according to a set of criteria. "Pattern" is taken here in its broader sense; it can be applied to anything. The molecular pattern or shape can be extended from the molecular geometrical contour to the set of parameters describing all useful molecular properties. In practice, since 1974, a tendency among authors to have increased the number of molecular parameters that are "tried" then "accepted" to be influential on odorousness has been noticed.

Thus, Döving (1974) started from 19 physicochemical characters to finally keep 11 of them in order to represent a relatively small olfactory similarity matrix (11 × 11). Dravnieks (1974) started from 11 parameters to finally keep 3, 4, or 5 of them according to the considered olfactory data sets. In 1977, he examined 118 indexes to finally keep 14 of them in order to describe supraliminal odorous intensities. Recent investigations by Jurs and co-workers (Ham and Jurs, 1985; Narvaez et al., 1986; Edwards and Jurs, 1989) respectively keep 13 molecular descriptors out of 18, 14 out of 47, and 7 out of 14.

What is the nature of these indexes? Physicochemical, quantic, geometric parameters used in Hansch-like methods are found, but in a greater number. Molecular fragments and topologic indexes are added to the list. An exhaustive analysis of these latter indexes have been recently proposed by Loukianoff (1988). Molecular fragments were introduced in olfaction by Dravnieks (1974,1977); their use in pharmacology have been widespread since Wissevesser line notation (WLN) of chemical formulas (Ash and Hyde, 1975). Stuper and Jurs (1976) designed a computer program named ADAPT that automatically generates sets of fragments called substructures starting from a given molecular structure. These substructures are such that pattern recognition techniques can be applied to them. Among molecular topology approaches, connectivity indexes have a special place. These indexes are related to the way the atoms are connected to each other within a given molecule. Thus, in a cyclic or in a branched molecule, an atom will be assigned a strong or weak connectivity index, respectively. These indexes, often strongly correlated to pharmacologic properties, were the object of several definitions, especially by Wiener, Balaban, Randic, Hosoya, Altenburg and Kier and Hall (1986), the latest possibly being the most interesting.

What can be thought of this tendency to increase the number of parameters in the establishment of structure-activity relationships? First of all, it must be noted that under these conditions, the adequacy between calculated and experimental olfactory properties is often extremely high. For instance, in Edwards and Jurs' publication (1989), a 0.93 correlation coefficient was found for 55 random substances, which is exceptional as compared to other methods. There are statistical tests such as the *F*-ratio to verify that the correlation remains significant despite the increase in the number of parameters. However, one might feel somewhat perplexed by this "steamroller" method for resolving the problem: the parameters "retained" by the program vary according to the sample under consideration (the olfactory threshold of this group of substances, the musky character of that family of compounds, etc.) and the "predictive" properties of the suggested model often only apply to molecules that resemble those used to "train" the computer. On the other hand, these models are not at all explicative, unlike "theories" that are, no doubt, excessively rigid in their dogma. Intermediate solutions such as Hansch's lead to greater intellectual support. In fact, the solution may consist of considering a vast number of parameters regrouping those that are recognized as significant in various types of olfactory properties and then reducing the number by only taking their mutually independant components. The number of thus defined independant parameters (certainly less than ten) could perhaps be used to establish a universal profile of the totality of olfactory nuisances.

6.4.4.3 Experimental Solubility Parameters

Beginning in 1972, Laffort and co-workers started from what appeared to them to be a double necessity: the use of "independant" parameters and their reliance on experimental bases. The totality of physicochemical and geometric properties used in the Hansch-like approach reflect low intermolecular forces that do not correspond to those brought into play in chemical links but that could be the same as those encountered in the solute-solvent phenomenon. This was the wager formulated at the start; all that remained was to model the phenomenon. The process included several steps, spread out between 1972 to 1982, with an evaluation (temporary?) of the method in 1987. Some of the publications are more concerned with olfaction (Dravnieks and Laffort, 1972; Laffort et al., 1974; Laffort, 1976; Laffort, 1977) and the others with physical chemistry (Laffort and Patte, 1976; Patte et al., 1982; Laffort and Patte, 1987). The procedure started with the construction of a program known as "Robin-Laffort", which is related in some ways to factorial analysis, but which makes it possible to "set the axis" of the reconstructed space. [This program has recently been more fashionably written under the name of POLY-STRU, as in "*poly*nomial *struc*turation analysis" by Callegari (1991).] This program was then applied to gas chromatographic data (an ideal solute-solvent model) provided by McReynolds (1970) from which four, then five, solute and solvent factors were taken, sufficient to take the totality of the chromatographic phenomenon into account (with a precision comparable to that of experimental measurements, whatever the chosen solutes and solvents). This approach was confronted with another semitheoretical one (Karger et al., 1976, 1978), leading to a mutual enrichment and corresponding to the change from four to five factors (Laffort and Patte, 1976). Then an experimental device was constructed, including a chromatograph with five columns of different types of polarity arranged in a series (diagram in Figure 6.5) linked to a calculator. Each pure solute injected into the apparatus generates six peaks (Figure 6.6), thus allowing the calculator to deduce five partial solubility parameter values: dispersion or nonpolar (α), orientation (ω), electronic or volumic polarizability (ε), acidity or proton donor (π), and basicity or proton acceptor (β). The experimental

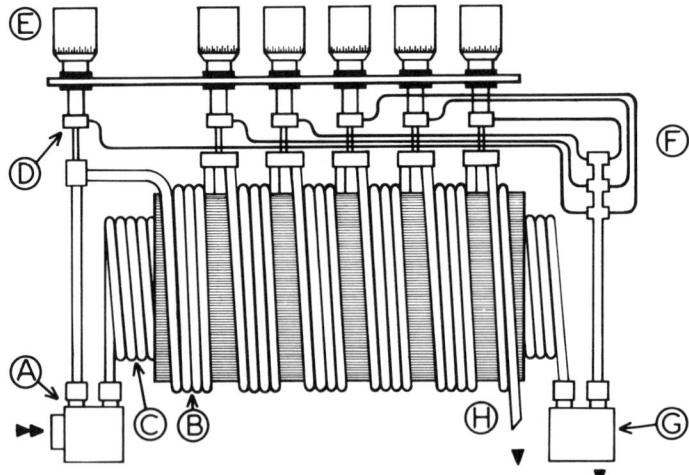

Figure 6.5. Diagram of the chromatographic device used. It includes five columns in series allowing the determination of five partial solubility parameters of pure substances. A = inlet block, B = measurements column, C = reference column, D = derivations, E = micrometric valves, F = manifold, G = detector cell (FID), H = exhaust (after Patte et al., 1982).

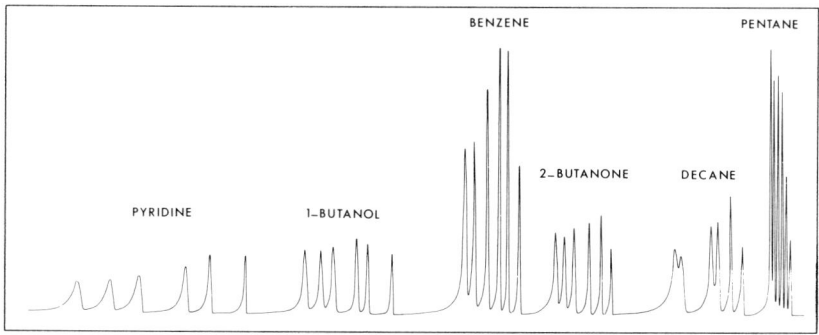

Figure 6.6. Chromatographic recordings obtained for different substances from the device described in Figure 6.5. Each compound generates six spikes whose retention times makes it possible to deduce their solubility parameters (after Laffort, 1981).

device made it possible to determine solubility factors for 240 solutes and 207 solvents (Patte et al., 1982).

What evaluation can be established from this approach? First of all, the accuracy of the physicochemical tool thus perfected can be underlined. Table 6.1 gives two examples: first, it includes ε values for different hydrocar-

Table 6.1. Several Examples of Solubility Parameters, after Patte et al. (1982). The Volumic Polarizability Factor ε is Clearly Linked to the Degree of Connectivity for Hydrocarbons. The Acidic Character of a Series of Alcohols, as Reflected by the Factor π, is Modulated in a Rather Precise Way by the Steric Hindrance and Double Bonds in α.

		ε			π
C2	Butane	0.00	C3	1-Propanol	0.75
	iso-Butane	−0.27		iso-Propanol	0.61
				Allylic alcohol	0.92
C6	Hexane	0.00	C4		0.75
	1-Hexene	0.50		1-Butanol	0.52
	Cyclohexane	1.16		2-Butanol	0.36
	Benzene	1.50		ter-Butylic alcohol	
C7	Heptane	0.00	C5	1-Pentanol	0.73
	2,4-Dimethylpentane	−0.50		ter-Amylic alcohol	0.33
	Toluene	1.37			
C9	Nonane	0.00	C6	1-Hexanol	0.77
	2,2,5-Trimethylhexane	−0.60		2-Hexanol	0.59
	1,3,5-Trimethylbenzene	0.99		3-Hexanol	0.47
C10	Decane	0.00	C8	1-Octanol	0.75
	Pinene	0.62		trans-1-Octene-2-ol	0.82
	trans-Decaline	1.57		2-Methylheptan-2-ol	0.35
				3-Methylheptan-3-ol	0.28

bons, that is, for substances without π electrons. The narrow link between ε values and the availability of *sigma* electrons can be noticed, that is, by what was called connectivity in the preceding paragraph: positive values for multiple links and cycles, negative values for branched hydrocarbons, and zero values for linear saturated hydrocarbons (for substances including heteroatoms—oxygen, sulfur, nitrogen, halogens—the phenomenon is more complicated). The second example concerns π values (proton donor) for a series of alcohols, which reflects whether the alcohol is primary, secondary, or tertiary, branched or not, cycled or with a double link. As far as the physicochemical angle is concerned, the approach seems rather satisfactory. It was even thought to use this device at the outlet of a classic chromatograph as a substance identifier, in the manner a mass spectograph is used, but that would have the advantage of being much more economic. In fact, certain confusions remain, notably between aldehydes and ketones on one hand and sulfurous, halogenic and benzenic compounds on the other hand (Laffort and Patte, 1987).

From the point of view of structure-activity relationships in olfaction, the evaluation is more reserved. Highly significant correlations have certainly been obtained, in the intensive as well as the qualitative domain, but the best are at a maximum of $r = 0.80$, which quite certainly means that the totality of the phenomenon is not taken into account by these five parameters (Laffort et al., 1974; Laffort, 1976; Laffort, 1977; Schiffman et al., 1977; Laffort and Patte, 1987). Experimental improvement of such an approach would most likely be accomplished by a two-dimentional physicochemical model and no longer by a three-dimensional model as in partition chromatography, that is, to be more precise, by the study of artificial membranes, which we will return to later.

Taking the preceding discussion into account, it goes without saying that this approach has nothing to do with the previously exposed "chromatographic theory"; however, since the two types of work have sometimes been confused, it is perhaps useful to underline the distinction.

A development of these solubility parameters, also often called "solvatochromic," consists of deriving them from the structure by using semiempirical calculations (Abraham et al., 1990, 1991). This is particularly interesting for substances for which experimentation cannot be done. In a recent study of QSAR in olfaction, a similar method has been performed, leading to a correct discrimination of nine clusters of olfactory (biological) origin with only 4% of misplaced odorants by using three out of five solubility parameters (Laffort, 1992, 1993). This result not only presents an interest by itself, but also because it seems to indicate that the receptor proteins "recognize" odorous molecules, mainly on the basis of their own overall properties (their size, for example) rather than their receptor site characteristics. The author's arguments for this assessment will not be summarized here. This totally new concept could also be applied to other biological activities of chemical products.

6.5 Conclusion and Perspectives

The impression given by the different chapters of the first part of this book is the great complexity of olfactory phenomenon. This is not surprising, since olfaction comes from both pharmacology (chemical captors) and sensorial physiology, two fields that are already very complex in themselves. This observation is particularly valid for the present chapter where the difference in approaches to the same problem is finally quite vast. Because of this fact, this chapter has been constructed somewhat differently than the others: technical precisions have been systematically avoided (absence of equations, among others) but, on the other hand, bibliographical references are more complete, so that readers who wish to deepen their knowledge on this or that point may do so without difficulty. Perspectives can be of two different natures: in the experimental field, the study of artificial membranes should certainly be increased and in the computer science field a certain number of trails could be followed.

6.5.1 Artificial Membranes

Classically, three types of phospholipidic membranes can be distinguished, constituting as many possible models of neuronal membranes.

1. *Bilayer membranes, like soapy water films.* When this kind of film loses its iridesence to become black, there are only two layers of phospholipids. This kind of model, although close to natural membranes, is but little used because of its fragility.
2. *Monolayer membranes, easy to obtain at the surface of a water body.* There, one can measure either surface potential or surface tension.
3. *Liposomes or bilayered microspheres, also quite easy to obtain.* The variation of their surface potential is measured through changes in the fluorescence of voltage-dependant dyes.

The second of these methods was applied to a study on tastes by Aiuchi et al., 1976 and the third method to a study on olfaction by Nomura and Kurihara (1987a and 1987b) of the same laboratory.

This type of experimentation remains to be developed as a physicochemical model that is relatively close to natural neuronal membranes even when one does not entirely subscribe to the interpretation made by Kurihara and co-workers of their work. The obtained correlations between olfactory properties and surface potentials of different types of liposomes lead these authors to put forward the hypothesis that olfactory discrimination could result from variable phospholipid and nonspecific protein membrane compositions, depending on the receptor cells. Insofar as this coding hypothesis relies on only two "letters," it is difficult to accept. On the other hand, more complete knowledge of this type of adsorption on pure lipidic membranes, then enriched with different types of proteins, would no doubt make it possible to progress noticeably in understanding the olfactory stimulation coding and mechanism.

6.5.2 Chemical Data Processing

One could include an entire group of disciplines having the use of computers and the problem of chemical message processing in common in this category.

6.5.2.1 Molecular Imagery

Molecular graphism, that is, reproduction molecular models on a polychromatic computer screen, has progressed considerably in the last few years (Doucet et al., 1987; Mornon, 1989). It supplies chemists and pharmacologists with a precise, rapid, and comfortable tool to study "structures" and "shapes" in the restricted sense of the term and thus to establish structure-activity relationships in the perspective previously called "organic chemistry."

Combining the resources of this technology and those of pattern recognition, Chastrette and co-workers have recently reactivated the old concept of osmophoric groups that they call "motifs" and that characterize the part of the molecule responsible for a given odorous note. In this way, these authors have defined the characteristics, in terms of distances, angles, and dihedral angles, that must exist between the different groups of atoms of a molecule for that molecule to have a musky odor (Chastrette and Zakarya, 1988) or the odor of sandalwood (Chastrette et al., 1990). The first of these works, which, for the moment does not include steroid and macrocyclic substances, takes the musky odor of 104 compounds into account and the second work takes the sandalwood odor of 57 odorous substances.

6.5.2.2 Molecular Topology

Earlier in this chapter, connectivity indexes according to different authors have been evoked, which should be made independant of each other (they are all related to molecule size, while taking the degree of "connection" of each atom with its neighbors into account). These indexes are not the only approach to molecular topology, that is, structural descriptions by means of algorithms which then make it possible to make calculations on these structures (for example, evaluation of molecular resemblance according to certain criteria). The principal methods are Morgan's algorithm (Choplin, 1985), Dubois' and co-workers' DARC system (Dubois and Viellard, 1971), and Moreau and collaborators' autocorrelation vector (Moreau and Broto, 1980; Moreau et al., 1989). The last method cited could have an interesting application in olfaction.

6.5.2.3 *Neuromimetic Networks Applied to Chemical Stimuli*

Neuromimetic or connectionist networks, which date from less than 10 years ago, constitute a new computer science tool in full expansion in different fields. For instance, in those fields related to sensoriality, one could cite recognition of speech or handwriting by machines (Alkon, 1989; Jutten, 1989; Bochereau and Bourgine, 1989). These tools, generally used to manipulate interconnection networks, are a particular category of computers ("transputers") composed of a large number of parallel microprocessors. Inspired by the tridimentional structure of living nervous systems, these devices allow an important amount of time saving, indispensable in certain applications. The first application of neuromimetic networks to the relationship between olfactory variables and physical parameters has been performed recently and seems to indicate a very promising use of this tool in this particular field (Callegari et al., 1992,1993).

Bibliography

Abraham, M. H., Whiting, G., Doherty, R. M., Shuely, W. H., 1990. Hydrogen bonding. Part 13. A new method for the characterisation of GLC stationary phases—The Laffort data set. *J. Chem. Soc. Perkin Trans. 2*, 1451–1460.

Abraham, M. H., Whiting, G., Doherty, R. M., Shuely, W. H., 1991. Hydrogen bonding. XVI. A new solute solvation parameter, -2, from gas chromatographic data. *J. Chromatogr.*, **587**, 213–228.

Acree, T. E., Barnard, J., Cunningham, D. G., 1984. A procedure for the sensory analysis of gas chromatographic effluents. *Food Chem.* **14**, 273–286.

Aiuchi, T., Kamo, N., Kurihara, K., Kobatake, Y., 1976. Physicochemical studies of taste reception. VI: Interpretation of anion influences on taste responses. *Chem. Senses Flavor* 2, 107–119.

Alkon, D., 1989. Mémorisastion et neurones. *Pour la Science* 143, 38–46.

Amoore, J. E., 1952. The sterochemical specificities of human olfactory receptors. *Perf. Essent. Oil Rec.* **43**, 321–330.

Amoore, J. E., 1968. Specific anosmias and primary odors. In *Theories of Odor and Odor Measurement*, edited by N. N. Tanyolaç. Robert College, Instanbul, pp. 71–85.

Amoore, J. E., 1970. *Molecular basis of odor.* C. C. Thomas, Springfield, IL, 200 pp.

Amoore, J. E.. 1977. Specific anosmia and the concept of primary odors. *Chem. Senses Flavor* 2, 267–281.

Amoore, J. E., Forrester, L J., 1976. Specific anosmia to trimethylamine: the fishy primary odor. *J Chem. Ecol.* 2, 49–56.

Amoore, J. E., Palmieri, G. Wanke, E., 1967. Molecular shape and odour: Pattern analysis by PAPA. *Nature* **216**, 1084–1087.

Amoore, J. E., Venström, D., 1967. Correlations between stereochemical assessments and organoleptic analyses of odorous compounds. In *Olfaction and Taste II*, edited by T. Hayashi. Oxford, Pergamon, pp. 3–17.

Anholt, R. R. H., 1987. Primary events in olfactory reception. *TIBS* **12**, 58–62.

Arcis, A., Charron, M., 1975. Un appareil original pour le dosage sélectif de composés sulfurés présents dans les gaz: le MEDOR-S. Cas pratiques d'utilisation. *Congrès Ass. Tech. Gaz,* DETN-Gas de France, Paris, 20 pp.

Ash, K. O., 1968. Chemical sensing: an approach to biological molecular mechanisms using difference spectroscopy. *Science* **162**, 452–454.

Bayonove, C., Cordonnier, R., 1970. Recherches sur l'arôme de muscat. *Ann. Techn. Agric.* **19**, 79–105.

Beauchamp, G., Yamazaki, K., Boyse, E., 1985. La reconnaissance olfactive de l'identité génétique. *Pour la Science* **95**, 79–86.

Beck, L. H., 1964. A quantitative theory of the olfactory threshold based upon the amount of the sense cell covered by an adsorbed film. *Ann. N.Y. Acad. Sci.* **116**, 448–456.

Beets, M. G. J., 1968. Odor and molecular structure. *Olfactologia* **1**, 77–92.

Beets, M. G. J., 1978. *Structure-Activity Relationships in Human Chemoreception.* Applied Science Publishers Ltd., London, 408 pp.

Bernuzeau, A., 1986. Les anosmies spécifiques et leurs applications. Thèse Paris 5, 45 pp.

Berton, A., 1959. Piles galvaniques sensibles à des traces de substances gazeuses, liquides ou solides. *Chim. Anal.* **41**, 351–358.

Bignett, E., Damiani, G., DeNegri, P., Ramoni, R., Avanzini, F., Ferrari, G., Rossi, G. L., 1987. Specificity of an immunoaffinity column for odorant-binding protein from bovine nasal mucosa. *Chem. Senses* **12**, 601–608.

Bochereau, L., Bourgine, P., 1989. Implémentation et extraction de traits sémantiques sur un réseau neuro-mimétique: exemple de la première annonce au bridge. Document interne. CEMAGREF, Parc de Tourvoie, B. P., 121, 92164 Antonny Cedex, 17 pp.

Boelens, H., 1976. Molecular structure and olfactive properties. In *Structure-Activity Relationships in Chemoreception,* edited by G. Benz. IRL press, London, pp. 197–206.

Buck, L., Axel, R., 1991. A novel multigene family may encode odorant receptors: a molecular basis for odor recognition. *Cell,* **65**, 175–187

Callegari, P., 1991. Traitements logiciels et propriétés des modèles matriciels multiplicatifs. Internal report at Laboratoire de Physiologie de la chimioréception, CNRS, Avenue de la Terrasse, 91190 Gif-sur-Yvette, France, 58 pp.

Callegari, P., Laffort, P., Rouault, J., 1992. OLFANET: a connectionist program simulating olfactory recognition. *Chem. Senses,* **17**, 855.

Callegari, P., Laffort, P., Rouault, J., 1993. Réseaux neuro-mimétiques et signifiant olfactif. In Les Sciences Cognitives, 4th Conference of AIDRI, Lyon, 11–12 of June, 1992 (in press).

Chanel, J., 1987. The olfactory system as a molecular descriptor. *NIPS* **2**, 203–208.

Chastrette, J., 1981. An approach to a classifiction of odours using physicochemical parameters. *Chem. Senses* **3**, 157–163.

Chastrette, M., Zakarya, D., 1988. Sur le rôle de la liaison hydrogène dans l'interaction entre les récepteurs olfactifs et les molécules à odeur de musc. *C.R. Acad. Sci. Paris* **307**, 1185–1188.

Chastrette, M., Zakarya, D., Pierre C., 1990. Relations structure-odeur de bois de santal: recherche d'un modèle d'interaction basé sur le concept d'hypermotif santalophore. *Eur. J. Med. Chem.* (in press).

Choplin, F., 1985. L'ordinateur en chimie. *Pour la Science* **95**, 50–59.

Davies, J. T., 1953. L'odeur et la morphologie des molécules. *Industr. Parf.* **8**, 74–79.

Davies J. T., 1965. A theory of the quality of odors. *J. Theoret. Biol.* **8**, 1.

Davies, J. T., 1971. Olfactory theories. In *Handbook of Sensory Physiology IV/1, Olfaction*, edited by L. M. Beidler. Springer-Verlag, Berlin, pp. 322–350.

Davies, J. T., Taylor, F. H., 1959. The role of adsorption and molecular morphology in olfaction: the calculation of olfactory thresholds. *Biol. Bull.* **117**, 222–238.

Delaleu, J. C., Holley, A., 1980. Modification of transduction mechanisms in the frog's olfactory mucosa using a thiol reagent as olfactory stimulus. *Chem. Senses* **5**, 205–218.

Delsaux, Y., 1985. Odeurs: une affaire de molécules. *Phosphore* (no. de février).

Devos, M., Rouault, J., Laffort, P., 1991. Standardized Olfactory Power Law Exponents in Man (en préparation).

Doré, J. C., Gordon, G., Jaubert, J. N., 1984. Approche factorielle des relations entre structure chimique et note odorante. *C.R. Acad. Sci. Paris* **299**, 315–320.

Doucet, J. P., Dubois, J. E., Weber, J., 1987. Représentations et simulations moléculaires. *Le Courrier du CNRS* **66–67–68**, 68–70.

Döving, K. B., 1974. Odorant properties correlated with physiological data. *Ann. N.Y. Acad. Sci.* **237**, 184–192.

Dravnieks, A., 1965. Contact potentials in detection of airborne vapors. In *Surface Effects in Detection*, edited by J. I. Bregman and A. Dravnieks. Spartan Books Inc., Washington D.C., pp. 103.

Dravnieks, A., 1974. A building-block model for the characterization of odorant molecules and their odors. *Ann. N.T. Acad. Sci.* **237**, 144–163.

Dravnieks, A., 1977. Correlation of odor intensities and vapor pressures with structural properties of odorants. In *Flavor Quality: Objective Measurement*, edited by R. A. Scanlen. ACS Symposium Series 51, American Chemical Society, Washington D.C., pp. 11–28.

Dravnieks, A., Laffort, P., 1972. Physico-chemical basis of quantitative and qualitative odor discrimination in Humans. In *Olfaction and Taste IV*, edited by D. Schneider. Wissens-Verlag-MBH, Stuttgart, pp. 142–148.

Dravnieks, A., O'Donnell, A., 1971. Principles and some techniques of high-resolution headspace analysis. *J. Agr. Food Chem.* **19**, 1049–1056.

Dravnieks, A., Trotter, P. J., 1965. Polar vapor detector based on thermal modulation of contact potential. *J. Sci. Instr.* **42**, 624.

Dubois, J. E., Viellard, H., 1971. Système DARC. Théorie de génération-description. *Bull. Soc. Chim. Fr.* 839–848.

Duchamp, A., Revial, M. F., Holley, A., MacLeod, P., 1974. Odor discrimination by frog olfactory receptors. *Chem. Senses Flavor* **1**, 213–233.

Dyson, G. M., 1937. Raman effect and the concept of odour. *Perf. Essent. Oil Rec.* **28**, 13–19.

Edwards, P. A., Jurs, P. C., 1989. Correlation of odor intensities with structural properies of odorants. *Chem. Senses* **14**, 281–291.

Eminet, B. P., Chastrette, M., 1983. Discrimination of camphoraceous substancces using physico-chemical parameters. *Chem. Senses* **7**, 293–300.

Etcheto, M., 1975. Mise au point d'un appareil de mesure des propriétés odorantes de substances pures sur des bases physiques. Diplôme de l'École Pratique des Hautes Études, 77 pp.

Etievant, P. X., Issanchou, S. N., 1987. Le goût du vin. *La Recherche* **18**, 1344–1353.

Etievant, P. X., Issanchou, S. N., Bayonove, C. L., 1983. The flavour of Muscat wine: the sensory contribution of some volatile compounds. *J. Sci. Food Agric.* **34**, 497–504.

Fabre, J. H., 1923. Souvenirs entomologiques. *L'Odorat,* Vol. 7. Delagrave, Paris, pp. 403–423.

Ferguson, J. F., 1939. The use of chemical potentials as indices of toxicity. *Proc. Roy. Soc. B* **127**, 387–404.

Friedman, H. H., Mackay, D. A., Rosano, H. L., 1964. Odor measurement possiblities via energy changes in cephalin monolayers. *Ann. N.Y. Acad. Sci.* **116**, 602–607.

Fuller, G. H., Steltenkamp, R., Tisserand, G. A., 1964. The gas chromatograph with human sensor: perfumer model. *Ann. N.Y. Acad. Sci.* **116**, 711–724.

Gavaudan, P., Poussel, H., Brebion, G., Schutzenberger, M. P., 1948. L'étude des conditions thermodynamiques de l'excitation olfactive et les théories de l'olfaction. *C.R. Acad. Sci. Paris* **226**, 1395–1396.

Getchell, M. L., Gesteland, R., 1972. The chemistry of olfactory reception: stimulus-specific protection from sulfhydrylreagent inhibition. *Proc. Nat. Acad. Sci.* **69**, 1494–1498.

Greenberg, M. J., 1979. Dependence of odor intensity on the hydrophobic properties of molecules. A quantitative structure odor intensity relationship. *J. Agric. Food Chem.* **27**, 347–352.

Greenberg, M. J., 1981. The dependence of odor intensity on the hydrophobic properties of molecules. In *Odour quality and chemical structures,* ACS Symposium series 148, edited by H. R. Moskowitz and C. B. Warren. American Chemical Society, Washington D.C., pp. 177–194.

Guillot, M., 1948a. Anosmies partielles et odeurs fondamentales. *C.R. Acad. Sci. Paris* **226**, 1307–1309.

Guillot, M., 1948b. Sur la relation entre l'odeur et al structure moléculaire. *C.R. Acad. Sci. Paris* **226**, 1472–1474.

Ham, C. L., Jurs, P. C., 1985. Structure-activity studies of musk odorants using pattern recognition: monocyclic nitrobenzenes. *Chem. Senses* **10**, 491–505.

Higashino, S., Takagi, S. F., Yagima, M., 1961. The olfactory stimulating effectiveness of homologous series, of substancies studied in the frog. *Jpn. J. Physiol.* **11**, 530–544.

Hodgkin, A. L., Katz, B., 1949. The effect of sodium ions on the electrical activity of the giant axon of the squid. *J. Physiol. (London)* **108**, 37–77.

Jaubert, J. N., Gordon, G., Doré, J. C., 1987. Une organisation du champ des odeurs. *Parf. Cosm. Arômes* **77**, 53–56; **78**, 71–82.

Johnston, J. W., 1953. Infrared loss theory of olfaction untenable. *Physiol. Zool.* **26**, 266–273.

Jutten, C., 1989. Réseaux neuromimétiques. Algorithmes, applications et machines. Document interne, Laboratoire de traitement d'image et de reconnaissance des formes, Institut National Polytechnique de Grenoble, 59 pp.

Karger, B. L., Snyder, L. R., Eon, C., 1976. An expanded solubility parameter treatment for the classification and use of different chromatographic solvents and adsorbents. I. General retention theory. *J. Chromatogr.* **125**, 71–88.

Karger, B. L., Snyder, L. R., Eon, C., 1978. An expanded solubility parameter treatment for the classification and use of different chromatographic solvents and adsorbents. II. *Anal. Chem.* **50**, 2126–2136

Kier, L. B., Hall, L. H., 1986. *Molecular Connectivity in Structure-Activity Analysis.* Research Studies Press, Letchworth (U.K.)

Laffort, P., 1963. Mise en évidence de relations linéaires entre l'activité odorante des molécules et certaines de leurs caractéristiques physico-chimiques. *C.R. Acad. Sci. Paris* **256**, 5618–5621.

Laffort, P., 1965. Efficacité odorante et activité thermodynamique. *Rev. Laryngol. Bordeaux Suppl.* 860–879.

Laffort, P., 1968. Some new data on the physico-chemical determinants of the relative effectiveness of odorants. In *Theories of Odor and Odor Measurement,* edited by N. N. Tanyolaç. Robert College, Istanbul, pp. 247–270.

Laffort, P., 1969. A linear relationship between effectiveness and identified molecular characteristics, extended to fifty pure substances. In *Olfaction and Taste III,* edited by C. Pfaffman. Rockefeller University Press, New York, pp. 150–157.

Laffort, P., 1976. A model of the olfactory mechanism based on chromatographic data: hypothesis, results and perspectives. In *Structure-Activity Relationships in Chemoreception,* edited by G. Benz et al. IRL Press London, pp. 185–195.

Laffort, P., 1977. Some aspects of molecular recognition by chemoreceptors. In *Olfaction and Taste VI,* edited by J. LeMagnen. IRL Press London, pp. 17–25.

Laffort, P., 1981. Mesure physico-chimique des odeurs. Limites et perspectives. *Parf. Cosm. Arômes* **39**, 39–44.

Laffort, P., 1992. Graphical representation of olfactory quality based on molecular parameters applied to experimental data from the Holley group. *Chem. Senses* **17**, 855

Laffort, P., 1993. Structuration graphique de la qualité odorante sur la base de paramètres moléculaires. Cas des données expérimentales du group de André Holley. *C.R. Acad. Sci. Paris* **316**, 105–111

Laffort, P., Callegari, P., Devos, M., 1991. The role played by the polarizability of solutes in their retention on pure GCL hydrocarbon phases. In *Advanced Study Institute on Theoretical Advancement in Chromatography and Related Techniques,* 18–30 August 1991. Ferrara, Italy.

Laffort, P., Dravnieks, A., 1973. An approach to a physico-chemical model of olfactory stimulation in vertebrates by single compounds. *J. Theoret. Biol.* **38**, 335–345.

Laffort, P., Patte, F., 1976. The solubility factors in gas-liquid chromatography. Comparison between two approaches and application to some biological studies. *J. Chromatogr.* **126**, 625–639.

Laffort, P., Patte, F., 1987. Solubility factors established by gas-liquid chromatography. A balance-sheet. *J. Chromatogr.* **406**, 51–74.

Laffort, P.,Patte, F., Etcheto, M., 1974. Olfactory coding on the basis of physico-chemical properties. *Ann. N.Y. Acad. Sci.* **237**, 193–208.

Laing, D. G., 1987. Coding of chemosensory stimulus mixtures. *Ann. N.T. Acad. Sci.* **510**, 61–66.

Laing, D. G., 1988. Relationship between the differential adsorption of odorants by the olfactory mucus and their perception in mixtures. *Chem. Senses* **13**, 463–471.

Lancet, D., 1984. Molecular view of olfactory reception. *TINS* 35–36.

Lancet, D., Pace, U., 1987. The molecular basis of odor recognition. *TIBS* **12**, 63–66.

LeMagnen, J., 1949. Étude d'un phénomène de sensibilisation olfactive. *C.R. Acad. Sci. Paris.* **228**, 122–124.

LeMagnen, J., 1952. Les phénomènes olfacto-sexuels chez l'Homme. *Arch. Sci. Physiol.* **6**, 125–160.

Loukianoff, M., 1988. Théorie des graphes et topologie moléculaire. Internal report at Laboratoire de Physiologie de la chimioréception, CNRS, Avenue de la Terrasse, 91190 Gif-sur-Yvette, France, 100 pp.

Mackay-Sim, A., Shaman, P., Moulton, D. G., 1982. Topographic coding of olfactory quality: odorant specific patterns of epithelial responsivity in the Salamander. *J. Neurophysiol.* **48**, 584–596.

Marin, A. B., Acree, T. E., Barnard, J., 1988. Variation in odor detection thresholds determined by charm analysis. *Chem. Senses* **13**, 435–444.

McReynolds, W. O., 1970. *Personal communication.* Celanese Chemical Co., Bishop, Texas.

Menevse, A., Dodd, G. H., Poynder, T. M., 1978. A chemical modification approach of the olfactory code. *Biochem J.* **176**, 845–854.

Miles, W. R., Beck, L. H., 1947. Infrared absorption in field studies of olfaction in bees. *Science* **106**, 512.

Miles, W. R., Beck, L. H., 1949. Infrared absorption hypothesis of olfaction. *Proc. Nat. Acad. Sci.* **35**, 292–310.

Moncrieff, R. W., 1961. An instrument for measuring and classifying odors. *J. Appl. Physiol.* **16**, 742–749.

Moreau, G., Broto, P., 1980. The autocorrelation of a topological structure: a new molecular descriptor. *Nouv. J. Chimie* **4**, 359–360.

Moreau, G., Broto, P., Turpin, C., Fortin, M., 1989. Réalisation sur ordinateur d'un screening de structures moleculaires de substances potentiellement anxiolytiques à l'aide de la technique d'autocorrélation. *Eur. J. Méd Chem.* (in press).

Mornon, J. P., 1989. *Personal communication.* Universite Paris VI, Tour 15-2e. 2, place Jussieu 75005 Paris.

Morvan, R. G., 1972. Olfaction. In *Encyclopédie Internationale des Sciences et des Techniques.* Presses de la Cité, Paris, Vol. **8**, 504–505.

Mozell, M. M., 1964. Evidence for sorption as a mechanism of the olfactor analysis of vapors. Nature **203**, 1181–1182.

Mozell, M. M., 1970. Evidence of a chromatographic model of olfaction. *J. Gen. Physiol.* **56**, 46–63.

Mozell, M. M., 1973 (personal communication). Departments of Physiology and Preventive Medicine, SUNY, Health Science Center at Syracuse, Syracuse, N.Y. 13210.

Mozell, M. M., Jagodowicz, M., 1973. Chromatographic separation of odorants by the nose: retention times measured across *in vivo* olfactory mucosa. *Science* **181**, 1247–1249.

Mozell, M. M. Sheehe, P. R., Hornung, D. E., Kent, P. F., Yougentob, S. L. Murphy, S. J., 1987. "Imposed" and "inherent" mucosal activity patterns. *J. Gen. Physiol.* **90**, 625–650.

Mullins, L. J., 1955. Olfaction. *Ann. N.Y. Acad. Sci.* **62**, 247–276.

Narvaez, J. N., Lavine, B. K., Jurs, P. C., 1986. Structure-activity studies of musks odorants using pattern recognition: bicyclo and tricyclo-benzoïds. *Chem. Senses* **11**, 145–156.

Nomura, T., Kurihara, K., 1987a. Liposomes as a model for olfactory cells: changes in membrane potential in responses to various odorants. *Biochemistry (Wash.)* **26**, 6135–6140.

Nomura, T., Kurihara, K., 1987b. Effects of changed lipid composition on responses of liposomes to various odorants: possible mechanism of odor discrimination. *Biochemistry (Wash.)* **26**, 6141–6145.

Ohloff, G., 1971. L'odorat et la forme des molécules. *La Recherche* **18**, 1068–1070.

Ottoson, D., 1956. Analysis of the electrical activity of the olfactory epithelium. *Acta Physiol. Scand. Suppl. 122*, **35**, 1–83.

Ottoson, D., 1958. Studies on the relationship between olfactory stimulating effectiveness and physico-chemical properties of odorous compounds. *Acta Physiol. Scand.* **43**, 167–181.

Ottoson, D., Von Sydow, E., 1964. Electrophysiological measurements of the odour of single components of a mixture separated in a gas chromatograph. *Life Sciences* **3**, 1111–1115.

Patte, F., Etcheto, M., Laffort, P., 1982. Solubility factors for 240 solutes and 207 stationary phases in gas-liquid chromatography. *Anal. Chem.* **54**, 2239–2247.

Pauling, L., 1946. Analogies between antibodies and simpler chemical substances. *Chem. Eng. News* **24**, 1064–1065.

Pelosi, P., Pisanelli, A. M., 1981. Specific anosmia to 1,8-cineole: the camphor primary odor. *Chem. Senses* **6**, 87–93.

Pelosi, P., Viti, R., 1978. Specific anosmia to 1-carvone: the minty primary odour. *Chem. Senses Flavor* **3**, 331–337.

Persaud, K. C., Pelosi, P., Payne, P. A., 1991. A multi-element gas and odour detector based on conducting organic polymers. *Chem. Senses* **16**, 402.

Pietras, R. J., Moulton, D. G., 1974. Hormonal influences on odor detection in rats: changes associated with the estrous cycle, pseudo-pregnancy, ovariectomy and administration of testosterone propionate. *Physiol. Behav.* **12**, 475–491.

Polak, E. H., Shirley, S. G., Dood, G. H., 1989. Concanavalin A reveals olfactory receptors which discriminate betwen alkane odorants on the basis of size. *Biochem. J.* **262**, 475–478.

Price, S., 1977. Specific anosmia to geraniol in mice. *Neurosci. Lett.* **4**, 49.

Price, S., Willey, A., 1988. Effects of antibodies against odorant binding proteins on electrophysiological responses to odorants. *Biochim. Biophys. Acta* **965**, 127–129.

Schiffman, S. S., 1974. Contributions to the physicochemical dimensions of odor: a psychophysical approach. *Ann. N.Y. Acad. Sci.* **237**, 164–183.

Schiffman, S. S., Robinson, D. E., Erickson, R. P., 1977. Multidimensional scaling of odorants: examination of psychological and physico-chemical dimension. *Chem. Senses Flavor* **2**, 375–390.

Shirley, S. G., Persaud, K. C., 1990. The biochemistry of vertebrate olfaction and taste. *Semin. Neurosc.* **1** (in press).

Shirley, S. G., Polak, E. H., Mather, R. A., Dodd, G. H., 198a. The effect of concanavalin A on the rat electro-olfactogram. Differential inhibition of odorant response. *Biochem. J.* **245**, 175–184.

Shirley,, S. G., Polak, E. H., Mather, R. A., Dodd, G. H., 1987b. The effect of concanavalian A on the rat electro-olfactogram at various odorant concentrations. *Biochem. J.* **245**, 185–189.

Shkapenko, G., Gerebtzoff, M. A., 1951. Critique expérimentale de l'intervention des radiations dans le mécanisme de l'olfaction. *Arch. Inter. Physiol.* **59**, 423–429.

Shurmer, H. V., 1987. Development of an electronic nose. *Phys. Technol.* **18**, 171–176.

Sicard, G., Holley, A. J., 1984. Receptor cell responses to odorants: similarities and differences among odorants. *Brain Res.* **292**, 283–296.

Southwick, E., 1989. *Personal communication.* Philip Morris Research Center, P.O. Box 26583, Richmond, Virginia, 23261-6483.

Stoll, M., 1965. De l'effet important de différences chimiques minimes sur la perception de l'odeur. *Rev. Laryngol. Bordeaux suppl.* 972–981.

Stoll, M., 1968. Discussions. *Olfactologia* **1**, 106.

Stuper, A. J., Jurs, P. C., 1976. ADAPT: a computer system for automated data analysis using pattern recognition techniques. *J. Chem. Inf. Comput. Sci.* **16**, 99–105.

Talou, T., 1992. Un "nez electronique" pour l'industrie. *La Recherche* **23**, 1058–1059.

Tanyolaç, N. N., 1965. The electro-odocell of odor measurement and surface effect. In *Surface Effects in Detection,* edited by J. I. Bregman and A. Dravnieks. Spartan Books Inc., Washington D.C., pp. 89–102.

Tanyolaç, N. N., Eaton, J. R., 1950. Study of odors. *J. Am. Pharm. Assoc.* **39**, 10.

Thompson, H. W., 1957. Some comments on theories of smells. In *Molecular structure and organoleptic quality, Monograph no. 1.* Society of Chemical Industry, London, pp. 103–115.

Venard, R., Pichon, Y., 1984. Electrophysiological analysis of the peripheral response to odours in wild type and smell-deficient of C mutant of Drosophila Melanogaster. *J. Insect. Physiol.* **30**, 1–5.

Vernet-Maury, E., LeMagnen, J., Chanel, J., 1968. Comportement émotif chez le Rat: influence de l'odeur d'un prédateur et d'un non-prédateur. *C.R. Acad. Sci. Paris* **267D**, 331–334.

Wilkens, W. F., Hartman, J. D., 1964. An electronic analog for the olfactory processes, *Ann. N.Y. Acad. Sci.* **116**, 608–612.

Wolkowski, Z. W., Moccat, D., Heymans, F., Godfroid, J. J., 1977. A quantitative structure-activity approach to chemoreception: importance of lipophalic properties. *J. Theoret. Biol.* **66**, 181–193.

Wright, R. H., 1954. Odour and molecular vibration. Quantum and thermodynamic considerations. *J. Appl. Chem.* **4**, 611–615.

Wright, R. H., 1981. Odor and molecular vibration. Redundancy in the olfactory code. In *Odour Quality and Chemical Structure,* edited by H. R. Moskowitz and C. B. Warren. ACS Symposium Series, 148, American Chemical Society, Washington D. C., pp. 123–141.

Wright, R. H., 1982. Odour and molecular volume. *Chem. Senses* **7**, 211–213.

Wright, R. H., Burgess, R. E., 1971. Molecular mechanisms of olfactory discrimination and sensitivity. In *Gustation and Olfaction,* edited by G. Ohloff and A. F. Thomas. Academic Press, London, New York, pp. 134–146.

Wright, R. H., Michels, K. E., 1964. Evaluation of far infrared relation to odor by a standards similarity methods. *Ann. N.Y. Acad. Sci.* **116**, 535–551.

Wysocki, C. J., Dorries, K. M., Beauchamp, G. K., 1989. Ability to perceive androstenone can be acquired by ostensibly anosmic people. *Proc. Natl. Acad. Sci.* **86**, 7976–7978.

Wysocki, C. J., Whitney, G., Tucker, D., 1977. Specific anosmia in the laboratory mouse. *Behav. Genet.* **7**, 171–188.

CHAPTER

7

Synergy and Inhibition in Olfaction

P. Laffort

7.1 Introduction

Throughout this chapter, only the problem of physiological interactions will be considered, which means that the case of reactional mixtures such as that of oxidizing or oxidizable substances, for example, will not be discussed even though it represents one of the more interesting ways of fighting odorous pollution [vaporized essential oils used in the interior of habitations to fight disagreeable odors contain oxidants in the same way as combustion products used in "Berger" (A Berger lamp is an antiodor device intended for use in the home that is still quite popular in France, whose action consists of catalytic combustion of aromatized alcohol) lamps]. At the same time, this implies that the influence of a solution's components on the volatility of some of them would not have been studied either; although this problem is also of great importance, it has already been discussed in Chapter 4. Therefore, the phenomena exposed below will concentrate on mixtures that are completely determined in the gaseous phase.

First of all, one should be reminded how much synergy and inhibition (in the chemical senses and, notably, in olfaction) differ from that observed in pharmacology, where the phenomena are very clear. There are known "antagonist" substances, for example, those of acetylcholine in the nervous synapses, which can completely block the receptors to the point where acetylcholine itself will be totally inactive: the phenomenon of total inhibition. Moreover, everyone is aware of the important potentiality effects between certain medications or with ethanol, which one should be careful of. Very

185

strong inhibitions, reminiscent of those in pharmacology have been found in the specialized olfactive receptors of the lobster (Derby et al., 1985; Ache et al., 1988), but nothing like it has been observed in generalist olfaction, which is of interest here: the phenomena are very subtle and therefore call for modeling. The contrast observed between the two types of olfaction is related to the difference in receptor specificity, as high in specialist olfaction as in pharmacology and as low in generalist olfaction as seen in Chapter 5.

The first studies on olfactory interactions were carried out at the same time as early experiments on olfaction by Passy (1895), Backman (1918), and Zwaardemaker (1907, 1925). From the beginning, the necessity of models has been felt; an experiment carried out by Cain and Drexler (1974) makes it possible to comprehend this in a more concrete way. These authors have studied many binary mixtures of pyridine and a second pleasant component (see Figure 7.1). For some mixture proportions, the overall perceived intensity was less intense than the perceived intensity of one of the two components; this situation is often called "subtraction" (Berglund et al., 1976). For other mixture proportions, the overall sensation was more intense than for each of the components but lower than their sum; this situation, which is most often observed, is called "hypo-addition." Finally, the rule of odor addition known as "the strongest component model" is observed for several intermediate cases (according to this rule, the perceived intensity of the mixture is equal to the more intense of the two components). In other words, no general trend can be definitely given for each of the odorous substance pairs studied: the result depends on the respective concentrations. This observation relates to the first comment of Figure 7.1. The other comments of this figure will be developed later.

Similar results were observed in electrophysiological experimentation (Etcheto et al., 1982; Laffort et al., 1989): the hypo-addition phenomenon, even though widely observed, does not bring very relevant information by itself. As a matter of fact, it has also been observed when a substance is added to itself (the olfactive sensation related to a given substance increases less than its concentration and is expressed by a power law whose exponent is less than 1 as has been seen in Chapter 5).

For a given experimental procedure and a given experimental subject, an entirely satisfactory model should make it possible to characterize each pair of odorous substances by only one number: its interaction index, regardless of the relative concentrations and of the perceived intensity levels. Additionally this number should have a reference value (for instance, 0, 1, or 100) when a substance is added to itself. Such indexes could then be used for prediction purposes in studies of the structure-activity relationship or even for the comparison of different integration levels within the same animal species. As will be seen, this objective has been reached recently.

The qualitative aspects of olfactive mixtures have been studied by a small number of investigators: Ekman et al., 1964; Laing and Willcox, 1983; Laing

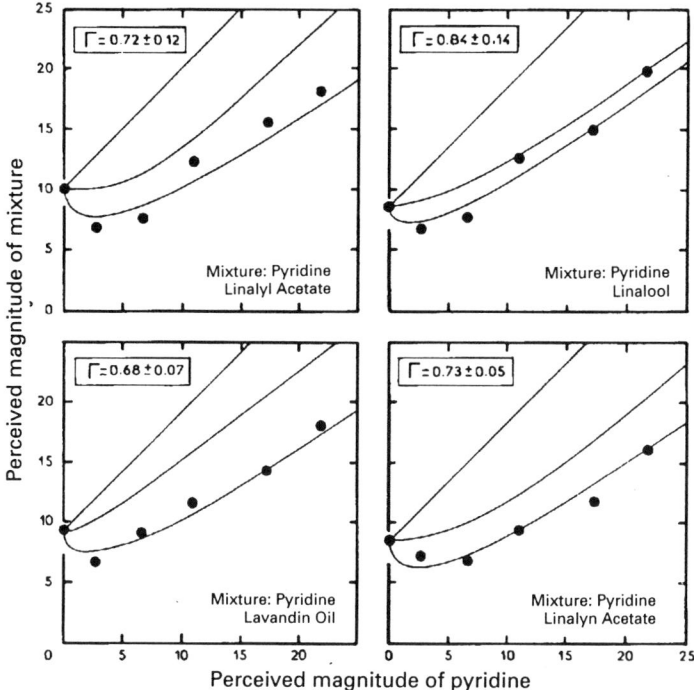

Figure 7.1. Intensity of olfactive sensations generated by four mixtures containing pyridine, expressed as a function of that produced by pyridine alone (after Cain and Drexler, 1974). Three theoretical curves have been drawn, corresponding to three models: additivity on the top, U at the bottom and UPL2 in the middle. Comments that will be developed throughout the chapter. (1) Experimental points exhibit a subtraction phenomenon on the left of each diagram and a hypo-addition phenomenon on the right. (2) The U model curve (as well as the vector model curve not represented here) is in agreement with the above observation. (3) Γ reflects the distance between the two lower curves, and cos $\alpha_{\shortparallel}$ reflects the distance between the two extreme curves. (4) Γ values of less than 1, reflecting a *true* inhibition, are clearly observed in at least three diagrams.

et al., 1984; Gregson, 1980, 1986. The conclusions drawn from these works were not clear. By contrast, a recent study by Olsson (1992) could provide a beginning of the key to this problem. According to this author, the maximum blending of odorous quality for a given pair of odorants occurs when relative concentrations correspond to equally strong odorous sensations smelled separately. It still remains to standardize a "blending index" and to understand which molecular characteristics of the compounds govern the values of such an index. However, the present study will be focused on quantitative aspects only.

7.2 Generation of an Appropriate Interaction Index

For pedagogic purposes, a chronological approach of intensity models having yielded a definition for an odorous interaction index has been adopted,
as each model was formulated to alleviate the inadequacies of the preceding
one. The increasing complexity of the mathematical formulations has gone
along with the progression of the unity of the observed phenomena (therefore, the conceptual simplification).

7.2.1 First Period: 1895–1971

7.2.1.1 Principal Works

7.2.1.1.1 Psychophysical Experiments

The following publications can be added to those already mentioned:
Rosen et al. (1962), Guadagni et al. (1963), Jones and Woskow (1964), Borelli
and Angleraud (1965); Kendall and Neilson (1966), Laffort (1968) and Köster
(1968, 1969).

7.2.1.1.2 Electrophysiological Experiments

Only one study carried out during this period is known (MacLeod, 1968) to
which two others can be added, published after 1971, but intellectually related to this period: Köster and MacLeod (1975) and Müller (1979).

7.2.1.2 Characteristics of this Period: Concentrations are Taken as Variables

Works from this long period before 1971 commonly express the intensity of
olfactive sensations generated by odorous mixtures as a function of component concentrations. At first glance, this reasoning, in direct relation with
the "stimulus-response relationship," seems to be the most logical. However, it generated hardly extrapolable results as exemplified by Figure 7.2
from Köster (1969). In these diagrams, isointensity curves have been drawn.
First of all, it can be noted that they are not homothetic for a given pair of
odorous compounds at different sensation levels. Furthermore, according
to synergy and inhibition criteria chosen by this author as well as others of
the same era, each of the curves passes unpredictably from one domain to
the other without the possibility of characterizing the pair of substances by
a sole interaction number. Two main tendancies of the curves can be noted:
shaped as a Poisson's distribution curve (curves *b, c,* and *d*) and shaped as
bimodal distribution curves (curves *a, e,* and *f*).

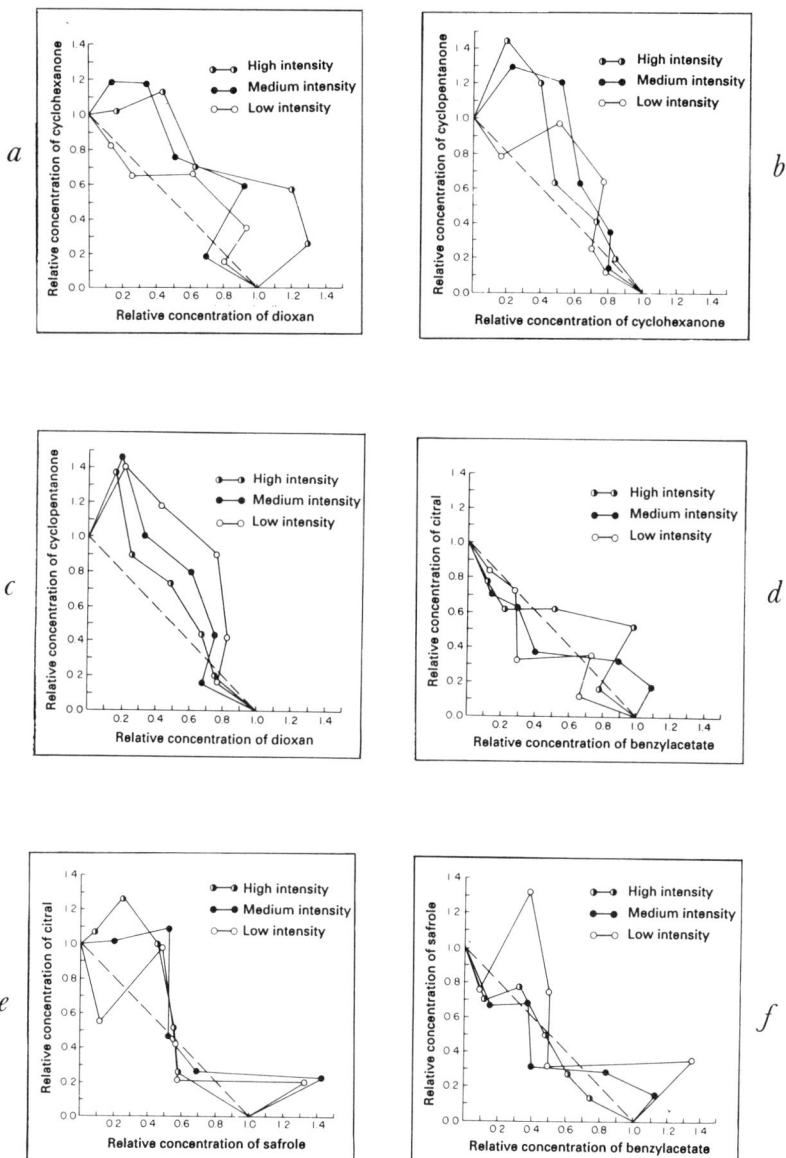

Figure 7.2. Iso-intensive curves of six binary mixtures after Köster (1969). The three levels "low", "medium" and "high" correspond to 30%, 50% and 70% of positive responses respectively. The dotted lines represent no-interaction references (for example, when a substance is added to itself). The areas under these lines are synergistic and those above the lines are inhibitory (see text).

7.2.2 Second Period: 1971–1979

7.2.2.1 Perceived Intensities of Components are Taken as Variables from Now On

For the first time, in 1971, Berglund et al. suggested that the perceived intensity (R_{AB}) of a mixture should be expressed as a function of perceived intensities R_A and R_B of components A and B and no longer as a function of the component concentrations. Since these authors were only using human verbal response, it was somehow a question of proposing a "psychopsychological" model rather than a "psychophysical" one. At this stage, Berglund et al. proposed a regressive method of R_{AB} as a function of R_A and R_B to study a mixture of two sulfides.

7.2.2.2 The Vector Model

The second really new event during this period was the proposition of Berglund et al. (1973) to submit perceived intensities to the vector addition rule (or the "parallelogram of forces") as shown in Figure 7.3. According to these authors, a unique number (cos α) characterizes a given pair of odorous substances (for a subject and a given procedure) whatever their respective concentrations and intensity levels (strong or weak). The concept of an angle characterizing a given pair of odorous substances was previously applied by

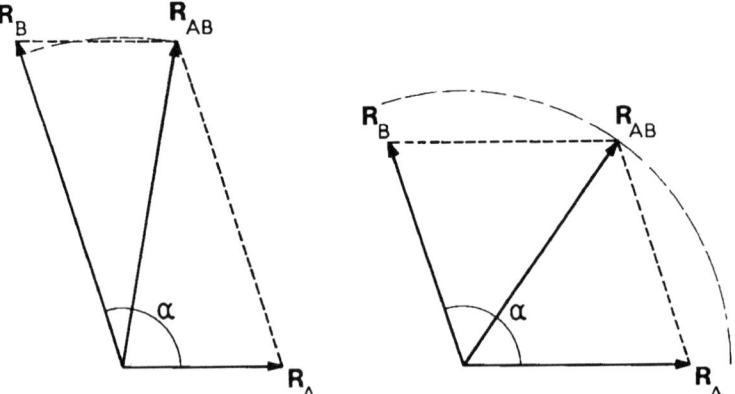

Figure 7.3. According to Berglund et al. (1973), the olfactory perception intensity of binary mixtures (R_{AB}) can be expressed as a function of perceived intensities R_A and R_B of components A and B using a vector model. In the two diagrams with the same angle of 110°, it can be seen that the vector model may explain, according to the respective values of the vectors, both hypo-addition (on the right) and subtraction (on the left) in agreement with experimental data (see second comment of Fig. 7.1) (after Laffort, 1989).

Ekman et al. (1964) to describe qualitative odorous similarity, but its application to intensity problems was only introduced by Berglund et al. in 1973.

The vector model was immediately found to be a good predictive tool (Cain and Drexler, 1974; Cain, 1975) not only because of high correlations observed between values calculated by this method and experimental values, but also because it took observational diversity into account, such as those reported in Figure 7.1 (second comment). Figure 7.3 illustrates how, according to the different values of vector components R_A and R_B, the vector sum R_{AB} can either be smaller than one of them (left) or, on the other hand, larger than each of them but smaller than their sum (right). In other words, these apparently different hypo-addition and subtraction situations can be expressed, at least in some cases, by the same angle α of the vector model. Consequently, α can be qualified as an intrinsic characteristic of the substance pair under consideration.

7.2.3 Third Period: 1979–1982

7.2.3.1 Concepts of τ (Tau) and σ (Sigma)

When two weak odors are are added, the resulting odor will most often be weak as well; in the same way, when two strong odors are added, the odor mixture will generally be strong. Therefore, most odor addition models are likely to provide relatively strong correlations between predicted and perceived odor strength. In order to overcome this difficulty when comparing models, two concepts were defined by Patte and Laffort (1979): σ (sigma) or synergy, which evaluates the deviation from additivity, and τ (tau), which can be defined as a proportion of perceived intensity. They are respectively expressed by the relationships

$$\sigma = R_{AB}/(R_A + R_B) \tag{1}$$
$$\tau = R_B/(R_A + R_B) \tag{2}$$

7.2.3.2 Definition of U and V Models

In the same publication, Patte and Laffort suggested two possible alternative models, called U and V, whose curve groups encompass those obtained from the vector model, as will be seen later. The U and V models are defined by Equations (5) and (7) and, along with the vector model, cover all the possibilities in which olfactive responses to mixtures are expressed as a function of response to the components. This is notably the case of the most simple models, often used precisely because of their simplicity. For example, the strongest component model is the V model in which $\cos \alpha = 0.5$; the simple additivity model is equivalent to the vector model when $\cos \alpha = 1$ or to the

U model when cos $\alpha = 0$. Finally, the euclidian additivity model is the vector model when cos $\alpha = 0$. By using Equations (1) and (2), the three general models can be expressed as functions of σ versus τ as follows:

Vector model

$$R_{AB} = \sqrt{R_A^2 + R_B^2 + 2\,R_A R_B \cos \alpha} \tag{3}$$

$$\sigma = \sqrt{1 + 2\,\tau\,(\tau - 1)\,(1 - \cos \alpha)} \tag{4}$$

V model

$$R_{AB} = R_A + R_B + 2 \ln f\{R_A, R_B\} \cos \alpha \tag{5}$$

(The mathematical catenary curve function or cosh function)

$$\sigma = 1 + 2 \ln f\{\tau, (1 - \tau)\} \cos \alpha \tag{6}$$

U model

$$R_{AB} = R_A + R_B + 2 \cos \alpha \sqrt{R_A R_B} \tag{7}$$

$$\sigma = 1 + 2 \sqrt{\tau\,(1 - \tau)} \cos \alpha \tag{8}$$

Figure 7.4 represents curve groups of σ as a function of τ for different cos α values in each of three general models. In order to represent eventual synergistic situations in the vector model, curves have been drawn for cos $\alpha >$ 1 which are called k in this particular case. The names U and V of the two new models have been inspired by the shape of the curves in Figure 7.4. The vector model forms curves reminiscent of a chain that are intermediate between the other two curves (in the strict sense of the term, expressed by another function [Function ln f signifies "the smallest value of the terms under consideration" (here, R_A and R_B)], "chain" refers to the form of a cable suspended by its two extremities, such as a high voltage electric cable between two pylons). By comparing these theoretical curves and the experimental data available at the time (that of Cain and Drexler, 1974 and Cain, 1975), Patte and Laffort (1979) concluded that the V model was not suitable at all and that the U model was better than or equivalent to the vector model.

Further studies (Laffort and Dravnieks, 1982; Olsson, 1986) showed that the experimental data fitting differences between U and vector models could be considered negligible. However, a difficulty encountered with the vector model should be emphasized (this is one of the reasons it is not used by Laffort and co-workers): dispersion of cos α values established by using Equation (3) are much higher than those established by using Equation (7) (Patte and Laffort, 1979; Laffort and Dravnieks, 1982). To overcome this difficulty, authors using the vector model generally limit the experimental data allowing calculation of cos α to equally strong components, the chosen criteria being $0.8 < R_A/R_B < 1.25$ (Olsson, 1986). In this particular case,

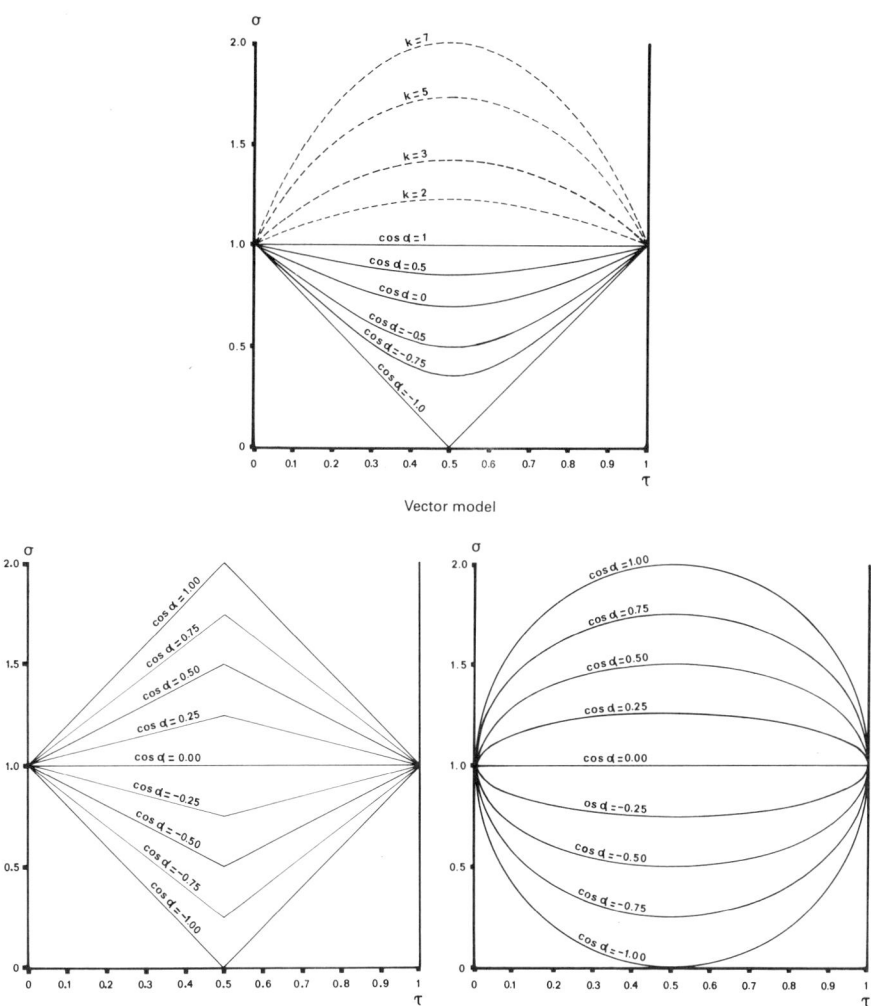

Figure 7.4. Representation of families of curves corresponding to three interaction models in a plot of σ versus τ for several hypotheses of cos α, after equations (4), (6), and (8) [in the upper diagram, the values of cos $\alpha > 1$ according to Equation (4) have been replaced by k]. Comparison of these theoretical curves with experimental data expressed in the same plot makes it possible to compare models in a more suitable way than direct comparison between predicted and experimental perceived intensities (after Patte and Laffort, 1979).

Equation (3) may be simplified as follows, using trigonometric transformation rules:

$$R_{AB} = (R_A + R_B) \cos (\alpha /2) \tag{9}$$

The advantage of Equation (9) over Equation (3) is that it no longer contains square or square root terms and thus does not increase experimental uncertainties. On the other hand, it presents the inconvenience of placing the determination of $\cos \alpha$ on a more limited set of data. Another difficulty inherent to the vector model having to do with mixtures having more than two components will be discussed later.

7.2.4 Fourth Period: 1982–1989

According to the specifications for an optimal model discussed in the introduction, there was still a considerable amount of work to be done at the beginning of this period. The main objection that could be made about vector and U models is that their respective $\cos \alpha$ values do not correspond to reference constants when a substance is added to itself. This experimental observation is a direct consequence of the power law discussed in Chapter 5 that governs stimulus-response relationships in sensorial physiology for low and medium sensation levels in electrophysiology as well as psychophysiology. Taking the equation from Chapter 5;

$$R = (C/C_0)^n \tag{10}$$

In olfaction, the exponent is generally smaller than 1. Therefore, the rule of additivity of perception cannot be applied when a substance is added to itself. From Equations (1), (2), and (10) and letting C and C' stand for two different concentrations of the same substance A, the rule can be written

$$R_{AA} = [(C_A + C_A')/C_A^0]^{n_A} \tag{11}$$
$$\sigma_{AA} = [\tau^{1/n_A} + (1 - \tau)^{1/n_A}]^{n_A} \tag{12}$$

Figure 7.5 shows families of σ versus τ curves for different values of the power law exponent n. In other words, when the experimental points of true mixtures do not generate a horizontal line in a σ versus τ diagram, it is not necessarily due to an interaction, but can be interpreted at least partially as a direct consequence of the power function.

7.2.4.1 UPL and γ Models

In order to overcome this difficulty, Laffort and Dravnieks (1982) suggested that the power law exponents and perceived intensities of the components be combined in a model called UPL (as in U model and power law). By comparing U and UPL models, these authors were able to define and index the true interaction γ (gamma), which has a value of 1 when a substance is

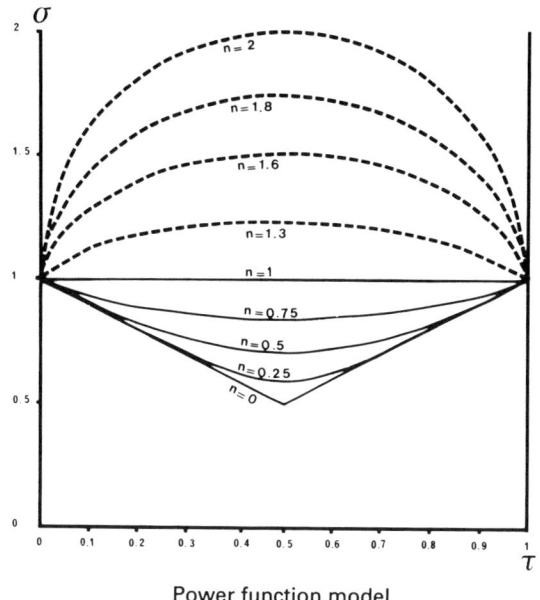

Power function model

Figure 7.5. Same representation of families of theoretical curves as in Figure 7.4 when one substance is added to itself with the hypothesis of the power function alone for different values of the n exponent according to Equations (10)–(12). When experimental points of true mixtures are not placed along the horizontal line (i.e., additivity) this may be due, at least in part, to the power function (after Laffort, 1989).

added to itself. By extension, in true mixtures a γ value equal to 1 was interpreted to show an absence of interaction and a value different from 1 as a true interaction.

The formulation of these two concepts will not be discussed in detail here since their generalization will be examined in the following step. The following is limited to showing how families of curves of a substance added to itself can be equally drawn from a model derived from a U model in which the value of cos α is deducted from exponent n.

Taking Equation (7)

$$R_{AB} = R_A + R_B + 2 \cos \alpha_{AB} \sqrt{R_A} \, R_B \ (U \text{ model})$$

Supposing that a substance is mixed with itself, that is to say, that $A = B$, the equation becomes

$$R_{AA} = R_A + R'_A + 2 \cos \alpha_A \sqrt{R_A \, R'_A}$$

Finally, supposing that responses R_A and R'_A to the two mixture components are identical ($R_A = R'_A$), one gets

$$R_{AA} = 2R_A + 2R_A \cos \alpha_A = 2R_A(1 + \cos \alpha_A)$$

According to (10), the above equation can be written

$$(2C_A/C_A^0)^{n_A} = 2(C_A/C_A^0)^{n_A}(1 + \cos \alpha_A)$$

or

$$2^{n_A} = 2(1 + \cos \alpha_A) \text{ or } \cos \alpha_A = 2^{n_A - 1} - 1 \tag{13}$$

From this equation, the $\cos \alpha_{UPL}$ values of an $A + B$ mixture were obtained by weighing the proportions of components R_A and R_B, with the help of an equation analogous to (19), and reintroducing them into an equation similar to (7), in place of $\cos \alpha_U$ values.

7.2.4.2 UPL2 and Γ Models

Taking into account the way that Equation (13) has been established, it seems obvious that it is not rigorously exact except when the two responses are identical. Laffort and Dravnieks (1982) showed that when one of the responses is nine times larger than the other ($R_A = 9R_B$), errors made in this way remain acceptable. However, limit conditions being impossible, a generalization of Equation (13) was suggested by Laffort et al. (1984, 1989) and Laffort (1989) in the form of Equation (17). Taking this modification into account, model UPL became UPL2 (as in "second version") and the γ index became Γ (capital gamma). The principle of the model is given by Equation (14) and its details are given by Equations (15)–(19):

$$\Gamma = \frac{1 + \cos \alpha_U}{1 + \cos \alpha_{UPL2}} = f \nearrow \begin{array}{l} R_{AB}, R_A, R_B \\ \\ \searrow n_A, n_B, R_A, R_B \end{array}$$

$$\Gamma = 1 \text{ (no interaction)} \tag{14}$$
$$\Gamma > 1 \text{ (synergy)}$$
$$\Gamma < 1 \text{ (inhibition)}$$

The numerator in (14) is directly deduced from Equation (7), which defines the U model:

$$\cos U = \frac{R_{AB} - R_A - R_B}{2\sqrt{R_A R_B}} \tag{15}$$

The establishment of the denominator in (14) necessitates the previous definition of P (proportion), $\cos \alpha_A$, $\cos \alpha_B$, and $\cos \alpha_{UPL2}$. One successively gets

$$P = \frac{R_B^{1/n_B}}{R_A^{1/n_A} + R_B^{1/n_B}} \tag{16}$$

$$\cos \alpha_A = \frac{1 - P^{n_A} - (1 - P)^{n_A}}{2 P^{n_A/2} (1 - P)^{n_A/2}} \tag{17}$$

$$\cos \alpha_B = \frac{1 - P^{n_B} - (1 - P)^{n_B}}{2 \, P^{n_B/2} \, (1 - P)^{n_B/2}} \tag{18}$$

$$\cos \alpha_{UPL2} = \frac{R_A \cos \alpha_A + R_B \cos \alpha_B}{R_A + R_B} \tag{19}$$

Equation (16) could have been written more simply by taking (10) into account. In fact, it is preferable that denominator variables in Equation (14) be the same as numerator variables (the responses, not the concentrations).

Equations (17) and (18) are given a priori and are not derived from previous relationships. They are justified by their validity in particular cases, which can thus be easily verified. In this way, (13) is a particular case of (17) in which $R_A = R_B$. The general nature of (17) has been verified for other relationship hypotheses of R_A and R_B values (Laffort et al., 1989).

Finally, Equation (19) is identical to the equation previously used in the UPL model (Laffort and Dravnieks, 1982); this is a weighing of $\cos \alpha_A$ and $\cos \alpha_B$ values by response proportions R_A and R_B.

As it is and without prejudging eventual later improvements, the Γ index fulfills the two conditions of the "specifications" discussed in the introduction: a unique number to characterize a given pair of odorous substances (in a determined experimental procedure and subject context), this number having a constant reference value when a substance is added to itself. We still need to describe several examples of application and extension of this concept and to compare it to several other models suggested in the recent past.

7.3 Several Cases of Γ Interaction Index Application

In the examples cited hereafter, it is sufficient to take the simplified equation (14) into account to approach the principle.

7.3.1 Visualization of Γ Significance

In figure 7.1, where the already cited Cain and Drexler's (1974) experimental data is drawn, three theoretical curves, each representing a model, have been drawn: that which is best fitted to the experimental points (U model), that which is least fitted (additivity), and that which is situated between the two others, which corresponds to the UPL2 model. As an example, Table 7.1 represents the calculations that have made it possible to draw these three curves for one of the four mixtures: that of lavandin essence and pyridine.

It could be said that classic criteria of interaction evaluation (the σ concept, for instance) somehow represent a "distance" between the two extreme curves, whereas Γ represents a "distance" between the intermediate curve and the bottom curve. It has been previously seen (14) that this second point of view is more appropriate for defining a criteria of true interaction (comment 3 of Figure 7.1).

Table 7.1. Application of Equations (14)–(19) to the Mixture of Lavandin Oil (A) and Pyridine (B) After Data from Cain and Drexler (1974). Olfactory Perceived Intensities R_A, R_B, and R_{AB} are Expressed in Arbitrary Units (Direct Estimates). The Smooth Lines in Figure 7.1 are Obtained from Equation (7) in Which the Value of R_A is Fixed at 9.4 Units, R_B Varies Continuously from 0 to 25 Units, and cos α is Taken for Each of Three Curves at Zero, Respectively: cos α_U and cos α_{UPL2}. The Shape of the Curves are R_{AB} as a Function of R_B. (cos α and Γ Represent Average Values.)

Experimental Data					Calculations					
Lavandin R_A	Pyridine R_B	Mixture R_{AB}	Exponents n_A	n_B	Cos α_U	P	Cos α_A	Cos α_B	Cos α_{UPL2}	Γ
9.4	2.5	6.8	0.18	0.40	−0.53	4×10^{-5}	−0.20	−0.07	−0.17	0.57
9.4	6.3	9.1	"	"	−0.43	4×10^{-4}	−0.25	−0.10	−0.19	0.70
9.4	10.6	11.5	"	"	−0.43	1×10^{-3}	−0.28	−0.13	−0.20	0.72
9.4	16.5	14.2	"	"	−0.47	4×10^{-3}	−0.31	−0.17	−0.22	0.68
9.4	21.0	18.0	"	"	−0.44	8×10^{-3}	−0.32	−0.19	−0.23	0.72
				Averages	$\cos\alpha_U$ −0.46				$\cos\alpha_{UPL2}$ −0.20	$\overline{\Gamma}$ 0.68

In this figure, it should equally be noted that the line of the intermediate curve (UPL2) does not allow one to be aware of the subtraction phenomena. Reciprocally, it could be said that any subtraction phenomena observed indicates a true inhibition; in the same way in the line in Figure 7.5, all experimental points situated below the triangle of coordinates (0.0,1.0), (0.5, 0.5), and (1.0, 1.0) would also indicate a true interaction. However, in both cases, only Γ (or another index constructed according to a similar principle) makes it possible to quantitatively estimate the degree of true interaction.

7.3.2 Electrophysiological Responses

Γ values for series of pairs of odorous substances can be established from global or unitary electrophysiological responses. Comparison at different integration levels for a given animal species could provide information about the evolution of the degree of interaction during olfactive information processing. At the present time, the only Γ values of electrophysiological origin that have been published concern electroantennographic (EAG) responses of honeybees to 18 pairs of substances (Laffort et al., 1984, 1989). These authors found $\Gamma = 1$ values for 12 of these pairs of odors and Γ significantly lower than 1 (inhibition) for the 6 others. Because simultaneous depolarization and hyperpolarization potentials were observed only for these 6 remaining pairs, Laffort et al. suggested that an electrical neutralization may occur during measurement, inaccurately reflecting information transmitted to the brain. Consequently, these authors concluded that there is probably no true interaction at the periphery in the honeybee. This hypothesis could be verified by single unit recording.

Other similarly directed information [Kuiper and Köster, 1988 (personal communication)] concerns an experiment in progress on EOG (electroolfactogram) response of frogs to three pairs of odorous substances (three combinations of Pentanone-3, Pentanone-2, and ethyl acetate). In the three cases studied, only classic negative potentials were observed and all three Γ values were found to be not significantly different from 1. It would be interesting to submit the numerous electrophysiological works carried out on the subject of mixtures to this relatively recent type of processing in such a way as to verify the hypothesis that seems to predominate for the moment: in general olfaction, there would not be true interaction at the periphery; the stimuli would be in competition with regard to the receptors, but each with its own effectiveness (threshold, power law exponent). The very weak specificity of the receptors (the blurred part of the periperic olfactive image) is consistent with this hypothesis. On the other hand, at higher levels of integration, true inhibitions appear (in olfaction, it does not seem that cases of $\Gamma > 1$—that is to say, true synergy—have been observed). For the moment, the only published Γ values concerning high integration levels have been established from psychophysical human subject responses, which will be discussed later.

7.3.3 Invariability of the Γ Index

In vector and U models, the supposed invariability of cos α necessarily leads to a symmetrical trace of σ as a function of τ. In fact, the experimental points are often placed asymmetrically in this type of diagram. Figure 7.6, established from psychophysiological responses obtained by Olsson (1986) for 41 different pyridine-dimethyl disulfide (DMDS) mixtures, gives an example. The hypothesis that Γ invariability makes it possible to draw a predictive curve, as well as an asymmetrical curve (asymmetrical?), can be seen as well.

In practice, equations (14)–(19) have been applied to experimental data in the following way.

1. First, individual Γ values are established for each of 41 points. An average value of 0.94, which will be taken as characteristic of this pair of odors, is found.
2. Next, with each particular mixture then having its own cos α_{UPL2} value, the denominator of Equation (14) is multiplied by 0.94, which leads to a predicted numerator value and thus a cos α_{U} value, this last value having been used to calculate a predicted value of R_{AB}.

The succession of R_{AB} values constitute the continuous trace of Figure 7.6. The correlation coefficients obtained between the experimental and predicted R_{AB} values are equivalent (r = 0.98, 0.99) when vector, U, or Γ models are used. On the other hand, correlations for σ_{AB} values are only 0.61 with vector and U models but they are 0.72 with the Γ index. It should be

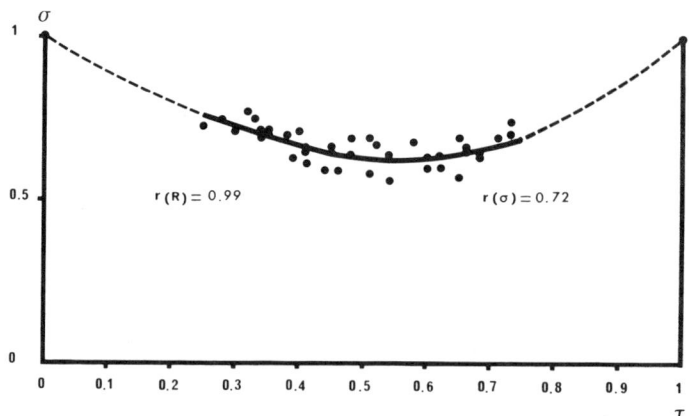

Figure 7.6. Representation in a plot of σ versus τ, psychophysical responses to 41 different mixtures of pyridine and dimethyl disulfide (DMDS), reported by Olsson (1986). The solid curve was drawn using Equations (14)–(19) for the Γ value to this experimental data; the dotted curves have been extrapolated. The theoretical curve as well as the experimental points distribution are asymmetrical, unlike the U or vector model curves (after Laffort, 1989).

noted that if $\Gamma = 1$ is used instead of 0.94, the correlations are practically unchanged ($r = 0.71$ for σ_{AB}).

This recent example, which is interesting because of the high number of experimental points, confirms the interest of an asymmetrical model, as well as the higher degree of Γ invariability in comparison with cos α vector and U models.

7.3.4 A Priori Construction of Isointensity Curves

Before the γ and Γ indexes were defined, several authors had already mentioned that apparent synergies and inhibitions could result from the power law when perceived intensities were expressed as a function of component concentration (Cain, 1975; Moskowitz and Barbe, 1977). According to Gregson (1986), this phenomenon has been known for over a century in other sensorial modalities and has been called "Fechner's paradox." (In fact, "Fechner's paradox" has been observed in vision and hearing only in symmetrical situations of true interaction. The influence of differences in the power law exponents of the components of mixtures has never been observed in these sensorial modalities; on the contrary, it occurs only in olfaction.) Be that as it may, once Γ was defined and its invariability was demonstrated for pairs of substances, it was tempting to express a priori perceived intensities of mixtures as a function of component concentrations using Equations (10) and (14)–(19). Unfortunately, by doing this, one ends up with inextricable equations known as implicit equations. Laffort et al. (1984, 1989) have surmounted this difficulty by using iterative processing for two Γ values equal to 0.75 (true inhibition) and 1 (absence of inhibition), respectively. Four families of curves with the latter hypothesis are illustrated in Figure 7.7. The form of the curves are shown to be dependant on the exponent values and the distance between them and the dotted diagonal as they come closer to it when the exponent values approach 1. Sometimes an absence of homothesis is observed inside the same family of curves.

Above all, the large variety of obtained curves can be noticed, which are reminiscent of some of the experimental curves in Figure 7.2: those called "Poisson's curves," for example (*b, c,* and *d*). Families of curves for $\Gamma = 0.75$, not given here, bring to mind experimental curves shaped as "bimodal distribution curves" (*a, e,* and *f*). It is easier to understand why it was so difficult to see coherence in experimental results obtained before 1971. This example makes it equally possible to confirm the synthetic character (integrating) of the Γ index.

7.3.5 Several Γ Index Values Obtained by Psychophysical Response

In a long-term predictive perspective of human olfactive perception of mixtures, it would be desirable to progressively accumulate Γ index values. At

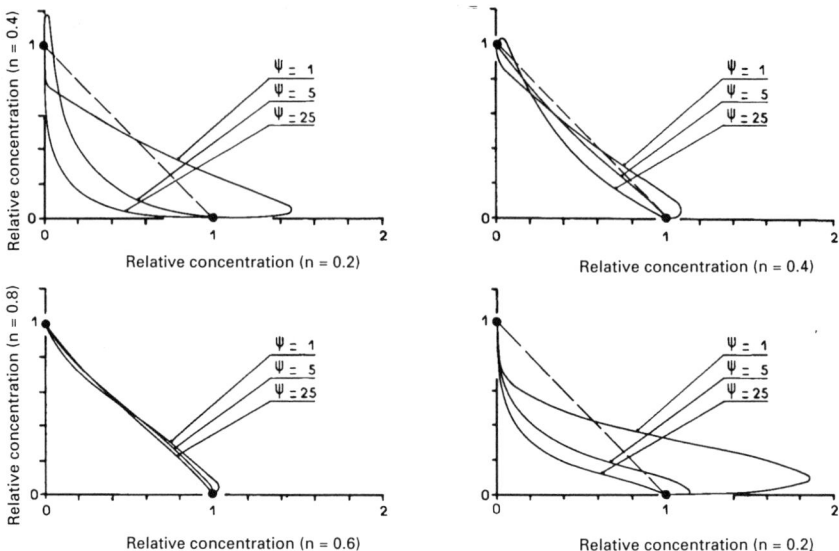

Figure 7.7. Theoretical olfactory isointensity curves obtained by computer simulation with the hypothesis that $\Gamma = 1$. The three levels of perceived intensity correspond respectively to 1, 5, and 25 times the perceived intensity at threshold. The diagonal dotted line corresponds to $n_A = n_B$, which is the case of a substance added to itself. The shape of the curves are similar to some experimental curves in Figure 7.2 (*b, c,* and *d*). One can note, as in this experimental figure, the absence of homothesis for families of curves at different sensation levels and the fact that the same curve sometimes passes succesively through the areas of so called synergy and inhibition (below and above the diagonal dotted line, respectively). The low value of the exponents and the space between them accentuates the phenomenon (after Laffort et al., 1989).

the moment, the situation is similar to that of the 1960s concerning power law exponents of pure substances: they were known for less than 10 products as opposed to the nearly 200 presently known (Devos et al., in preparation). On one hand, only an accumulation of Γ values will make it possible to cross check data from different sources and, on the other hand, to test cause-effect relationship hypotheses with the concerned molecular structures.

Most of the known data have already been cited in this chapter. The restricted list that, at this time, may be considered as exhaustive is given in Table 7.2. These results are too few to draw hypotheses on the relationships between the nature of the components and the Γ values of a mixture. It is already known that odor significance is implicated in higher brain feedback control of bulb activity (cf. Chapter 1). It is certainly also involved in relationships in the central brain, either specifically olfactive or implicating other tracts, notably the trigeminal nerve. Thus, one can expect true interactions to be associated with qualitative or hedonic distances and hence with

Table 7.2.   Several Values of the Γ Index Obtained by Psychophysical Responses. (The Calculations are Those of Laffort and Co-Workers; Data Between Brackets Corresponds to Standard Deviations and to the Number of Experimental Points.)

Substances	Γ	(σ ; N)	References
Pyridine-Linalyl acetate	0.72	(0.12; 6)	Cain and Drexler (1974)
Pyridine-Linalool	0.84	(0.14; 6)	"
Pyridine-Lavandin oil	0.68	(0.07; 6)	"
Pyridine-Dimethyl disulfide	0.94	(0.06; 41)	Olsson (1986)
Propanol-Amyl butyrate	0.78	(0.17; 30)	Cain (1975)
Geraniol-Eucalyptol	0.65	(0.14; 30)	De Wijk (1989)

the molecular properties that govern them. The significant increase in the number of reliable Γ measurements, which make it possible to verify and refine these hypotheses on numerical experimental bases, should be effective in the near future on the basis of experiments in progress in four or five laboratories and implementation of their previous works.

7.4 Other Models Suggested in the Recent Past

7.4.1 "Simple" Models

Apart from the additivity model, which is totally inadequate in olfaction, the strongest component model and the euclidean additivity model (already discussed in Section 7.2.3.2.) are included in this category. They are respectively expressed by the following equations:

Strongest component

$$R_{AB} = \text{Sup} \{R_A, R_B\} \tag{20}$$

[The function Sup signifies "the highest value of the terms under consideration" (here, R_A and R_B)]
Euclidean additivity

$$R_{AB} = \sqrt{R_A^2 + R_B^2} \tag{21}$$

On the basis of different experimental data, these models have been tested against the more complicated ones discussed above (Laffort and Dravnieks, 1982). The results show that they are less sensitive and less predictive, but taking their simplicity into account, they could be useful for estimating the order of magnitude of expected experimental results in a predictive perspective. In other words, the values that they make possible to predict are never absurd and can provide useful guidelines when more complex models are applied.

Another model that can be included in this category is the one suggested by Guadagni et al. (1963) for subliminal levels. The principle of this model is the addition of odor units (OU):

Guadagni et al. model

$$OU_{AB_A} = C_A/C_B^0 + C_B/C^0 \tag{22}$$

The mixture would be odorless when OU_{AB} is smaller than 1 and odorous in the opposite case ($OU_{AB} > 1$). This empirical rule has been applied in the food industry with success in some cases and failure in others (Guadagni et al., 1963, 1966; Buttery et al., 1974). At the present time this model appears to have been abandoned.

7.4.2 Schutte's Model (1985)

This model, like the Γ index, in its principle takes the exponent from the power law into account as well. However, to derive its formulation the author assumes a constant value for the exponents regardless of the nature of the concerned substances. Such a restrictive assumption, contrary to experiments, removes most of the interest that this model could present.

To be more specific, Schutte's model implies two assymetical interaction indexes S_{AB} and S_{BA} that represents a regression with respect to the vector, U and Γ models. S_{AB} and S_{BA} values are equal to 1 when a substance is added to itself but in the case of true mixtures, they do not allow one to discriminate the apparent interaction (as resulting from the exponents) from the interaction proper.

According to its author, this model presents another difficulty due to its non-applicability to levels close to threshold value.

7.4.3 The Constant Ratio Model ("Equiratio")

This model was suggested under the name "equiratio" by Frijters et al. (1983, 1984, 1987) who successfully applied it to sapid mixtures. Its principle is as follows: the concentration-response straight line in log-log coordinates for a series of dilutions of a mixture would be in between the straight lines corresponding to the complements of the mixture, based on the ratio of the mixture's components. Although the name is new, the method was already suggested more than 30 years ago by Sales (1958) for odorous mixtures and used in the gas industry over a period of many years (Borelli and Angleraud, 1965; Blanchard, 1976). However, Angleraud (1968) then Blanchard (1976) demonstrated that the assumption was not experimentally verified as shown by Figure 7.8. The nonapplicability of this model to olfaction was recently confirmed by Schiet and Frijters (1988).

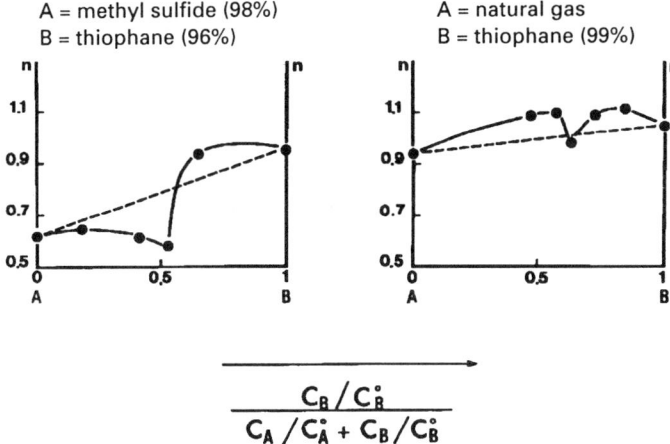

Figure 7.8. Variation of the exponent of the power law for two types of odorous mixtures containing thiophane (or tetrahydrothiophene) (after Angleraud, reported by Laffort, 1968). Each point corresponds to six or seven dilutions of the same mixture (consequently all having the same ratio C_A/C_B). The nonlinear line of the points, for the exponents as well as the thresholds, was confirmed by Blanchard (1976) for three binary thiol mixtures. These results showed, well before their successful application to tasting from 1983 on, that the model known as "equiratio" could not be applied to olfaction.

7.5 Extension to Mixtures with More than Two Components

The authors of all the previously cited models have suggested formulas allowing them to be extended to include more complex mixtures. This section will be limited to vector and U model general formulas. The justification of these extensions has been more or less verified according to the case.

7.5.1 The Vector Model

In 1974, Berglund suggested the equation

$$R'(p) = \sqrt{\sum_{i=1}^{p} R_i^2 + \sum_{i=1}^{p-1} \sum_{j=i+1}^{p} 2\, R_i R_j \cos \alpha_{ij}} \tag{23}$$

in which p is the number of components of a mixture. Moskowitz (1979) then Laffort and Dravnieks (1982) have shown, from experimental data, that this equation sometimes generates imaginary predicted values (i.e., square roots of negative values) for tertiary and quaternary mixtures. More precisely, it can be shown that this can happen each time the following condition is not verified:

$$\cos \alpha \geq \frac{1}{1 - p} \tag{24}$$

For example, for an infinite number of components, $\cos \alpha$ must not be smaller than 0; in other words, α must not be larger than 90°, which is contrary to the usual experimental data (most often one finds α values that are equal or larger than 110°). Of course, there is an exact and more complex formulation of the addition of more than two vectors that does not present this inconvenience, but it has not been checked against experimental data.

7.5.2 U, UPL2, and Γ Models

Laffort and Dravnieks (1982) have suggested the following formulations:

$$R_{(p)} = \frac{1}{p} \left(\sum_1^p R_{(p-1)} + \sum_1^p R_{(1)} \right) + \frac{2 \cos \alpha \, (p)}{p} \sum_1^p \sqrt{R_{(p-1)} R_1} \tag{25}$$

or $\cos \alpha$ is defined by

$$\cos \alpha \, (p) = \frac{\displaystyle\sum_{BIN\,=\,1}^{p(p-1)/2} \cos \alpha_{(BIN)} \, R_{(BIN)}}{\displaystyle\sum_{BIN\,=\,1}^{p(p-1)/2} R_{(BIN)}} \tag{26}$$

In Equation (26), R_{BIN} is what was previously called R_{AB} and $\cos \alpha_{(BIN)}$ is what was called either $\cos \alpha_U$ or $\cos \alpha_{UPL2}$, depending on the case.

These authors have tested the compared validity of several models (on two groups of tertiary and quatenary psychophysical data), from which those with occuring power law exponents were excluded, these exponents not being known. For one of the data groups, the vector and U models were found to be equally good and the "simple" models were slightly less effective. For the second data group, the U model was by far the best.

7.5.3 "Simple" Models

These models do not present any difficulty in generalizing their formulation. The comments that concern them are the same as for binary mixtures.

7.5.4 Schutte's Model

This author has suggested an extended formula of his model, but has not verified its validity on experimental bases.

7.5.5 The Frijter et al. Model

Extension of the constant ratio model does not present any particular difficulty either, but its validity has not been tested on odorous mixtures with more than two components.

7.5.6 General Comments

The overall impression that emerges from this brief survey on multiple odorous mixtures is that an experimental effort could be developed on this subject in such a way so as to better test (and possibly improve) existing models. It has not been possible to apply the Γ model, in particular, at the present time.

7.6 Models Applied in Other Modalities

7.6.1 The Models Proposed in Taste by Hyman and Frank (1980)

These authors propose two indices.

1. The independant component index (ICI), defined as follows:

$$ICI = R_{AB} \tag{27}$$

This index is, of course, equivalent to σ, defined in (1). As has already been seen, σ is a useful tool in taste models (comparison between predicted and experimental σ values), but it cannot be considered as a suitable interaction index since, when a substance is added to itself, its value is not constant.

2. The mixture discrimination index (MDI), also called *substitution model*. It is defined by

$$MDI = R_{AB}/R' \tag{28}$$

in which R' is "the response to either component at a concentration equal to the sum of two concentrations of one chemical equivalent in effect to the components of the mixture." Applying the power law [Equation (10)] to this definition, it becomes two equations:

$$R' = (R_A^{1/n_A} + R_B^{1/n_A})^{n_A} \tag{29}$$

and

$$R' = (R_A^{1/n_B} + R_B^{1/n_B})^{n_B} \tag{30}$$

These two equations are only equivalent when the exponents for the two components are the same. That is, as has already been seen, a restrictive view of the reality (even in taste, but perhaps less dramatically than in olfaction). In the particular case where the two exponents are the same, it can be easily demonstrated that MDI is equivalent to the γ index, which is in turn a particular case of the Γ index for equally strong components. In other words, the Γ model can be considered as a generalization of the substitution model.

7.6.2 The Isobole Approach Applied in Pharmacology

According to the authors using this method (for example, Berembaum, 1981) an index of interaction I can be calculated as follows:

$$I = \frac{C_A}{C'_A} + \frac{C_B}{C'_B} \tag{31}$$

in which C'_A and C'_B are the concentrations of A and B that would individually elicit the same response as the mixture. Taking the hypothesis of the power function for single components [Equation (10)], it becomes

$$I = \left(\frac{R_A}{R_{AB}}\right)^{1/n_A} + \left(\frac{R_B}{R_{AB}}\right)^{1/n_B} \tag{32}$$

Recently, Sühnel (1992) applied this model to olfaction and taste.

Without entering into details, it can be said that this interaction index satisfies one of the two conditions defined as suitable at the beginning of this chapter: if one odorant is added to itself, the interaction index I remains constant, equal to 1. By contrast, I values do not characterize given pairs of odorants each time the exponents are different (one can estimate values of I by adding x and y coordinates of isointensity curves in Figures 7.2 and 7.7). This concept does not allow any prediction; only a description. This is a return to "the period before 1971" in olfaction. Pharmacologists could possibly benefit from the investigations carried out on this topic in olfaction during the last few years.

7.7 Conclusion

In the light of recently exposed results, an assessment of synergy and inhibition in olfaction can be established as follows.

1. At the periphery of the olfactory system there would not be an interaction, strictly speaking, but only olfactory stimuli competition with regards to the receptors. This competition, which is called apparent inhibition, is quite well reflected by the expression called $\cos \alpha_{UPL2}$, even if it is established from data integrated by the central nervous system (psychophysical). This expression makes it possible to bring forward the synergistic character of substances with weak power law exponents and the

inhibiting character of substances with strong exponents. Works in progress on establishing standardized banks of power law exponents (Chapter 2) and on predictive attempts from threshold values (Chapter 5) will be immediately applicable in this field.

2. At the totally integrated level, a *stricto sensu* inhibition will appear, more or less strong depending on the substance pairs, well characterized by the Γ index. At this time, the too restricted number of known Γ values does not allow hypothesis formulation on the connection between degrees of thus characterized true interaction and other olfactory or molecular properties.

Bibliography

Ache, B. W., Gleeson, R. A., Thomson, H. A., 1988. Mechanisms for mixture suppression in olfactory receptors of the spiny lobster. *Chem. Senses* **13**, 425–434.

Angleraud, O., citée par Laffort, 1968, *op. cit.*

Backman, E. L., 1918. The olfactology of the methyl-benzol series. *Onderz. Physiol. Lab. Utrecht* **5**, 349–364.

Baker, R. A., 1964. Response parameters including synergism-antagonism in aqueous odor measurement. *Ann. N.Y. Acad. Sci.* **116**, 495–503.

Berembaum, M. C., 1981. Criteria for analyzing interactions between biological active agents. *Adv. Cancer Res.* **35**, 269–335.

Berglund, B., 1974. Quantitative and qualitative analysis of industrial odors with human observers. *Ann. N.Y. Acad. Sci.* **237**, 35–51.

Berglund, B., Berglund, U., Lindvall, T., 1971. On the principle of odor interaction. *Acta Psychologica* **35**, 255–268.

Berglund, B., Berglund, U, Lindvall, T., 1976. Psychological processing of odor mixtures. *Psychol. Review* **83**, 432–441.

Berglund, B., Berglund, U, Lindvall, T., Svensson, L. T., 1973. A quantitative principle of perceived intensity summation in odor mixtures. *J. Exper. Psychol.* **100**, 29–38.

Blanchard, J., 1976. Relations entre l'odeur d'un gaz et sa teneur en mercaptans. *Congrès Assoc. Tech. Industr. Gaz en France (AIG)*, 62, rue de Courcelles, Paris, 1988.

Borelli, F., Angleraud, O., 1965. Studi pratici sull'odorizzazione intensita di odore di differenti gas. *Gas Roma* Suppl. 7-8, 27 pp.

Buttery, R. G., Black, D. R., Guadagni, D. G., Ling, L. C., Connolly, G., Teranishi, R., 1974. California bay oil. I. Constituents, odor properties. *J. Agric. Food Chem.* **22**, 773–777.

Cain, W. S., 1975. Odor intensity: mixtures and masking. *Chem. Senses Flavor* **1**, 339–352.

Cain, W. S., Drexler, M., 1974. Scope and evaluation of odor counteraction and masking. *Ann. N.Y. Acad. Sci.* **237**, 427–439.

Derby, C. D., Ache, B. W., Kennel, E. W., 1985. Mixture suppression in olfaction: electrophysiological evaluation of the contribution of peripheral and central neural components. *Chem. Senses* **10**, 301–316.

Devos, M., Patte, F., Rouault, J., Laffort, P., Van Gemert, L. J., 1990. *Standardized human olfactory thresholds in air.* IRL Press, Oxford, 165 pp.

De Wijk, R. A., 1989. Temporal factors in human olfactory perception, Thesis of Utrecht University (the Netherlands), 255 pp.

Ekman, G., Engen, T., Kunnapas, T., Lindman, R., 1964. A quantitative principle of qualitative similarity. *J. Exper. Psychol.* **68**, 530–536.

Etcheto, M., Pichon, Y., Laffort, P., 1982. Sensibilité antennaire de l'abeille à des mélanges binaires de substances odorantes. *J. Physiol. (Paris)* **78**, 266–269.

Frijters, J. E. R., De Graaf, C., 1987. The equiratio taste mixture model successfully predicts the sensory response to the sweetness intensity of complex mixtures of sugars and sugar alcohols. *Perception* **5**, 615–628.

Frijters, J. E. R., Ooude, Ophuis, P. A. M., 1983. The construction and prediction of psychophysical power functions for the sweetness of equiratio sugar mixtures. *Perception* **12**, 753–767.

Frijters, J. E. R., De Graaf, C., Koolen, H. C. M., 1984. The validity of the equiratio taste mixture model investigated with sorbital-sucrose mixtures. *Chem. Senses* **9**, 241–248.

Gregson, R. A. M., 1980. A model of paradoxical odour mixture perception. *Chem. Senses Flavor* **5**, 257–269.

Gregson, R. A. M., 1986. Quantitative and qualitative intensity components of odour mixtures. *Chem. Senses Flavor* **11**, 455–470.

Guadagni, D. G., Buttery, R. G., Harris, J., 1966. Odour intensities of hop oil components. *J. Sci. Fd. Agric.* **17**, 142–144.

Guadagni, D. G., Buttery, R. G., Okano, S., Burr, H. K., 1963. Additive effect of subthreshold concentrations of some organic compounds associated with food aromas. *Nature* **200**, 1288–1289.

Hyman, A. M., Frank, M. E., 1980. Effects of binary taste stimuli on the neural activity of the hamster chorda tympani. *J. Gen. Physiol.* **76**, 125–142.

Jones, F. N., Woskow, M. H., 1964. On the intensity of odor mixtures. *Ann. N.Y. Acad. Sci.* **116**, 484–494.

Kendall, D. A., Neilson, A. J., 1966. Sensory and chromatographic analysis of mixtures formulated from pure odorants. *J. Food Sci.* **31**, 268–274.

Köster, E. P., 1968. Relative intensity of odour mixtures at supra-threshold level. *Olfactologia Suppl.* **2**, 29–41.

Köster, E. P., 1969. Intensity in mixtures of odorous substances. In *Olfaction and Taste II,* edited by C. Pfaffmann. The Rockefeller University Press, New York, pp. 142–149.

Köster, E. P., Mac Leod, P., 1975. Psychophysical and electro-physiological experiments with binary mixtures of acetophenone and eugenol. In *Methods in Olfactory Research,* edited by D. G. Moulton, A. Turk, and Johnston Jr., Academic Press, London-New York-San Francisco, pp. 431–444.

Laffort, P., 1968. Interactions quantitatives dans un mèlange d'odeurs: niveau liminaire. *Olfactologia Suppl.* **1**, 95–104.

Laffort, P., 1989. Models for describing intensity interactions in odor mixtures: a reappraisal. In *Perceptions of Complex Smells and Tastes,* edited by D. G. Laing, W. S. Cain, R. L. McBride, and B. W. Ache. Academic Press, Sydney, pp. 205–223.

Laffort, P., Dravnieks, A., 1982. Several models of suprathreshold quantitative olfactory interaction in humans applied to binary, ternary and quaternary mixtures. *Chem. Senses* **7**, 153–174.

Laffort, P., Etcheto, M., Patte, F., Marfaing, P., 1989. Implications of power law exponent in synergy and inhibition of olfactory mixtures. *Chem. Senses* **14**, 11–23.

Laffort, P., Patte, F., Etcheto, M., 1984. Inférence de la "loi de puissance" dans les phénoménes de synergie et d'inhibition de la perception olfactive. *J. Physiol. (Paris)* **79**, 63A–64A.

Laing, D. G., Willcox, M. E., 1983. Perception of components in binary odor mixtures. *Chem. Senses* **7**, 249–264.

Laing, D. G., Panhuber, H., Willcox, M. E., Pittman, E. A., 1984. Quality and intensity of binary odor mixtures. *Physiol. Behav.* **33**, 309–319.

Mac Leod, P., 1968. Interactions quantitatives dans un mélange d'odeurs. Étude électrophysiologique. *Olfactologia, Suppl.* **1**, 23–27.

Moskowitz, H. R., 1979. Utility of the vector model for higher-order mixtures: a correction. *Sensory Process* **3**, 366–369.

Moskowitz, H. R., Barbe, C. D., 1977. Profiling of odor components and their mixtures, *Sensory Process* **1**, 212–226.

Müller, W., 1979. Effectiveness of mixtures of odorants in the frog's olfactory receptors. *Chemoreception Abstracts* **7**, 953.

Olsson, M., 1986. Perceptual models of odor interaction for binary mixtures. National Institute of Environmental Medicine, S-104 Stockholm, Report 8/1986, 22 pp.

Olsson, M. J., 1992. Evaluation of interaction in olfactory and taste mixtures. *Chem. Senses* (in press).

Passy, J., 1895. Revue générale sur les sensations olfactives. *Année Psychologique* **2**, 363–410.

Patte, F., Laffort, P., 1979. An alternative model of olfactory quantitative interaction in binary mixtures. *Chem. Senses Flavor* **4**, 267–274.

Rosen, A. A., Peter, J. B., Middleton, F. M., 1962. Odor thresholds of mixed organic chemicals. *J. Water Pollut. Control Fed.* **34**, 7–14.

Sales, M., 1958. Odeur et Odorisation des gaz. *7ème Congr. Inter. Ind. Gaz, Roma*, **36**, 166 pp.

Schutte, L., 1985. A new model for describing interactions in odour mixtures. In *Progress in flavour research 1984*, edited by J. Adda. Elsevier, Amsterdam (The Netherlands), pp. 3–13.

Shiet, F. T., Frijters, J. E. R., 1988. An investigation of the equiratio-mixture model in olfaction psychophysics: a case study. *Percep. Psychophys.* **44**, 304–308.

Zwaardemaker, H., 1907. Uber die Proportion der Geruchskompensation. *Arch. Anat. Physiol. (Leipzig) Suppl.* **31**, 59–70.

Zwaardemaker H., 1925. *L'Odorat*. G. Doin, Paris, 305 pp.

8

Sources of Volatile Organic Compounds and Study of Odorous Pollution

P. LeCloirec, M. Gueux, H. Paillard, and C. Anselme

8.1 Generalities-Classification of Odorous Sources

Many agricultural, industrial, and even domestic activities are sources of olfactive nuisances. Arbitrarily, it is possible to group them in two categories according to their origin (Cahiers Techniques/Technical Notebooks,1984).

1. *Odors due to fermentation.* Many types of waste (solid or liquid) or mineral or vegetable substances are likely to evolve and produce volatile odorous material. These transformations can take place in an aerobic or anaerobic environment.
2. *Odors from processing plants.* This is the case of the chemical industry, but also that of the food and fragrance industries.

Raw material from these activities is sometimes odorous, but their synthesis often produces volatile by-products responsible for odors. It should be noted that treatment of manufactured products (drying, handling, packing, etc.) is one possible source of odorous gas effluents.

Bouscaren (1984) gives another classification that has to do with chemical or biological reactions that cause odorous emissions. Four categories are given.

1. *Thermal decomposition of organic compounds.* This is the case of solid waste incineration, foundries, petrochemical industries, thermal power stations, etc.
2. *Anaerobic decomposition of organic material.* For example, yeast or food manufacture.

3. *Anaerobic decomposition of animal products.* This situation is found in meat quartering plants, gelatin manufacturing, and fish meal factories. These types of factories have been the object of numerous analyses of concentrations of various odor-producing products, as will be seen during the course of this chapter.
4. *Animal excrements.* Intensive animal farming produces this type of nuisance; this is the case of pig manure and chicken droppings.

These classifications attempt to regroup the activities that are possible sources of olfactive nuisances. By concentrating on them, we will give examples of several activities that create malodorous effluents:

- Industrial emissions
- Odors due to the food industry
- Nuisances due to wastes (solid or liquid)
- Odors in potable water

Due to the large range of sources of olfactive nuisances, this presentation cannot be exhaustive.

8.2 Industrial Gaseous Emissions

Multiple industrial activities can be the source of olfactive nuisances because of raw materials or the transformation of basic products. Table 8.1 shows the classification of odorous compound emissions. They are nearly always complex mixtures of various molecules. Cheremisinoff and Young (1975) give the principal molecules encountered in gaseous emissions from the chemical and petrochemical industry (Table 8.2). One can note the omnipresence of NH_3, SO_2, and H_2S in the discharges.

Table 8.1. Classification of Industrial Volatile Compound Emissions (CITEPA, 1986).

Activities	% Emissions
Combustion	3
Refineries	7
Iron and steel industries	2
Basic chemical industries	10
Metal part degreasing	18
Printing	10
Automobile manufacturing	6
Painting, other than automobile	22
Fermentation	10
Waste elimination	1
Other	11

Table 8.2. Examples of Odorous Products in Chemical and Petrochemical Industries (Cheremisinoff and Young, 1975).

Industry	Source of Odors	Types of Odor
Refineries	Gas and gas recycling systems	SO_2, H_2S, NH_3, organic hydrocarbon acids, aldehydes, mercaptans
	Catalytic cracking	SO_2, NH_3, aldehydes
	Fluid catalysis	SO_2, NH_3, aldehydes, hydrocarbons
	Boilers	SO_2, NH_3, H_2S, aldehydes, hydrocarbons
	Warehouses	hydrocarbons
Inorganic chemical industry	Fertilizer/phosphate production	NH_3, aldehydes, SO_2
	Phosphoric acid	SO_2, H_2S, aldehydes, and other odorants
	Soda	NH_3
	Nitric and sulfuric acids and lime	NH_3, aldehydes, SO_2
Organic chemical industry	Organic chemistry	Mercaptans, NH_3, hydrocarbons, SO_2, organic acids, and other odorants
	Paints	NH_3, hydrocarbons, SO_2, aldehydes, organic acids, and other odorants
	Plastics	NH_3, hydrocarbons, SO_2, aldehydes, organic acids
	Rubber products	NH_3, hydrocarbons, SO_2, aldehydes, organic acids
	Soaps, detergents	NH_3, hydrocarbons, SO_2, aldehydes, organic acids, and other odorants
	Textiles	NH_3, hydrocarbons, SO_2, aldehydes, organic acids, and other odorants

8.2.1 Activities Related to Energy

In the case of industries related to energy production, it is necessary to understand activities such as complex petrochemicals and gas, coal, or petroleum combustion. In the petrochemical industry, three categories of odorant sources can be distinguished: diffused sources (warehouses, desulfurization units, water purification units), canalized sources (Claus units, ovens, and combustion reactors) and sources of lesser importance (steam crackers, tar settlers) (Cahiers Techniques/Technical Notebooks, 1984). Table 8.3 gives examples of several sulfurous products and their concentrations in various emissions.

Other families of products are found as well. Thus, ammoniac, some amines, and phenolic products are found at levels of several 100 mg/m^3 in coal gasification plants and factories (Bombaugh, 1988) and in petrochemical industries (Persson et al., 1987; Bouscaren, 1984). Cokework residues produce a large variety of odorous products at high concentrations (many tens of mg/m^3). The flow rates are very low. Lafaye and Gillard (1984) give several values: NH$_3$, 175 mg/m^3; benzene, 48 mg/m^3; toluene, 13 mg/m^3; styrene, 2 mg/m^3; and napthalene, 11 mg/m^3. More precise analyses of odorous molecules have been done on diesel motor exhaust gases (Partbridge et al., 1983). A large range of products is observed, divided into two large fractions: oxygenated molecule extracts and aromatic compounds. The most significant product is bezaldehyde with levels ranging from 0.002 to 0.175 ppm; other values given range from 0.009 ppm (Swarin and Lipari, 1983) all the way to 3.5 mg/m^3 for Johnson et al. (1981). Product types such as pyrene, fluorene, idanone, and cinnoline are found at trace levels.

8.2.2 Wood, Paper and Viscose Industries

In the paper industry, the strongest odors are given off by wood shaving degasification, recycling, drying, and pyrolysis of black liquors and eventually lime kiln vents. Emissions range from 1 to 10 kg/h for dimethyl, diethylsulfur, and methyl mercaptan for firing condensers and from 10 to 50

Table 8.3. Odorous Compound Levels Emitted by Energy-Linked Activities.

Activity	Odorous Compound (mg/m^3)			References
	H$_2$S	EtSH	SO$_2$	
Petrochemicals		3×10^{-7}		Dorgelo, 1978
Natural gas	19.5			Detrie, 1969
Combustion gas	300			Reissner, 1971
Boilers			100–3400	Vicard, 1985
Coal power stations			578	Detrie, 1969
Oil power stations			1598	Detrie, 1969

kg/h for evaporator-condensers (Cahiers Techniques/Technical Notebooks, 1984; CNGE Report, 1987).

Köhler et al. (1981) studied the odor emissions due to wood working industries (panels, small logs, . . .). Concentrations ranging from 50 to 60 mg C/Nm^3 are present in the raw gas. The portion of formic aldehyde is important and varies from 30 to 70% of the total carbon.

Table 8.4 shows several sulfurous product values in paper and viscose industries. The high concentrations of H_2S and mercaptans in paper pulp should be noted. In the viscose industry or sponge maufacture, one notices the high CS_2 concentrations occuring in these industrial processes.

8.2.3 Paints and Polymers

Gaseous effluents due to the manufacture or application of paints are loaded with solvents and volatile compounds occuring in the formulation of these paints. The emission flow rates are generally important and can be situated between 40,000 and 100,000 Nm^3/h according to the ventilation system. A study on paint and varnish industry data (Lemasle and Martin, 1980) shows the existence of pentoxone, diethyl or triethylamine, and isopropanol as well as products belonging to the glycol family (ethyl, butyl, and hexyl glycol) in gaseous emissions. Jensen (1984) studied paint firing cabins in the automobile industry. The levels of volatile organic compounds varied from 400 to 800 mg C/m^3 with an average of 600 mg C/m^3. Table 8.5 gives the composition of these gases. The products mentioned are generally found in this type of emission. Köhler (1984) and Laplanche et al. (1986) note, moreover, butyl acetate, ethyl glycol acetate, xylene, and butanol in concentrations situated

Table 8.4. Flow Rates and Volatile Odorant Compound Content in Paper and Viscose Industries.

Activities	Paper Pulps	Paper	Viscose
Flow Rate (m^3/h)		$100-10^{-5}$	0.6
Odorous compounds (mg/m^3)			
H_2S	>1000	$300-10^{-5}$	2,500
EtSH			
MeSH		100	
Me_2S		50	
Me_2S_2		40	
CS_2			1,250
SO_2		216	
References	Roberts et al., 1971	Wilson, 1973 Electric Co., 1980	Countaulds, Ltd., 1975

Table 8.5. Compounds Contained in One Emission of a Firing Cabin in the
Automobile Industry (Uno Jensen, 1984).

Compounds	Quantity
Total carbon	600 mg/m³
Ethyl glycol	8%
Butyl glycol	4%
Hexyl glycol	6%
Aliphatic alcohols	75%
Others	7%

between 10 and 60 mg/m³ in odorous emissions from paint workshops in the
automobile industry.

In the development or adaptation of polymers, one finds monomers in
emissions. A study on styrene emissions (Laplanche et al., 1985) shows that
during the adaptation of polystyrene, concentrations of about 19 mg/m³ are
found at working area levels, 20 mg/m³ close to the operator and 18 mg/m³
at floor level. During impregnation, the atmosphere can contain amounts
from 120 to 200 mg/m³.

8.2.4 Food Industries

8.2.4.1 Yeast Manufacture

Alimentary yeast manufacturing, taking into account fermentation processes, implicates the emission of odorous gas. The flow rates are quite variable, ranging between 5,000 and 15,000 m³/h. Large families of odorous molecules are present. Table 8.6 shows the amounts of several analyzed
products. Because of fermentation phenomena, large proportions of alcohols (methanol, ethanol, n-butanol, isobutanol) and ketones (acetone,
methyl ethyl ketone, diethyl ketone) are found. Organic acids in concentrations of several tens of μg/m³ are encountered; butyric, isobutyric, valeric,
propionic, and caproic acids can be analyzed. Total organic acidity represents 0.2–1 mg/m³. Wastewaters pose an important odor problem, whether
it is in the air surrounding the collectors, the residual waters themselves, or

Table 8.6. Nature and Amounts of Odorous Compounds in Yeast Maufacture

Compounds	Concentration (mg/m³)	
	Fabrication	Wastewaters
H₂S and mercaptans	0–32	15–180
NH₃	0–6	1–2
Amines (methyl, ethyl, triethyl)	0.1–2	2–14
Alcohols	2–20	5–700
Ketones	1–30	10–800

in the vapors emitted by the condensations of liquid wastes produced during fabrication.

8.2.4.2 Human and Animal Alimentation

Several studies on the nature and concentrations of odorous molecules have been conducted in Japan concerning the fabrication of dishes based on fish or vegetables. Koizumi et al. (1987) studied the odors emitted during warehousing or while the fish was being cooked. These results are given in mg of malonaldehyde emitted per kg of fish. Depending on the type of fish, values varied from 0.5 to 3 mg/kg for the cooked primary material, but could go up to 400 mg/kg for fish stocked at 5 °C for six days. When working on bean cooking and preparation, Tokimoto and Kobayashi (1988) identified numerous products. Thus, one finds the family of alcohols (methanol, butanol, propanol, . . .), ketones (acetones, butanedione, heptanone, octonone, acetophenone), hydrocarbons of the terpene type, organic acids (acetic acid), phenols, lactones, and nitrogenous compounds of the pyridine type.

Food preparation for domestic animals also generates odorous effluents. If warehousing of raw materials can induce several nuisances, it is due most of all to dryers. The flow rates are generally important, from 10 to 100,000 Nm^3/h and the gases can be hot. The analysis of such emissions shows the presence of ammoniac from 0.1 to O.3 mg/m^3, sulfurous products from 1 to 5 mg/m^3, alcohols (ethanol, isobutanol) from 0.1 to 3 mg/m^3, and aldehydes and ketones. Galé (1984) also notes the presence of organic acids in low concentrations of several tens of $\mu g/m^3$.

8.2.4.3 Sugar Manufacturing

In conducting a survey in nine sugar manufacturing plants, Huismann et al. (1987a) noted that 25–30 odor sources were identified. Nevertheless, an average of six per plant are responsible for 90–95% of the total emanations produced. The principal odorous emissions are due to the following fabrication steps:

Extraction: 2.3%
Pulp drying: 30.7%
Carbonation: 10.8%
Evaporation of gas: 12.3%
Building ventilation: 8.5%
Condenser: 30.7%
Other: 4.7%

The percentages given are based on a flux, that is to say, on the product of the gaseous flow rate and the odor concentration determined by olfactometric analysis.

Qualitative and quantitative analyses have identified 38 products that are responsible for odor emissions (Huisman et al., 1979; 1987b). The identified constituents include furane-type compounds as well as a certain number of pyrazines at rates of several hundreds of $\mu g/m^3$. Nitrosamines and polycyclic hydrocarbons are found in concentrations of several tens of $\mu g/m^3$ at emission in gases qualified by the authors as unusable. Ketones and aldehydes (100–500 $\mu g/m^3$) and sulfurized hydrogen (several hundreds of $\mu g/m^3$) were also identified as well as phenols and aliphatic alcohols (several hundreds of $\mu g/m^3$). The toxicity aspect of effluents is elaborated in regard to data given by the literature, which is accessible for only certain products.

8.3 Olfactory Nuisances due to Wastes

Wastes or their processing implicate olfactive nuisances. It is necessary to understand the term "wastes" in a broad sense. This section will be concerned with odors due to household wastes, but also with animal carcasses, fish debris, manure, or their vaporization.

8.3.1 Animal By-Product Industries

In activities linked with the treatment of animal by-products, from fish to meat quartering, fat processors generate very highly odorous gaseous effluents. Table 8.7 gives a scale of odorant molecule concentrations encountered in this type of industry. For the most part, these odors can be attributed to sulfurized and aminated compounds. The indicated amounts can vary depending on the type of processing being done in the packing plants: feather hydrolysis, meat cooking, fat removal from bones, etc. However, it is at this time that odorant emissions are the strongest. On the other hand, workshop air is more diluted. As for chicken dropping dehydration, the high levels of sulfurized products (H_2S and mercaptans) should be noted. In the specific case of fish meal fabrication, only amino products are present.

8.3.2 Animal Droppings and Manure

Concentrated intensive animal farming creates a disagreeable olfactive nuisance for the neighborhood. This odorant pollution can be due to the farm location, manure spreading, or manure treatment. Practically all types of animal farming are implicated: cattle, pig, fowl, etc. A list of more than 75 compounds has been inventoried in odorant emissions. Large families of molecules are present: volatile organic acids, aldehydes, ketones, esters, amines, sulfurs, mercaptans, and heterocyclic nitrogenous compounds (Barth et al., 1984; Sweeten, 1988). These products are present in manure or are due to biological degradation; the compounds influencing most of the odors are organic acids, ammoniac, and amines (Miner and Hazen, 1989). Emissions are subject to seasonal change. Thus, Meyer and Converse (1981) show that odor production is multiplied by 2.18 for H_2S and 1.18 for NH_3 when the temperature rises from 15.5 to 23 °C. Molecule quantification gives

Table 8.7. Odorous Product Content in Industries Processing Animal By-Product or Fish (according to Nominé, 1979; Martin et al., 1986; Le Cloirec et al., 1988).

Compounds Dropping (mg/m³)	Meat Quartering		Worm Farming	Fat Processing	Fish Processing	Chicken Dehydration
	Noncondensible by Cooking	Workshop Gas				
H₂S and mercaptans	100–1,000	0.1–2	26	low	<0.01	800–18,000
NH₃	500–1,500	2–20	90–130	2–75	140	180–1,000
Amines						
Methylamine	1–5	<0.05				
Trimethylamine	10–60	0.1–0.2		0.02–0.2	2.8	40–80
Ethylamine	0–10	0.1–0.3	0.5	3		
Dimethylamine	10–30	0.2–0.3			17	
Aldehydes						
Acetaldehyde	3–20	<0.02				20–60
Propionaldehyde	1–20					
Ketones						
Acetone	5–10					
Butyric acid	0–10					

Table 8.8. Several Compound Levels Emitted During Household Waste Fermentations (Adapted from Work by Anderson, 1984).

Compounds	Concentrations
H_2S	0.0047
Organic sulfur	0.07
Mercaptans	0.003
Amines	0.4
Organic acids	1.5
Scatole	0.22
Alcohol	0.3

H_2S and NH_3 a value of about 20 mg/m³; other molecules have a concentration of between 10 and 100 µg/m³.

8.3.3 Household Wastes

There are a variety of treatments in existence for household wastes, which often indicates an olfactive nuisance. Identification and qualification work on emitted odorant molecules has mostly been done on aerobic or anaerobic fermenters.

In the case of anaerobic processes, Anderson (1984) gives a list of malodorous products that are present in emissions (Table 8.8). Perry et al. (1984) identified the same product classifications during a study pilot on anaerobic digestion of household wastes. Mercaptans and H_2S are due to the degradation of amino acids such as methionine, cystine or taurine, and organic acids; alcohols or ketones are generated by carbohydrate fermentation. Amines and ammoniac are produced by amino acid degradation. Indole or scatole type compounds can be tryptophane metabolism products.

Recently, a study on an aerobic household waste fermenter has been undertaken (Le Cloirec, 1989). Sulfurized compounds were not detected and ammoniac was quantified (at several mg/m³). On the other hand, one finds amines at rather high levels from 30 to 50 mg/m³, alcohols from 20 to 30 mg/m³, and aldehydes from 1 to 3 mg/m³. These concentrations vary with temperature, mixing rate, and the quantity of air influx in the reactor, on the average of 10,000 m³/h.

8.4 Vault Gases

Airtight concrete vault implantation, either above ground or buried, allows the use of land unsuitable for inhumation in open ground. However, the cadaveric decomposition process that develops in these vaults generates important liquid and gaseous fluxes, which are a source of olfactive nuisances and thus need to be treated.

8.4.1 Post Mortem Processes

A living body consists of a large number of differentiated cells. It is essentially composed of proteins, lipids, glucides, water, salt, organic metallic compounds, and a stable inorganic rigid structure.

The maintainence of life requires a biological and physical-chemical equilibrium. Death is the rupture of this equilibrium under the action of physical, chemical, and microbial phenomena (Simonin, 1955; Lery and Isnard, 1984). In the first hours of death, one observes body cooling, the beginning of dehydration, and cadaveric rigidity. This is the result of biochemical modifications of glucogen, adenosine triphosphate (ATP), and muscular phosphocreatine (PC) (Rosset, 1988). Under the action of endogenic enzymes, after blood circulation and O_2 consumption stop, these constituents liberate acid substances. The initially neutral pH drops from 7 to 5.5. According to the nature of the muscles and their glycogen, ATP and PC reserve rate and microorganism proliferation is limited (Feodoroff, 1987; Rosset, 1988). At the end of this acidifying phase, the rupture of the digestive barrier leads to an intestinal germ invasion of muscle and skin tissue (Caspar and Burn, 1934; Guiraud and Galzi, 1981; Rosset, 1988). This is expressed by a gaseous infiltration or putrid emphysema that can lead to rupture of the cutaneous membrane. Tissue autolysis is expressed by progressive liquefaction (Rimoux et al., 1988). The previously latent microfloral activity intensifies with cadaveric conditions, particularly favorizing anaerobic germs.

These first steps of mineralization of organic material, the period during which nitrogen, carbon, oxygen, and hydrogen constitute the superficial and profound putrefaction that are the origin of odor and appearance alterations, This process can be favorized or, on the contrary, inhibited at each step according to environmental conditions, the atmospheric gaseous composition, and, in particular, the temperature (Rimoux et al., 1988).

8.4.2 Decomposition in a Vault

After a latent period during the muscular acidic phase and tissue autolysis, aerobic and anaerobic saprophytic flora establish themselves and develop according to the environmental conditions created. In a tight concrete vault that is more or less ventilated, these activities develop on the cadaver and in liquids produced by soft tissue and viscera hydrolysis. These rapidly form a favorable fermentation environment: their solubilized organic element composition, simplified by tissue autolysis, facilitates microorganism assimilation. Nitrogenous substance degradation causes discharge alcalinization by progressive ammonification and malodorant substance production.

In acid pH, amino acid decarboxylation discharges amines and CO_2; in alcaline pH, acidaminolytic activity is intense. The reactions, essentially caused by deamination, discharge ammoniac and fatty acids, some volatile amines, mercaptans produced by sulfurized residues (organic sulfur being in proteinic majority), and CO_2 (Lambin and Germain, 1969). For example,

- Glutamic acid (Lambin and Germain, 1969):

$$COOH—(CH_2)_2—CO—COOH + NH_3 \text{ to}$$
pH = 3
Ketoglutaric acid

$$COOH—(CH_2)_2—\underset{\underset{NH_2}{|}}{CH}—COOH \nearrow$$

$$\searrow COOH—(CH_2)_3—NH_2 + CO_2 \text{ to pH} = 5$$
Aminobutyric acid

- By decarboxylation, basic amino acids give off polyamides (ptomaines) that have a particularly fetid odor:

-Lysine   → Cadaverine + CO_2
$$NH_2—(CH_2)_3—\underset{\underset{NH_2}{|}}{CH}—COOH → NH_2—(CH_2)_5—NH_2$$

-Arginine → Ornithine + NH_3 → Putrescine + CO_2

$$HN = \underset{\underset{COOH}{\diagup}}{C}—NH—(CH_2)_3—\overset{\overset{NH_2}{|}}{CH} → H_2N—\underset{\underset{(CH_2)_3}{\diagdown}}{CH}—COOH → H_2N—(CH_2)_4—NH_2$$
$$\underset{\underset{NH_2}{\diagup}}{(CH_2)_3}$$

- Sulfurized amines allow the formation of H_2S, mercaptans, and organic disulfide (Lambin and Germain, 1969):

-Methionine   → Cysteine   → Taurine + CO_2

$$\begin{array}{ll}
CH_2—S—CH_3 & CH_2—SH \\
| & | \\
CH_2 & → CH_2—NH_2 → HSO_3—(CH_2)_2—NH_2 \\
| & | \\
CH—NH_2 & COOH \\
COOH &
\end{array}$$

or   or

  Dimethyldisulfide   Mercaptoethylamine + CO_2 + NH_3
   or
  Dimethylsulfide $HS—(CH_2)_2—NH_2$

Aromatic amino acids produce cyclic compounds with a mephitic odor at certain concentrations; transformations can be effectuated by numerous intestinal bacteria (Proteus, Clostridies) (Lambin and Germain, 1969).

-Tryptophan + $H_2O \rightarrow$ Indole + Pyruvic acid + NH_3

Glucidic and lipidic hydrolysis produce hydrocarbonated chains that ferment easily, the decomposition of which gives off acetic, propionic and butyric volatile fatty acids, aldehydes, ketones, and organic acids. The quantity and proportion of these soluble and volatile substances determine complex odors, certain compounds being particularly malodorous (butyric and caproic acids).

Under favorable conditions, compounds produced by hydrolysis undergo successive degradations leading to the formation of acetates, H_2 and CO_2. Starting from these substrates, methanogenic bacteria can then intervene.

8.4.3 Influential Factors

8.4.3.1 Gaseous Atmospheric Composition

Oxygen allows the development of aerobic germs on the cadaver (*Pseudomonas, Acinetobacter, . .*). Their rapid metabolism, exothermic, humid, and alcaline due to the consumption of nitrogenous substances is favorable to the extension of profound microbial proliferation developed in the abdominal cavity where the rH is low toward muscular tissues. In this absence, lactic fermentations become predominant and favorize acidification of the environment. Nitrogenous substance consumption is next to nothing, thus putrefaction is limited (Rozier et al., 1985; Rimoux et al., 1988).

CO_2 between 5% and 50% of the atmosphere, has a bacteriostatic action and inhibits superficial flora (Rozier et al., 1985).

Water activity (A_w): Environmental humidity determines the intensity of microbial multiplication. In a humid atmosphere, bacterial invasion is rapid, whereas in a dry atmosphere, there is a slow progression of mold (Rozier et al., 1985; Rosset, 1988).

8.4.3.2 pH

Alkaline pH favors the implantation of active proteolytic flora which is responsible for the putrefaction process (essentially, *Clostridium* and *Bacillus*).

8.4.3.3 Temperature

As a general rule, the germ multiplication is slower at lower temperatures.

8.4.4 Example of Gas Production of Animal Cadavers in Tight Concrete Vaults

An experimental protocol has been implemented on pig cadavers in tight concrete vaults (Lery et al., 1988); one with air circulation (Box V) and the other with simple gas decompression (Box NV). The study was conducted over more than 4 years time with monitoring on each box:

- Physical and chemical environment: O_2, temperature, liquid emission pH
- Macroscopic evolution of the bodies
- Quantitative and qualitative analysis of cadaveric liquids and gases

8.4.4.1 Observations on Animal Decomposition

In the applied experimental conditions, the ventilation of Box V (O_2 supply, humidity supply, CO_2 evacuation) favors the rapid implantation of superficial putrefaction flora (*Pseudomonas, Acinitobacter*). This proteolytic flora generates ammonia in the medium and allows for the in-depth development of anaerobic bacteria (essentially, *Clostridium*) that multiply very rapidly when the rH reaches a sufficiently low value after O_2 consumption by the tissues (Rosset, 1988; Rimoux et al., 1988). This same flora colonizes discharged liquids. These metabolisms are accompanied by a characteristic brown color of the bodies and a rapid aminolysis step of the liquids, accompanied by odorous sulfurous gas production (mercaptans) followed by ammoniac and several amines.

Unfavorable initial environmental conditions, like those of Box NV (without air circulation), delay the ammoniacal putrefaction process by a year. Dry atmosphere and an acidic to neutral pH allow a slow proliferation of molds on the surface (*Aspergillus, Cladosporium, Penicillium*) participating in lipid hydrolysis and oxidation reactions. These metabolisms are accompanied by a dry and wrinkled appearance of the body, rancid and sickly odor, and lead to a relative conservation of the liquids. Once started by high temperatures, the transformation evolves in the same manner as in the other box.

The gaseous production essentially comes from liquid fermentation. Its intensity and composition in sulfurous, nitrogenous, and carbonaceous

Table 8.9. Concentrations and Odor Thresholds of Produced Sulfurized Compounds

Produced Compound	Box V Maxi mg/m³	Box NV Maxi mg/m³	Odor Threshold mg/m³	× 100% Recognized mg/m³	Toxicity Threshold mg/m³
Methyl mercaptan			0.004	0.07	1
Ethyl mercaptan	192	206	0.0025		1.25
Sulfurized hydrogen	2.8	14.2	0.0007	1.4	14

chemical families varies with the course of degradation. The temperature intensifies the microbial metabolic rate and increases gasification.

8.4.4.2 Emission of Malodorous Compounds

In the presence of intense putrefaction activity, olfactive nuisances are reinforced by high temperature and by the interferences between the released compounds. Production fluxes are intermittent and limited to annual summer periods.

8.4.4.2.1 Sulfurized Compounds (Table 8.9)

A comparison of the NV and V profiles shows (Figure 8.1):

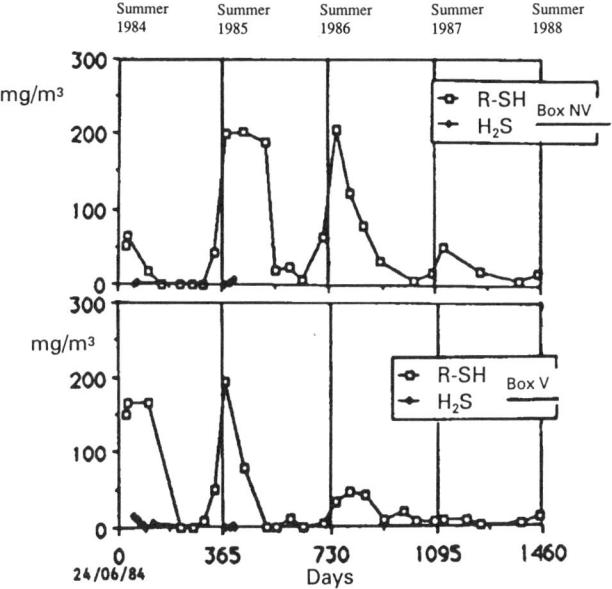

Figure 8.1. Sulfurized compound production.

1. An important production of sulfurized compounds in Box V during the first 2 years
2. A similar production in Box NV, but delayed by 1 year

Sulfurized compound identification reveals the marginal character of H_2S production. The short monthly phases of H_2S emission detected in the summer months during the first two years correspond to integument splitting on the bodies. Mercaptans were measured in the form of ethyl mercaptans in possible interference with other acetylated mercaptans of methyl, propyl and butyl types and with H_2S.

8.4.4.2.2 Nitrogenous Compounds (Table 8.10)

Methylamine, ethylamine, and isopropylamine type amines were found in both boxes during two and a half years of decomposition, though at much higher concentrations in the box where there was no ventilation (Figure 8.2).

In the ventilated box, ammoniac volatilization was characteristic at elevated temperatures and alcaline pH.

In the nonventilated box, this phenomenon did not become significant until after three years of decomposition and at much lower concentrations.

8.4.4.2.3 Carbonaceous Compounds (Table 8.11)

Carbonaceous compounds appeared simultaneously in both boxes, but in much lower concentrations in the presence of air (Figure 8.3). C_2 and C_3 chains constituted the essential part of the flux. Aldehydes alone were measured during a year and a half, then mixed with alcohols and ketones. After 2 years of decomposition, acetone was the principal constituent measured in Box NV. No methane emission was measured.

Table 8.10. Concentrations and Odor Thresholds of Produced Nitrogenous Compounds.

Produced Compound	Box V Maxi mg/m³	Box NV Maxi mg/m³	Odor Threshold mg/m³	× 100% Recognized mg/m³	Toxicity Theshold mg/m³
Ammoniac	1,030	131	33	38	18
Methylamine	20	25	0.30	4.5	12
Trimethylamine	37.8	19	0.000,5	10	
Ethylamine	31	28		1.6	
Diethylamine		69		0.6	
Triethylamine		2.5		1.3	
Isopropylamine	11.5	135		2.6	
Butylamine		4.14		1	

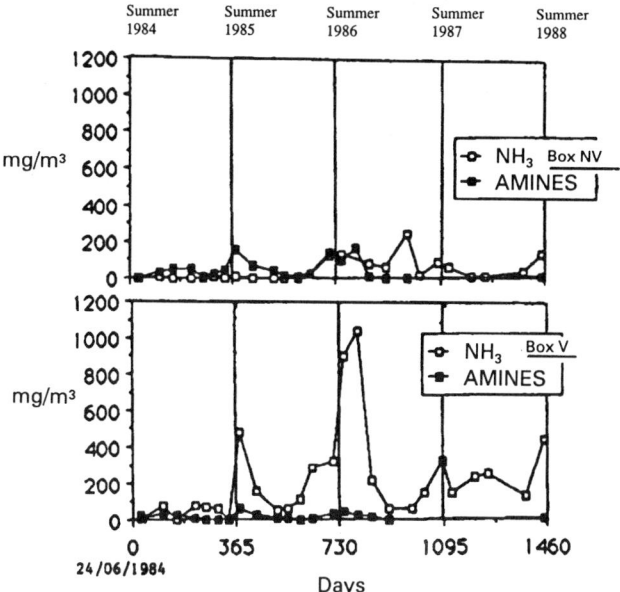

Figure 8.2. Nirogenous compound production.

Table 8.11. Concentrations and Odor Thresholds of Produced Carbonaceous Compounds.

Produced Compound	Box V Maxi mg/m³	Box NV Maxi mg/m³	Odor Threshold mg/m³	x 100% Recognized mg/m³	Toxicity Threshold mg/m³
Acetaldehyde	46	21	0.6		180
Propionaldehyde	107	192		0.2	
Butylaldehyde	4	12		0.0002	

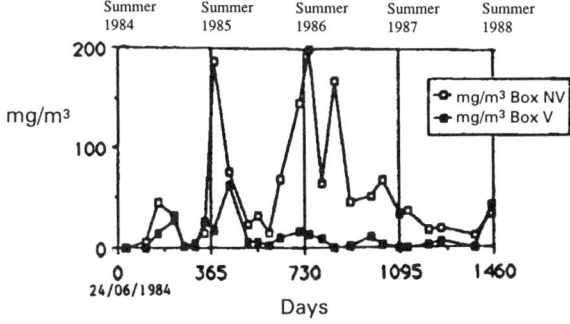

Figure 8.3. Carbonaceous compound production.

229

8.4.4.3 Quantitative Characteristics of Gaseous Emissions

The gaseous flux could only be measured in Box NV. The gas production kinetics are characterized by a latent phase during the first year, then by an alternation of exponential production phases in hot periods and plateau phases in cold periods during the following years. The accumulated quantity of measured gas at the end of 4 years of decomposition was in the neighborhood of 10 m³.

It should be noted that gaseous volumes measured in Box NV correspond to decomposition activity that was retarded by about 1 year in comparison to Box V. On the other hand, only decomposition gases were measured since there was no air entering Box NV.

8.4.5 Conclusion

The microbial developments that are responsible for the decomposition process are primarily affected by environmental composition in direct relation with pH, rH, and A_w parameters and by environmental conditions such as temperature and the amount of oxygen in the air. The ecosystem thus created determines the identity and volume of implanted flora whose successive actions generate soluble, volatile, and gaseous substances over a period of many years and are sources of complex and extremely varied odors.

8.5 Wastewater Treatment Plants

8.5.1 Causes of Bad Odors in Wastewater Treatment

The collection and treatment of urban and industrial wastewaters are frequently at the origin of olfactive nuisances. These wastewaters, which are loaded with organic particles and dissolved matter, nitrogenous compounds (including ammoniac), and phosphorus, can lead directly or indirectly to the formation of disagreeable odors by the intermediary of purification by-products (greases, sludges), following a well-known biological fermentation process that is activated in reduction conditions.

Furthermore, some industrial discharges contain, from the start, extremely volatile compounds used in manufacturing processes such as sulfides (Harkness, 1980), aldehydes, alcohols (Brandl and Stover, 1988), or even ammoniac, which can be at the origin of intense olfactive pollution. Besides these direct industrial discharges of odorous compounds, the other principal odorous compounds given off by wastewaters essentially belong to reduced sulfurized product and nitrogenous product families.

Sulfurized compounds constitute the majority of the olfactive molecules found in sewage plants and effluent collection networks. These include mercaptans, organic sulfurs and disulfurs and, above all, hydrogen sulfide. As shown in Figure 8.4, established by Harkness, when the effluent or organic deposit septicity is affected, anaerobic bacteria reduce sulfates and sulfurized organic compounds (amino acids, detergents) to mostly hydrogen sulfide and some organic sulfides (mercaptans, polysulfides), which are at the origin of bad odors (Kienov et al., 1982). When H_2S is present in the gaseous phase, it is oxidized as sulfuric acid that is responsible for concrete corrosion and attacks electromechanic equipment made of steel (Bowker et al., 1985). The reduction of sulfates by sulfate-reducing bacteria (*Desulphovibrio* and *Desulphotomaculum*) occurs within the normal redox potential domain ranging from -200 to -300 mV. In practice, it can be observed that odor emissions are related to the presence of H_2S and organic sulfides starting at a redox potential below -50 mV (Crook and Rbwen, 1987).

Nitrogenous compounds can also be at the origin of olfactive nuisances, mainly at sewage plant level and more rarely at network level. These compounds are essentially ammoniac, amines, and, to a lesser extent, indole and

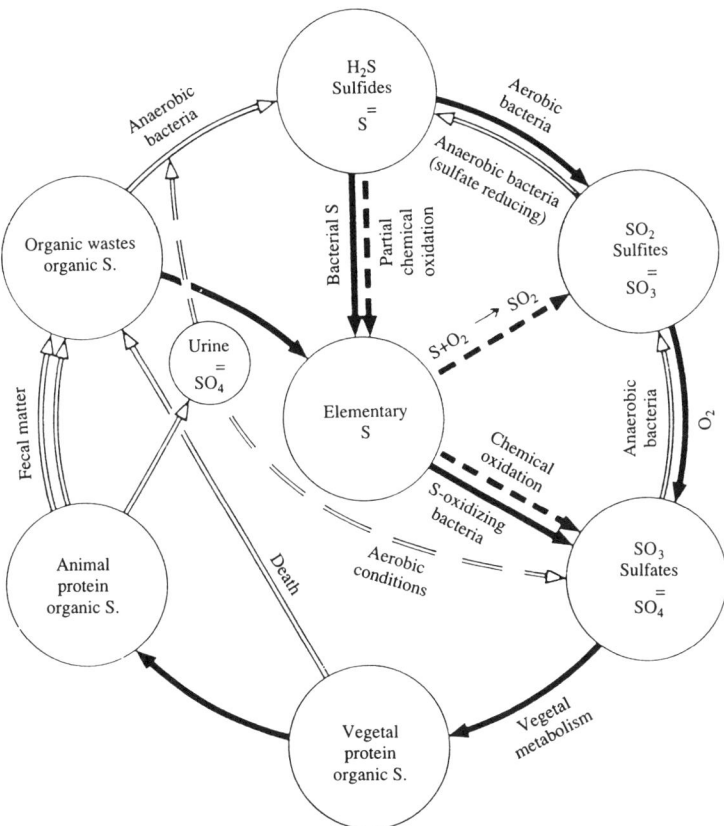

Figure 8.4. Sulfur cycle in sewerage (according to Bowker, 1985).

scatole. In sewerage, part of the nitrogen content comes from urine, whose average composition is as follows:

Urea	25 g/L
Uric acid	0.6 g/L
Creatinine	1.5 g/L
Ammoniac in N	0.6 g/L

The other part comes from biological protein and amino acid degradation. Ammoniac can be equally formed by hydrolysis of nitrogenous organic compounds in networks with a long residence time and when the temperature is elevated. Methylamine and dimethylamine are present in urine in low concentrations. Amines are generally a bacterial metabolic by-product of amino acids in anaerobiose (Harkness, 1980).

Lysine → Cadaverine (putrid odor)

$$NH_2—(CH_2)_4—CH(NH_2)COOH → NH_2—(CH_2)_4—CH_2—NH_2$$

Indole and scatole, which have a nauseating fecal odor, are formed by anaerobic fermentation starting from the amino acid tryptophan:

Other odorant compounds such as those belonging to the volatile fatty acid (VFA), aldehyde, alcohol, or ketone families are sometimes responsible for olfactive nuisances in specific cases:

1. Direct industrial aldehyde discharges, which present an agreeable fruity odor
2. Anaerobic sludge or effluent digestion, which favorizes volatile fatty acid, aldehyde, or ketone production
3. Thermic treatment of purification sludges after anaerobic digestion, which provokes degassing of VFA and aldehydes previously formed by a biological process

These compounds are by-products produced by bacterial fermentation of carbohydrates that are first transformed into acids (acidification phase) then into acohols, aldehydes, and ketones. In this class of compounds, acetic acid, butyric acid, valeric acid, ethanol, acetaldehyde, isovaleraldehyde, and acetone can be present in the atmosphere of purification systems. Industrial effluents containing large amounts of sugars and proteins (food industries) or fresh biological sludges can release this type of compound after

Table 8.12. Characteristics of Principal Compounds Responsible for Odors in Wastewater Treatment Plants.

Compound Type	Compound	Molar Mass	Chemical Formula	Odor Characteristics	Odor Threshold (mg/N m³ air)	Vapor Pressure (atmosphere)	Boiling Temperature (°C, 760 mm Hg)
Sulfurized	Hydrogen sulfide	34.1	H_2S	Rotten egg	0.0001–0.03	20 (25 °C)	.62
	Methylmercaptan	48.1	CH_3SH	Cabbage, garlic	0.0005–0.08	2 (26 °C)	8
	Ethylmercaptan	62.1	C_2H_5SH	Rotting cabbage	0.0001–0.03	0.53 (18 °C)	23
	Dimethylsulfide	62.13	$(CH_3)_2S$	Rotting vegetables	0.0025–0.65	0.53 (18 °C)	37
	Diethylsulfide	90.2	$(C_2H_5)_2S$	Ether	0.0045–0.31	0.05 (18 °C)	92
	Dimethyldisulfide	94.2	$(CH_3)_2S_2$	Putrid	0.003–0.014	0.078 (24 °C)	109
Nitrogenous	Ammoniac	17	NH_3	Very pungent, irritating	0.5–37	0.016 (20 °C)	.33
	Methylamine	31.05	CH_3NH_2	Rotting fish	0.021	2 (10 °C)	.7
	Ethylamine	45.08	$C_2H_5NH_2$	Pungent, ammoniacal	0.05–0.83	1 (16.6 °C)	17
	Dimethylamine	45.08	$(CH_3)_2NH$	Rotting fish	0.047–0.16	2 (25 °C)	7
	Indole	117.5	C_8H_6NH	Fecal, nauseating	0.0006	<0.001 (25 °C)	254
	Scatole	131.5	C_9H_8NH	Fecal, nauseating	0.0008–0.10	<0.001 (25 °C)	266
	Cadaverine	102.18	$NH_2(CH_2)_5NH_2$	Rotting meat		<0.001 (25 °C)	178
Acids	Acetic	60.05	CH_3COOH	Vinegar	0.025–6.5	0.001 (25 °C)	118
	Butyric	88.1	C_3H_7COOH	Rancid butter	0.0004–3	0.001 (25 °C)	163.5
	Valeric	102.13	C_4H_9COOH	Sweat, perspiration	0.0008–1.3	0.001 (35 °C)	186.5
Aldehydes & Ketones	Formaldehyde	30.03	$HCHO$	Acrid, suffocating	0.033–12	1 (-20 °C)	-19.5
	Acetaldehyde	44.05	CH_3CHO	Fruit, apple	0.04–1.8	1 (20 °C)	21
	Butyraldehyde	72.1	C_3H_7CHO	Rancid	0.013–15		74.8
	Isovaleraldehyde	86.13	$(CH_3)_2CHCH_2CHO$	Fruit, apple	0.072		92.5
	Acetone	58.08	CH_3COCH_3	Sweet/fruit	1.1–240	0.26 (23 °C)	56.5

fermentation in networks or in sludge thickeners. For example, butyric acid (rancid odor) is a biological degradation by-product of an amino acid: glutamic acid (Martin et al., 1986). The principal compounds responsible for bad odors in wastewater purification are given in Table 8.12, classified by family.

In practice, the principal causes of bad odors (Paillard and Blondeau, 1988) can be as follows:

1. The nature of effluents heavily laden with large quantities of highly biodegradable organic material or sometimes receiving malodorous industrial wastes
2. Networks that have a structure that favorizes fermentation (residence time longer than 3 h in flow inversion, large deposits of fermentable material, cascade flow inversion), which would be that much more important in the case of higher temperatures and larger loads of organic matter
3. Water treatment processes (aeration basins with shallow diffusers) and sludges (thermal conditioning) that can favor degassing of odorous compounds previously formed in the network or generated by these purification treatments
4. Badly adapted purification works that provoke turbulences or intense effluent mixing. This is the case of cascade at wastewater treatment plant inlets or a badly designed degreaser, for example

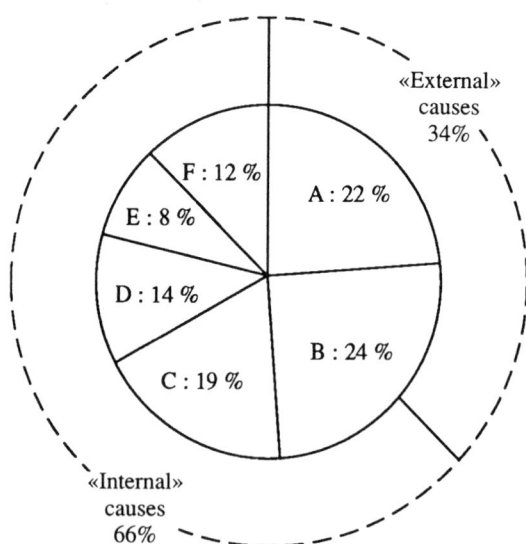

Figure 8.5. Causes of formation and degassing of odorous substances on over 100 wastewater treatment plants in Germany (Böhnke and Frechen, 1983). A: Odorous substances discharged in the effluent (industrial wastes). B: Odorous substances formed by anaerobic fermentation. C: Exploitation errors or insufficient maintainence. D: Undesirable degassing. E: design error/water or sludge treatment overload. F: Various causes.

5. Exploitation conditions of the network and wastewater treatment plant, which can be insufficient or unadapted (working with overly heavy loads in the biological reactor, for example, or a residence time in the thickener exceeding 48 h)

A survey done by the German Environmental Ministry on over 100 wastewater treatment plants (Böhnke and Frechen, 1983) shows that the causes of olfactive nuisances can be diverse (Figure 8.5) with, however, nearly a quarter of the nuisance cases due to anaerobic fermentation. The wastewater collection network is often one of the major sources of bad odors that are given off at pumping stations and principally at the wastewater treatment plant intake, which is generally not covered or ventilated. Other nuisance sources exist in a wastewater treatment plant, such as sludge thickeners, for example, or, more generally, all sludge treatment steps. These different purification treatment steps are represented in Figure 8.6.

Treatment Step		Average Odor Intensity in Odor Units/m³ of Air (Frechen, 1987)
RW:	raw water	45
IB:	intake basin + screening	85
P:	pretreatment	
	desilting + degreasing	60
PS:	primary settling	55
AD:	anaerobic digestion	
	(often non-existent)	
BT:	biological treatment	
	heavy load	60
	average load	50
	light load	30
SS:	secondary settling	30
SSW:	secondary sludge well	45
PSW:	primary sludge well	85
FST:	fresh sludge thickener	200
SDB:	sludge dewatering building	400

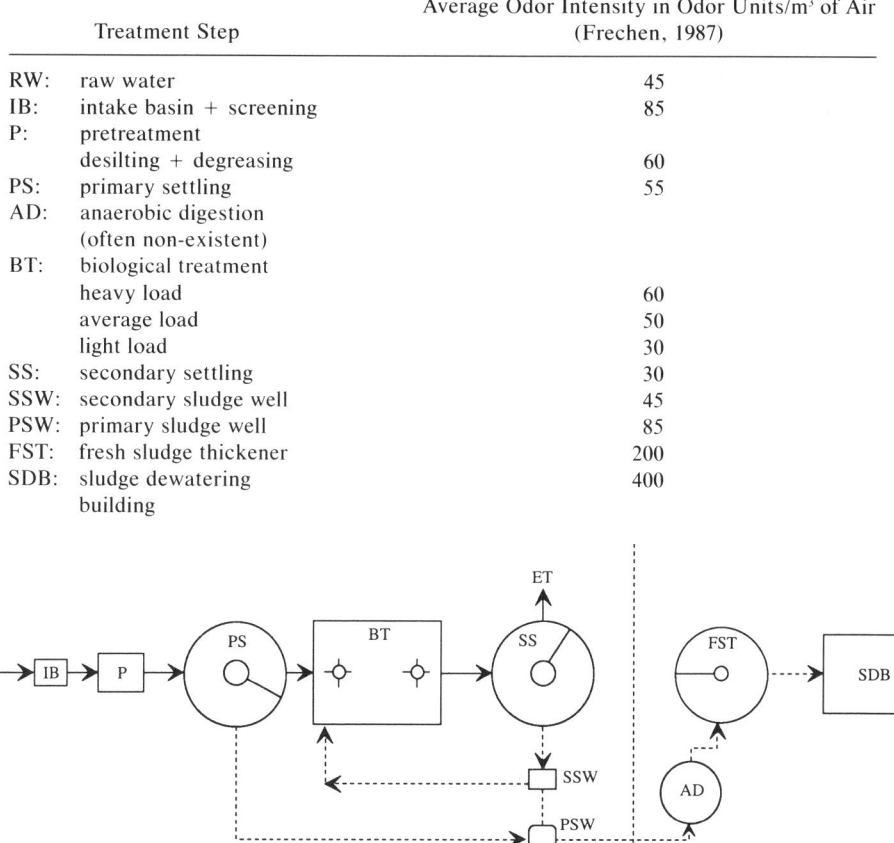

Figure 8.6. Scheme of a classic wastewater purification treatment line.

If it is relatively easy to pinpoint a source of bad odors simply with the human nose, it is more difficult to identify each olfactive nuisance source in a wastewater treatment plant and determine which are the major sources to be eliminated in order to lower the total odor level below the local resident's tolerance threshold. It is thus necessary to know how to quantify each source and determine if it creates a nuisance or not. Since there is currently no analytical method available to directly measure the nuisance, the only possible experimental procedure to qualify and quantify an odorous source consists of a combination of physical-chemical analysis and the olfactometric analysis.

8.5.2 Analysis of Bad Odor Sources at the Sewage Network Level

Bad odor problems are frequent at the pumping station level where the transiting effluent has a high upstream residence time in the piping (more than 3 h). Table 8.13 gives the average and maximum concentrations of diverse odorous compounds as analyzed on more than 10 pumping stations located upstream of French sewage treatment works. Hydrogen sulfide is the predominant compound. Organic nitrogen (amines, indole, scatole) may occur at non-negligible concentrations. Among the sulfurized organic compounds, only methyl mercaptan occurs at low concentrations. Ammoniac concentrations are generally below the olfactive threshold of 5 mg/Nm^3. These organic compound concentrations depend directly on the septicity of the effluent, which is itself related to the organic load, the temperature, and the residence time. In Finland, for example, where climatic conditions and effluent composition are different in comparison with France, hydrogen sulfide and methyl mercaptan concentrations are much lower (about 0.1 mg/m^3) but ammoniac concentrations are equivalent (Kangas et al., 1986). Some dimethyl sulfide, an oxidation product of methyl mercaptan, is also present in low concentrations (0.13 mg/Nm^3 average), although it has not been detected in France.

After numerous olfactometric and physical-chemical analyses on many pumping stations in Australia, (Koe and Brady, 1986) established a direct

Table 8.13. Characteristics of Raw Effluents on 10 Wastewater Purification Plants Included in an Odor Analysis Campaign.

	NTK (mgN/L)	NH$_4^+$ (mgN/L)	N organ. (mgN/L)	S^{2-} (mg/L)	pH	E/H$_2$ (mV)	T (°C)
Average	57.4	44.4	13.2	0.99	7.6	104	22.9
Maximum	80.5	60	20.5	3.90	8.15	353	26
Minimum	48.8	31	9	0.03	7.2	−260	17.9

relationship between the odor level and the H_2S concentration for a given site:

$$Cs = 41.(X)^{0.57}$$

with Cs odor level in standard odor unit per m^3 of air (SOU/m^3)
X mg/Nm^3 in hydrogen sulfide

This relationship is valid in the case of domestic effluents. If the effluent receives a large amount of industrial wastes, the constants of this equation are modified. Studies in progress in France show that the correlation between olfactometric measurements and that of H_2S alone is unsatisfactory. This correlation is clearly improved with the addition of the organic nitrogen parameter, which presents non-negligible concentrations in pumping stations (cf. Table 8.2). However, it seems hazardous to try to relate olfactometric analysis and several physical-chemical analysis parameters, taking into account the mutitude of compounds present in an odorous source and the rather large uncertainty of olfactometric analysis.

Otherwise, Koe and Brady (1986) have shown that a direct relationship exists between the quantity of odor emitted [QO in Standard Odor Unit (SOU) per s] and the pumping flow rate (Q in L/s) in pumping stations:

$$QO = 0.22.Q^{1.28}$$

with

$$QO = Rv \cdot Cs$$
$$QO = \text{quantity of odor emitted in SOU/s}$$
$$Q = \text{effluent flow rate in L/s}$$
$$Rv = \text{ventilation rate in } m^3/s$$
$$Cs = \text{odor level in SOU/}m^3$$

The inconvenience of olfactometric determination of odor levels expressed as a dilution rate to the perception threshold for 50% of the subjects (called K50) is to not fix a threshold over which there is a nuisance risk.

In practice, bad odor emissions are limited to the pumping station level and manholes when H_2S in the effluent has a concentration lower than 1 mg/m^3. For effluents with a lower than normal pH (the pH of a domestic effluent is from 7.5 to 8) and elevated temperatures (higher than 25 °C), less than 0.5–0.7 mg S^{2-}/L should be maintained in the effluent to limit olfactive nuisances related to sulfurized products.

When H_2S is in equilibrium between gaseous and aqueous phases, a theoretical relationship derived from Henry's Law exists that allows the calculation of [H_2S] in the air by the amount of [H_2S] in the water:

$$Y = HX \text{ with } Y = [H_2S] \text{ in the air in molar fraction}$$
$$X = [H_2S] \text{ in the water in molar fraction}$$
$$H = \text{Henry's constant}$$

H depends on the temperature and pH. For sulfides in H_2S form, the related Henry's constant H^* is only a function of temperature:

$$H^* = \exp(-2.035,2/T + 13,094)$$

In water, hydrogen sulfide is more or less disassociated in function with pH, and, in a lesser measure, in fuction with temperature (Morton and Card, 1986):

$$H_2S \rightleftharpoons H^+ + HS^-\quad K1 = ([H^+][HS^-])/[H_2S] = 1 \times 10^{-7}\ (20\ ^\circ C)$$
$$HS^- \rightleftharpoons H^+ + S^{2-}\quad K2 = ([H^+][S^{2-}])/[HS^-] = 1.3 \times 10^{-13}\ (20\ ^\circ C)$$

The actual Henry's constant can thus be expressed as follows:

$$H = H^*\, 1/[1 + (K1/10^{-pH} + K1\, K2/10^{-2\,pH})]$$

At the pumping station level, the transfer equilibrium is never affected. The actual transfer yields are from 1 to 60% in relation to theory according to the site. This variability can be explained mainly by the hydraulic systems and configurations particular to each pumping station. Thus, if it is difficult to calculate the quantity of hydrogen sulfide emitted into the air from sulfide concentrations in the water with precision in all cases, it is possible to do so for a given pumping station after having determined its H_2S transfer yield.

It is always necessary to know the effluent sulfide concentration in order to sustain this model. This information can be obtained by direct analysis which may be delicate depending on the method used, or estimated by using laboratory-determined empirical prediction equations.

8.5.3 Bad Odor Sources in Urban Wastewater Purification Plants

Figure 8.6 makes it possible to visualize the principal olfactive nuisance sources at different treatment steps in a purification plant. It has to do with an olfactometric analysis campaign done on numerous urban wastewater purification plants in Germany. The odor levels are average values expressed in SOU/m³ of air. It is obvious that sludge treatment is the major olfactive nuisance source. In certain cases, the odor level can reach very high values (Frechen, 1987); for example,

- Dewatering building, band filter working: 3,330–95,500 SOU/m³
- Thermal sludge thickener: 71,000 SOU/m³
- Discharged matter: 530–39,000 SOU/m³

Effluent intake and pretreatments are often at the origin of bad odors as well, especially in hot weather. This is the case, for example, in the south and southwest of France. This observation is less verifiable in a country like Germany where temperatures are lower. This phenomenon should be put in relation with the capacity of the effluent to become septic in the network and thus apt to generate reduced sulfurized compounds.

If the olfactometer gives interesting results as to the odorous level of each source, it does not allow identification of the compounds at the origin of odors or their concentrations. Nevertheless, this information is indispensable in order to define a deodorization treatment if necessary. An analysis campaign conducted by Anjou Recherche in collaboration with the Ministry of the Environment on many French purification plants is currently being done. The results obtained on 10 plants are summarized in Table 8.13. Most of these plants are covered and the buildings are ventilated. The sulfurized compound families are in the majority; H_2S is present in high concentrations compared to mercaptans and disulfides.

The sources of high concentrations in total sulfides are located at the head of the plant, at the effluent intake basin level, and throughout the sludge treatment line. Among the mercaptans, only methyl mercaptan is present, notably at the thickener level where its concentration can exceed 1 mg/m^3. The presence of dimethyl and diethylsulfide above the head works should be noted. These compounds are probably oxidation products of mercaptans (Kangas et al., 1986).

As far as nitrogenous compounds are concerned, ammoniac is present in low concentrations except at the sludge storage building level when sludges are stabilized with lime. Ammoniac concentration can reach levels of more than 20 mg/m^3, whereas the olfactive threshold is around 5 mg/m^3. Nitrogenous organic compounds (amines, indole, scatole) have low concentrations compared to sulfurized compounds. The highest concentrations are detected in the head works, during sludge thickening and during band press dewatering.

Lime addition during thickening or for dewatered sludge stabilization has a tendancy to favorize amine and especially ammoniac stripping by pH increase. Among the amines, traces of trimethylamine and diethylamine are essentially present most of the time at the water treatment works level. On the other hand, at the sludge thickening level, total amine concentration [in $(C_2H_5)_2NH$] can reach 3 mg/Nm^3 (Psarofaghish and Adam, 1975) and as high as 6 mg/Nm^3 in digestion gases. Indole and scatole, which have a pronounced fecal odor, can reach concentrations of 10–30 $\mu g/m^3$ (Belin, 1986).

The presence of sulfurized and nitrogenous compounds in the atmosphere of purification plant head works is in direct relation with the quality of the effluent entering the treatment line. Table 8.3 shows that the nitrogen content is stable enough though sulfide content is quite variable. A rather low pH (lower than 7.5) in comparison with the average, combined with high temperature and a low (negative) redox potential, corresponds to a high sulfurized compound content in the air. A high pH (more than 8.1) seems to favorize organic nitrogen stripping.

Besides reduced sulfurized compounds and nitrogenous compounds, the other families of compounds such as aldehydes, ketones, or acids are only present at trace levels . Concentrations are generally lower than olfactive thresholds for these compounds (cf. Table 8.1). In certain cases, one finds aldehydes, ketones, and acids in high concentrations when the sludge is thermally conditioned with thickening and dewatering. This is the case of the purification plants at Achères close to Paris or Le Mans. At Le Mans, for example, one finds mercaptans but, above all, acetone (2 mg/m^3), acetaldehyde (45 mg/m^3), isovaleraldehyde (18 mg/m^3), and organic acids (butyric acid: 3 mg/m^3, valeric acid) are found in thermal conditioning gases. Digested sludges tend to give off amines and organic acids as well. Digestion gases are loaded with odorous compounds (H_2S, mercaptans, ammoniac, amines, and organic acids). These odorous gases as well as thermal conditioning gases are generally burned with the methane produced in the boiler used for sludge conditioning or incineration.

Treatment works considered to be sources of important odors are now covered and ventilated. Ventilation rates used at the present time are from 3 to 12 times the volume of the room per hour. They are calculated in such a way that the dilution obtained allows the required toxicity thresholds concerning gaseous pollution for personnel to be met and desirable H_2S levels in order to limit concrete and equipment corrosion. When proper ventilation rates are applied, odorous compound contents present in the air collected on the total ventilated purification plant works are in the neighborhood of the contents illustrated in Table 8.14.

8.5.4 Bad Odor Sources in Industrial Wastewater Treatment Plants

Each odor problem related to industrial effluent purification is an individual case. However, one can consider that there are two main types of olfactive nuisances: odors related to direct volatile compound discharges and odors produced by anaerobic fermentation or the treatment of industrial effluents.

If the olfactometer is frequently used (Thal et al., 1978; North, 1980) to identify and quantify odorous sources, it is generally necessary to call on sophisticated physical-chemical analysis techniques by a gaseous chromatograph coupled with a mass spectometer in order to precisely identify and quantify the major odorous compounds (Van Lagenhove et al., 1985).

1. *Direct volatile odorous compound discharges.* Odorous compounds given off by aqueous industrial effluents can be diverse. Several examples are described in Table 8.15. Concentrations can be extremely variable in function with the industrial activity and the purification process in use. However, it can be observed that odorous levels are sometimes very high in comparison with those obtained for gaseous extracts in urban wastewater purification plants. Tannery discharges, for example, heavily loaded with sulfides used in tanning agent formulas, generate high H_2S and mercaptan concentrations by fermentation, which are given off at stocking and water purification works levels (Harkness, 1980).
2. *Volatile odorous compounds resulting from the fermentation of industrial wastewaters.* Food industry discharges, which are loaded with organic

Table 8.14.   Average and Maximum Concentrations of the Principal Odorous Compounds in the Polluted Gas Evacuation Vent of an Urban Wastewater Purification Plant.

		Maximum Concentrations	
	Average Concentrations	Without Thermal Treatment*	With Thermal Treatment*
Nitrogenous products			
Total nitrogen (mgN/m³)	2.37	21	2.15
Ammoniacal nitrogen (mgN/m³)	2.69	20.3	0.75
Organic nitrogen (mgN/m³)	0.26	0.8	1.23
Sulfurized products			
Total sulfides (mgH₂S/m³)	6.76	14.6	2.6
Dimethyldisulfide (mg/m³)	<0.15	<0.15	2.6
Diethyldisulfide (mg/m³)	0.36	0.73	2.5
Dimethylsulfide (mg/m³)	0.14	0.65	2.4
Ethylmercaptan (mg/m³)	<0.09	<0.09	2.3
Methylmercaptan (mg/m³)	0.89	1.64	5.5
Hydrogen sulfide (mg/m³)	3.19	3.04	7.5
Volatile solvents (mg/m³)			
Acetaldehyde	0.022	0.04	45.25
Propionaldehyde	0.006	0.007	1.64
Isobutyraldehyde	0.006	0.007	2.37
Butyraldehyde	0.006	0.007	3.55
Isovaleraldehyde	0.006	0.007	18.52
Acetone	0.013	0.08	1.89
Butanone-2	0.006	0.007	3.3
Olfactometer			
Olfactive threshold ($K50$) in SOU/m³	29,075	176,822	242,332

*Thermal sludge conditioning.

matter (carbohydrates, proteins), can generate odorous compounds by anaerobic fermentation, notably sulfurized compounds, nitrogenous compounds (ammoniac, amines), and volatile fatty acids. Moreover, these effluents are often treated in anaerobic digesters, which produce biogas, or in aerated or nonaerated ponds. If digester treatments are rarely bad odor sources because they are tight and biogas containing H_2S and mercaptans is burned, the same is not true of ponds. Slaughterhouse effluent treatment in aerated ponds followed by nonaerated ponds can provoke extremely intense olfactive nuisances (Hargett and Rybus, 1982).

From a general point of view, odorous fluxes produced by fermentable industrial effluent treatment are much more important compared to those

Table 8.15. Several Examples of Odorous Pollutants in Industrial Purification Plant Atmospheres.

Industry	Sulfurized Compounds	Nitrogenous Compounds	Aldehydes and Ketones	Acids and Alcohols	Hydrocarbons and Solvents	Olfactometer (SOU/m^3)	References
Chemical							
Pharmacy	—	Acrylonitryle	—	Phenols Acrolene	Aromatic solvents (benzene, toluene)	715,000	(Brandl and Stover, 1988) (North, 1980)
Insecticides	H$_2$S	—	—	Alcohols	Chlorobenzene, chlorine		(Ministry of the Environment, 1984)
Perfumes	—	—	Aldehydes, Ketones	—	—		
Petrochemical Industry							
Wastewaters	H$_2$S, Mercaptans	NH$_3$	—	—	Hydrocarbons		(Bouscaren, 1984)
Iron and Steel Industry							
Wastewaters	H$_2$S	—	—	—	Aromatic solvents (benzene, toluene)	1,070–207,400	(Ballaye, 1986)

		Amines	Formaldehyde		Solvents	1,200–18,000	(Bouscaren, 1984) (North, 1980)
Textile	—	—		—	—	1,200–18,000	(Bouscaren, 1984) (North, 1980)
Paper Making Industry	H₂S, Mercaptans Dimethylsulfide Dimethyldisulfide			—	—		(Bouscaren, 1984)
Food and Agricultural Industries							
Fish processing	—	Trimethylamine Cadaverine Putrescine NH₃	—	Fatty acids (Butyric acid)	—	150,000–400,000	(Bouscaren, 1984) (North, 1980)
Slaughterhouses	H₂S, Mercaptans	NH₃, Amines	Aldehydes	Fatty acids	—	6,000–1,350,000	(Bouscaren, 1984)
Pig farming	H₂S, Mercaptans	NH₃	Aldehydes	Fatty acids, alcohols (Propionic and butyric acid)	—		(Bouscaren, 1984)
Manure treatment	Disulfides	Trimethylamine					(Bouscaren, 1984)

243

given off by urban wastewater purification plants. One finds sulfurized compounds most frequently and, for certain industrial activities, such as fish processing or meat quartering, high amine concentrations.

8.6 Odorous Pollution Sources in Water Potabilization

Many different odorous compounds have been identified in water. These substances have extremely varied origins. They can be found in raw water before any treatment either because of the presence of pollutants due to industrial origin or because of an intensive microorganism development leading to a particularly odorous metabolite release in the water. These compounds can also be formed during potabilization treatment by the action of chemical agents (ozone, chlorine, . . .) being used on organic matter in the water. Finally, the evolution of water quality in networks can lead to the appearance of odorous molecules during distribution.

8.6.1 The Nature of Odorous Compounds

8.6.1.1 Inorganic compounds

For water to taste "neutral," the salt content should be approximately that of saliva to which the taste receptors are adapted (O'Mahoney, 1972; Bartoshuk, 1974). From data obtained from taste assessments of individual salts in water solutions, Zoeteman (1978) recommended maximum salt levels in order to obtain water with an acceptable taste quality. These levels are given in Table 8.16.

Among the metal ions that can be present in potable water, iron could be tasted in distilled water at 0.05 mg/L, copper at 2.5 mg/L, manganese at 3.5 mg/L, and zinc at about 5 mg/L (Water Research Center, 1981).

Table 8.16. Maximum Acceptable Levels of Salts in Water to Prevent Offensive Taste (Zoeteman, 1978).

Salts	Salt Concentration (mg/L)	Cation Concentration (mg/L)
NaCl	465	185
$MgCl_2$	47	12
$CaCl_2$	350	105
$NaHCO_3$	630	175
$Mg(HCO_3)_2$	58	10
$Ca(HCO_3)_2$	610	150
Na_2SO_4	—	—
$MgSO_4$	840	170
$CaSO_4$	1,020	300

8.6.1.2 *Organic Compounds*

Most raw and/or natural waters to be potabilized contain between 1 and 10 mg organic carbon per litre, which in fact covers many thousands of specific organic compounds. Two of the principal characteristics of these compounds are the large variety of chemical families called into question and their extreme concentration variability, both in time and place (seasonal and location variations). One can distinguish organic micropollutants (10–30% of the total organic carbon) and organic macromolecules, which represent the organic matrix of waters (70–90% of the TOC—Total Organic Compound).

In natural waters, there are many types of organic compounds that can be classified by order of importance into six major groups: humic substances, hydrophilic acids, carboxylic acids, peptides and amino acids, carbohydrates (sugars), and hydrocarbons (Thurman, 1985; Bruchet and coworkers, 1985, 1987). Except for volatile fatty acids, no data is available in water-related literature describing the effect of this organic matrix on tastes and odors of water.

Different oxidation processes used in potable water treatment plants modify the nature of this organic matrix. This modification leads to the formation of products that are responsible for bad tastes, as will be seen in the chapter concerning removal of tastes and odors from potable water (Chapter 13).

On a concentration scale going from 100 µL to 1 ng/L, organic compounds in waters can belong to numerous chemical families. These products are either naturally present in raw waters (microorganism metabolites, for example) or introduced by human activity. In general, they correspond to products most easily identified by mass chromatography/spectrometry.

8.6.2 Anthropogenic Origins

As far as specific organic pollutants are concerned, no one—including the polluters—possesses a complete inventory of compounds present in industrial or urban effluents. In most cases, the only known characteristics of these effluents are the global parameters such as total organic carbon or the chemical demand in oxygen.

It is nearly impossible to publish an exhaustive list of all the organic pollutants that can be detected in raw waters. For this reason, one can quote Keith (1976): "If only 5% of the more than two million known organic compounds eventually find their way into our waters, the number of identifiable organic compounds in the world's waters could be estimated at 100,000." There are many lists of organic compounds identified in different types of water in the literature (Keith, 1976; Keith, 1981). Only several examples are given in this chapter, assuming that the type of trace organic compounds identified in the water are the same for all industrialized countries. Taste and odor thresholds quoted as examples in the following paragraphs are taken from two principal compilations (Fazzalari, 1978 and Van Gemert and Nettenbreijer, 1977).

8.6.2.1 Dispersed Chronic Pollution in Urban Areas

Many hydrocarbons, halogenated solvents, aerosol propellants, and refrigerants, including trichloroethylene, tetrachloroethylene, carbon tetrachloride, dichloromethane, and freons, are produced at a rate of several billion kilograms per year. These products are the most frequently encountered synthetic pollutants in potable water because of their massive production, widespread use, chemical and biological stability, volatility, and negligible adsorption in soils and sediments. Cotruvo (1985) gives an example of potable water contamination of about 0.1 μ/L that could go as high as several milligrams per litre, notably for underground waters. Table 8.17 gives odor detection thresholds for some of these chlorinated solvents (Van Gemert and Nettenbreijer, 1977).

Water contamination from gasoline and other refined fuels and solvents is becoming an increasingly frequent occurence. This is especially true of gasoline and its components such as benzene, toluene, and xylenes, as well as several kinds of alkylbenzenes. The majority of these products can be responsible for taste and odor problems, as shown by odor detection threshold values of napthalene (0.000,5 mg/L) and t-butylbenzene (0.000,5 mg/L) according to Zoeteman et al. (1971).

8.6.2.2 Municipal Wastewater Effluents

Garrison et al. (1976) have established a list of specific compounds identified in wastewaters. Many carboxylic acids and alcohols as well as phenols, aromatic hydrocarbons, and chlorinated solvents are present in concentrations at the microgram per litre level. The presence of trichloro, tetrachloro, and pentachloro anisoles, compounds known to cause intense moldy odors, have been detected in both secondary and tertiary effluents (Burlingame et al., 1976a and 1976b). The potential effect of these latter compounds on the organoleptic quality of waters can be evaluated from the odor detection threshold value of 2,3,6-trichloroanisole, which is less than 0.1 ng/L (Van Gemert and Nettenbreijer, 1977).

Table 8.17. Odor Detection Thresholds of Several Chlorinated Solvents Identified in Urban Area Groundwaters.

Solvent	Odor Detection Threshold (mg/L)
1,4-dichlorobenzene	0.000,3
Trichloroethylene	0.5
Tetrachloroethylene	0.3
Carbon tetrachloride	0.2

8.6.2.3 Industrial Wastewater Effluents and Waste Disposal

The nature of specific compounds in industrial wastewaters depends, of course, on the type of industry planned. The destruction of chemical wastes is an ever-increasing problem in industrialized countries. Contamination of groundwaters due to improper solid waste storage conditions has been recognized as the most significant source of present and future pollution (Cotruvo, 1985). The odor and taste of water can indicate the presence of pollutants that present a health risk.

In the Parisian area, the Agence de Bassin Seine Normandie and other water distribution professionals have published a study in which they evaluated problems associated with the safety and security of water resources. Part of this study concerned the inventory of all chemical products that are stored in large quantities in the Parisian suburbs. These locations are upstream from pumping stations and potable water treatment plants that use the Seine river as a raw water source. More than 400 products have been listed, but the odor threshold concentration for only a few of them is known (Table 8.18) (Mallevialle and Suffet, 1987).

8.6.2.4 Miscellaneous Origin

It is extremely difficult to establish a complete list of all the specific organic pollutants present in water when one considers the fact that a biological or chemical transformation of these products can occur in the environment. Hundreds of pesticides totaling hundreds of millions of kilograms per year are in use worldwide. Herbicides such as triazines are extremely mobile in water and are frequently found in surface water; their odor thresholds are given in Table 8.19.

8.6.3 Biological Sources of Tastes and Odors

8.6.3.1 Algae and Actinomycetes

The organisms most often related to the development of tastes and odors (mainly musty or earthy types) are actinomycetes and different types of algae (notably blue-green), though protozoans, fungi, and other aquatic microorganisms have been implicated from time to time.

In 1965, Gerber and Lechevalier reported the isolation and identification of an earthy-smelling component of certain actinomycete cultures, a compound they named "geosmin" (odor threshold of less than 10 ng/L) from the Greek "ge" meaning earth and "osme" meaning odor. Later, Gerber (1969) isolated a musty-smelling compound from certain actinomycete cultures, which she identified as 2-methylisoborneol (MIB). In the same year, Medsker et al. (1969) isolated MIB (odor threshold of less than 10 ng/L) among 3 out of 28 cultures studied. Medsker et al. (1969) as well as Collins et al. (1970) referred to the compound as 2-exo-hydroxy-2-methylbornane. Two

Table 8.18. Taste and Odor Threshold Concentrations of Different Industrial Products.

Product	Threshold (mg/L of water)	Industry
Acetone	from 5 to 265 (O)[a]	Paints, varnishes, perfumery, plastics
Acetic acid	from 0.007 to 200 (O)	Printing, photography, textiles
Allyl chloride	14,700 (O)	Chemistry
Ammonium chloride	210 (T)[b]	Chemistry
Anilene	70 (O)	Rubbers
Anisole	0.05 (O)	Chemistry
Benzene	from 2 to 30 (O)	Inks, printing, paints, varnishes
Bromoform	0.3	Chemistry
Butyl acetate	0.043 (O)	Perfumery, plastics, paints
Butyric alcohol	from 0.5 to 40 (O)	Fungicides, insecticides, perfumery, paints, pharmaceutics, plastics
Chloroform	0.1 (O)	Plastics, pharmaceutics
Cyclohexane	200 (O)	Oils, rubbers, photoengraving
Diacetone alcohol	44.1 (O)	Paints, wood, photography
2,4-Dichlorophenol	from 0.002 to 0.21 (O)	Chemistry
1,3-Dichlorobenzene	0.02 (O)	Chemistry
Ethyl acetate	5 (T)	Food, paints, varnishes
Ethyl alcohol	from 100 to 2,400 (O)	Antifreeze, explosives, perfumery, pharmaceutics
Ethyl amine	10 (O)	Greases, oils, resins, perfumery, pharmaceutics
Formaldehyde	50 (O)	Disinfectants, insecticides, coloring, plastics, textiles
Formic acid	from 450 to 8,000 (O)	Galvanoplasty, textiles, metallurgy
Glucose	4×10^{-5} (T)	Chemistry
Glycerol	from 38 to 440	Chemistry
Heptane	50 (O)	Chemistry
Heptanoic acid	3	Chemistry
Isobutyl acetate	0.037 (O)	Paint solvents
Bleach	5×10^{-2} (T)	Paper mills
Lactic acid	40 (O)	Chemistry
Magnesium sulfate	1,000 (T)	Pharmaceutics
Methyl alcohol	from 10 to 1,600 (O)	Motor fuel, paints, varnishes
Methyl-ethyl ketone	50	Paint solvents, varnishes, glues, household cleaning products
Metronidazole	149 (T)	Chemistry
Nitric acid	65 (T)	Chemical fertilizers, explosives, metallurgy, textiles, pharmaceutics
Nitropropane	25 (O)	Chemistry
1-Octanol	0.13 (O)	Chemistry
Oxalic acid	120 (T)	Household cleaning products, wax
Phenol	from 1 to 6 (O)	Chemistry

(Continued)

Table 8.18. *(Continued)*

Product	Threshold (mg/L of water)	Industry
Pivalic acid	50 (O)	Chemistry
Propionic acid	20 (O)	Chemistry
Pyridine	from 0.003 to 4 (O)	Chemistry
Saccarine	4.4 (T)	Galvanoplasty
Salicylic acid	from 20 to 90 (O)	Chemistry
Sodium chloride	450 (T)	Chemistry
Sodium hydroxide	320 (T)	Chemistry
Stearic acid	20 (O)	Chemistry
Styrene	from 0.05 to 0.73 (O)	Polystyrene and rubber making
Sulfuric acid	13 (O)	Detergents, chemical fertilizers, galvanoplasty, metallurgy
Tartric acid	75 (O)	Chemistry
Tetralin	18 (O)	Resin solvents, greases, varnishes, household cleaning products
Toluene	1 (O)	Paint solvents, varnishes, glues, inks, coloring, insecticides, pharmaceutics
Trichlorethylene	0.5 (O)	Galvanoplasty, printing, textiles, industrial stain removers
Xylene	from 0.02 to 1.8 (O)	Paints, printing, insecticides
Isobutyl acetate	0.073 (O)	Paints, solvents

[a](O) Odor threshold concentration.

[b](T) Taste threshold concentration.

years after Gerber and Lechevalier's work, Safferman et al. (1967) isolated geosmin from blue-green algae.

Since 1969, taste and odor problems of biological origin have been widely studied in the literature and actinomycetes and blue-green algae (cyanophycea) have been the subject of numerous scientific reports. Blue-green algae that produce geosmin and MIB have been universally recognized as the principal cause of tase and odor problems in water resources.

8.6.3.2 Other Microorganisms (Bacteria, Fungi, Zooplankton)

Occasionally taste and odor problems in water resources are caused by other bacteria, fungi, zooplankton, and nemathelminthes. MacKenthun and Keup (1970) reported that ferro-bacteria could cause tastes and odors in deep water and in water distribution networks. Sulfate-reducing bacteria (*Desulfovibrio desulfuricans,* for example) can reduce sulfates to hydrogen sulfide, which has a characteristic odor of rotten eggs (Lin, 1976). Wajon et al. (1985a, 1985b, 1985c) attributed the "muddy" odor in an Australian water

Table 8.19. Odor Thresholds and Characteristics of Several Pesticides and Herbicides (according to Sigworth, 1965).

Herbicides or Pesticides	Odor Characteristics	Odor Threshold (mg/L)
Hexachlorocyclohexene		
delta isomer	Earthy/musty	0.000,13
beta isomer	Musty	0.000,32
alpha isomer	Musty	0.088
gamma isomer (lindane)	Chlorine	12.0
Chlordane (5% granulated)	Musty	0.000,32
(40% powder)	Musty & chlorinous	0.002,5
Aldrin	Musty/muddy	0.017
Endrin	Musty & chlorinous	0.018
Heptachlor	Musty & chlorinous	0.02
Parathion	Rotten onion	0.04
Dieldrin	Musty/camphor	0.041
Isopropyl 2,4-dichlorophenoxacetate	Chlorophenol or iodoform	0.055
Delnav	Kerosene	0.06
Malagran 5 (5% Malathion)	Garlic	0.081
Toxaphene	Musty/muddy	0.14
Dichlorodiphenyl Trichloroethane	Chlorophenol or iodoform	0.035
Rotenone	Musty & chlorinous	0.36
Malathion	Onion	1.0
2,4,5-T	Iodoform/musty	2.92
2,4-D	Chlorophenol/musty	3.13
Methoxychlor	Musty/chlorinous	4.7

distribution network to dimethylpolysulfides that were possibly produced by several species of the bacterial genus *Pseudomonas* or *aeromonas hydrophilia*. Bacterial decomposition of some blue-green algae results in the production of mercaptans and dimethylpolysulfides (Slater and Block, 1983; Jenkins et al., 1967; Krasner et al., 1986). Certain bacteria (*Pseudomonas*, for example) can convert sulfurized amino acids (methionine and cysteine) to hydrogen sulfide, methyl mercaptan, and dimethylpolysulfides (Whitfield and Freeman, 1983). Thus, bacterial degradation is an important source of bad odors in water.

Müller et al. (1982) have shown that these phenomena can occur either in the natural medium or on slow sand filters, thus affecting water quality. According to Wood et al. (1983), odor-producing fungi live in all habitats known to be potential sources of earthy tastes in potable water.

Zooplankton, including crustaceans such as *Cyclops* and *Daphnia* and the rotifer *Keratella* can be responsible for fishy odors if they are abundant (Lin, 1976). Nematodes can secrete odorous compounds as well; an oily compound giving an earthy and musty smell was isolated from nematode cultures (Chang et al., 1960; Cobb, 1918). Certain amoebae are also capable of causing tastes and odors (Chang et al., 1960).

8.6.4 Tastes and Odors Generated During Water Potabilization

All chemical products, such as coagulants, oxidants, or disinfectants used to remove particles, organic compounds, and microorganisms can react with organic compounds present in water and thus form by-products. Some of these by-products are well-known taste and odor compounds. This is particularly true of oxidant or disinfectant use (ozone, halogens, chloramines, chlorine dioxide).

8.6.5 Tastes and Odors in Distribution Networks

The phenomena of taste problems appearing in water distribution networks are still poorly understood and, since they are often transient, it is difficult to make a general evaluation of their mechanisms. However, it is possible to give a partial response to this question by listing the probable origins of these tastes and odors:

1. Biological origin
2. Disinfectant residual and oxidation by-products
3. Emissions from pipes and reservoir coatings
4. Diffusion of pollutants through synthetic pipes

8.6.5.1 Tastes and Odors of Biological Origin

In certain cases (high temperature, low water velocity), important bacteria regrowth can be observed in some parts of the water distribution network or certain reservoirs despite the presence of a disinfectant residual at the treatment plant outlet. A sudden increase in the number of microorganisms can lead to the appearance of tastes such as musty, earthy, and fishy. This phenomenon is due to the release of a certain number of metabolites from these microorganisms. In most cases, these tastes are linked to the presence of methylisoborneol and geosmin produced either by actinomycetes (Henly et al., 1969; Rosen et al.,1970; Silvey and Roach, 1954; Silvey 1954) or by blue-green algae (Izaguirre et al., 1981). However, the conditions underlying the appearance of these phenomena and the existing relationship between the number of microorganisms and odor intensity are not yet completely understood. These tastes can appear in networks containing low quantities of actinomycetes (Water Reasearch Center, 1981) but in which a large number of heterotrophic bacteria have been found. This means that a large variety of microorganisms can release metabolites that are responsible for tastes.

Generally, one can suppose that bacterial growth is favored by

1. A high assimilable organic carbon (AOC) concentration due either to the dissolved organic carbon residual contained in the distributed water or to the release of materials used in water distribution (Van der Kooij et al., 1982)
2. A high temperature
3. Low water velocities in certain parts of the network that are insufficiently cross connected
4. Important residence times (more than 48 h)

Several authors (Krasner et al., 1986; Wajon et al., 1985a, 1985b, 1985c; Giger and Schaffner, 1981) have described the development of muddy (swamp) and fishy tastes and odors in distribution networks in connection with the appearance of sulfurized compounds (dimethylpolysulfides). Some of these authors have detected dimethyltrisulfide concentrations from 5 to 250 ng/L (Wajon et al., 1985c), which is very important in view of the odor threshold of this molecule (10 ng/L) (Buttery et al., 1976). These sulfurized compounds are produced in distribution networks under anaerobic conditions. This bacterial degradation leads to the formation of compounds such as hydrogen sulfide, methylmercaptan, and dimethylpolysulfide, which all have a characteristic odor of rotten vegetables.

8.6.5.2 Tastes and Odors Caused by a Disinfectant Residual

The use of an oxidant with an important residual bactericidal potential can cause taste problems. These problems are either directly related to the presence of the oxidant residual itself or to the slow reaction of the oxidant residual with the organic compounds present in the distributed water.

In the first case, the problem is mainly olfactory and complaints are generally confined to consumers located at the beginning of the distribution network. These consumer complaints mention chlorine odors and even fishy odors (in the case of a chlorine dioxide disinfection or the presence of dichloramine) usually noticed in the shower.

The second case is much more complex and complaints frequently mention "pharmaceutical and medicinal" tastes: that is to say, tastes due to chlorophenols. These tastes are caused by slow kinetic reactions between the disinfectant residual and organic matter contained in the water or released by reservoir and pipe coatings or synthetic pipes. These reactions involve, for example, additive chlorine reactions on organic matter such as chlorine dioxide or free chlorine oxidation of organic matter. The resulting by-products generally harm the organoleptic qualities of distributed water.

8.6.5.3 Emissions from Pipe and Reservoir Coatings

The presence of iron in water is either natural, the result of metallic pipe corrosion, or corresponds to a ferrous salt residual used for coagulation and can give a disagreeable taste to water. The average taste threshold is about 3 mg/L for the majority of the population, but can reach 40 µg/L for those who are the most sensitive (about 5% of the population) (Water Research Center, 1981; Cohen et al., 1960).

Corrosion of pipes or fittings containing copper or zinc may give rise to a significant increase in concentrations of these metals in water. Both give an astringent taste to water. Taste threshold concentrations are approximately 3–7 mg/L for copper and 20 mg/L for zinc (Water Research Center, 1981; Cohen et al., 1960).

Tastes and odors may also be due to the release of organic compounds from lining and coating materials used to protect the internal surface of pipes and reservoirs from corrosion. In these cases, tastes are generally described as "pharmaceutical and medicinal" or "solvent and/or chemical-like." Various authors have reported naphtalene dissolution (Zoeteman, 1978) or polycyclic aromatic hydrocarbons (Alben, 1979a, 1979b) from certain bituminous materials (asphalt, coal tar) currently in use in the potable water field. Certain pipe and coating materials lead to a decrease in the water quality of the distribution network either by favoring bacterial growth or by introducing organic odorous pollutants into treated water. Many organic solvents are used during coating installation and can thus eventually contaminate the water. Several authors have noted releases of clorinated solvents such as trichloroethylene (Yoo et al., 1984) or tetrachloroethylene (Stinson and Carns, 1983), leading to the detection of "pharmaceutic" type tastes and odors. In other cases (Yoo et al., 1984; Krasner and Means, 1986; Alben et al., 1987) releases of important concentrations (exceeding 100 µg/L) of benzene derivatives (xylenes, for example) have been noted. Recent works (Bruchet et al., 1986a, 1986b) show the presence of different ketones such as methylethyl or methylisobutylketone in concentrations exceeding 1 mg/L (taste threshold = 2 mg/L), which are associated with the use of epoxy and vinyl materials as reservoir coatings.

In the water distribution field, the use of synthetic pipes (mainly PVC and polyethylene) has been increasing in the last few years. However, in certain cases, changes in the organoleptic quality of the water have been noted. The following three principal causes for these changes have been identified in the literature:

1. Dissolution of polymer additives
2. Oxidation of the internal surface during extrusion (manufacturing problem) and dissolution of the resulting polar compounds during contact with water
3. Migration of external contaminants through the pipe

Several authors have associated the appearance of numerous organic compounds released by synthetic pipes with intense tastes and odors of plastic or burned plastic.

Polyethylene pipes, particularly high-density polyethylene pipes (HDPE) generally have a very low external contaminant permeability but, on the other hand, they contribute to the release of a large number of additives, mainly phenolic additives used as antioxidants during the manufacturing process (Anselme and co-workers, 1986, 1986a, 1986b, 1985b; World Water, 1984). The presence of phenolic additives such as 4-methyl-2,6-di-t-butyl phenol as well as aldehydes in the water seems to be the cause of burned plastic odors noticed when new HDPE pipes are installed in a distribution network. Compounds released by PVC are generally tetrachloroethylene (Demond, 1985) and phtalates esters (Marshall et al., 1982; Marshall, 1985).

8.6.5.4 Diffusion of Pollutants Through Synthetic Pipes

Plastic pipes are permeable (Reich et al., 1985) to a range of substances, especially hydrocarobons and phenols. In order to avoid potential odor problems caused by external pollutant migration through the pipes, careful consideration should be given to their installation.

Gas penetration through polymeric materials used to make potable water pipes can be described as a solution diffusion. The gas is dissolved in the polymer on the outside surface, diffuses through the pipe wall, and desorbs from the inner surface. Table 8.20 gives a comparison of the permeation coefficients of several gases for PVC, HDPE, and LDPE (low-density polyethylene).

Low-density polyethylene has a permeation coefficient that is two to three times higher for most gases than high-density polyethylene. PVC is less porous to gases than either HD or LD polyethylene. LDPE is known to be a very permeable polymer to various hydrocarbons such as toluene, benzene (Marshall et al., 1982; Meheus et al., 1984), oil, or gasoline compounds as well as chlorinated solvents such as trichloroethylene (Marshall et al., 1982; Vonk, 1984; Kreft et al., 1981). PVC pipes are generally less permeable to external contaminants than low and high density polyethylene pipes, respectively (Kreft et al., 1981); they are notably very permeable to chlorinated solvents and benzene derivatives.

Table 8.21 gives several examples of interactions occuring between PVC and organic compounds. This table includes weight gains (%) observed on PVC samples brought into contact with different solvents.

Table 8.20. Permeation Coefficients of Several Materials by Different Gases (according to Flodstad, 1985).

Gas Material	Air	N_2	O_2	H_2	CO_2	CH_4	C_2H_4
PVC	0.4	0.1	1.2	29	2.4	—	—
HDPE	12	7.2	24	84	96	19	31
LDPE	24	19	62	180	240	72	120

Table 8.21. Equilibrium Swelling and Interaction Parameters for PVC Systems/ Organic Compounds (according to Berens, 1985).

	Equilibrium Adsorption in PVC	
Compounds	Weight Gain (%)	Time (h)*
CH_2CL_2	>800**	2
$CHCL_3$	227	6
CCL_4	~0	(280)
$1,1-C_2H_4Cl_2$	132	6
$1,2-C_2H_4Cl_2$ (EDC)	~700**	6
$1,1,1-C_2H_3Cl_3$ (TCA)	67	600
$1,1,2-C_2H_3Cl_3$	>400**	10
C_2HCl_3 (TCE)	70	5
C_2Cl_4 (PCE)	35	800
$1,2-C_3H_6Cl_2$	176	20
Benzene	50	12
Toluene	50–60	15
Xylene	42	100
Acetone	170	2
Methanol	~1.0	(300)
Ethanol	~0.1	(300)
i-Propanol	~0.5	(300)
n-Hexane	~0.6	
Vinylchloride	30	160

*Time to apparent equilibrium 24-mil (0.6 mm) PVC sheet or duration of experiment.
**Compound soluble in PVC.

The total of these examples show that the penetration of odorous pollutants through pipes in plastic material used for potable water distribution is an important potential risk. The greatest possible care should thus be used in the choice of land and installation areas for this type of pipe.

8.6.6 Tastes and Odors Created in Consumer Systems

When water leaves the treatment plant, it is generally free of all disagreeable tastes and odors, with the possible exception of chlorine problems. Some of these problems can reappear in the principal distribution network but also in the consumer's distribution network. In this case, certain complaints are difficult to address and resolve because the problem is localized and can simply be created by the consumer's plumbing installations. The two most common complaints are as follows:

1. Septic odors (rotten egg) of hydrogen sulfide and mercaptans, which can be found in networks that have high sulfate concentrations
2. Musty odors, which can be found in waters that have stood for long periods of time either in pipes or reservoirs, such as those in hotels or secondary residences

Another source of nuisances that can appear in houses, apartment buildings, or hotels is bacteria growth in water softeners (or other types of individual treatment equipment, such as those containing activated carbon).

Bibliography

Alben, K. (1979a). GC/MS analyses of potable water for evidence of contamination by coal tar compounds used in storage tank coatings. *Div. Environ. Chem. Am. Chem. Soc.,* 19th Conf. Honolulu, Hawaii, April.

Alben, K. (1979b). Chemical composition of acid base and neutral leachate from a commercial coal tar. *Abstracts of International Symposium on Aquatic Pollutants,* Jekyll Island, GA, October.

Alben, K., Shpirt, E., Bruchet, A. (1987). Rates of leaching organics from coating material. *American Water Works Association Annual Conference,* Kansas City, MO.

Anderson, L. (1984). *Proceeding, Caractérisation et Techniques de Réduction des Nuisances Olfactives dans les Industries de Procédés.* Société Belge de Filtration, Bruxelles, April.

Anselme, C. (1986a). Internal Report No. PM/MB 2878, Lyonnaise des Eaux, 78230 Le Pecq.

Anselme, C. (1986b). Internal Report, Lyonnaise des Eaux, 78230 Le Pecq.

Anselme, C. (1987). Internal Report Nos. CA/GP 264, CA/JD 317, and QAO CA/DC 107, Lyonnaise des Eaux, 78230 Le Pecq.

Anselme, C., et al. (1985a). Can polyethylene pipes impart odors in drinking water. *Environ. Technol. Lett.* **6**, 477.

Anselme, C., et al. (1985b). Characterisation of low molecular weight products desorbed from polyethylene tubings. *Sci. Total Environ.* **37**, 371.

Anselme, C., et al. (1986). Influence of polyethylene pipes on tastes and odors of supplied water. *American Water Works Association Annual Conference,* Denver, CO.

Barth, C. L., Elliott, L. F., Melvin, S. W. (1984). Using odor control technology to support Animal Agricultural. *Trans. ASAE* 859–864.

Bartoshuk, L. M. (1974). NaCl thresholds in man: thresholds for water taste or NaCl taste? *J. Compar. Physiol. Psych.* **83**, 2,310.

Belin, C. (1984). Measurements of trace organic gaseous compounds in the atmosphere. *Proceeding, Caractérisation et Techniques de Réduction des Nuisances Olfactives dans les Industries de Procédés.* Société Belge de Filtration, Bruxelles, April, 165–169.

Berens, A. R. (1985). Prediction of organic chemical permeation through PVC pipe. *J. Am. Wat. Works Ass.* **77**, 11, 57.

Böhnke, B., Frechen, F. B. (1983). Untersuchungen über die Gerushsprobleme auf Klärangen ursachen und behebung unter Einbeziehug der Kosten—unter besonderer Berücksichtigung der problem in Nordrhein—Westfalen, Rapport Minister für Ernährung, Landwirtchaft und Forsten des Landes Nordrhein—Westfalen.

Bombaugh, K. J., Rhodes, W. J. (1980). Discharges from Coal gazification plants. *Environ. Sci. Technol.* **22**, 12, 1389–1396.

Bouscaren, R. (1984). Les odeurs et la désodorisation. *Techn. Sci. Meth.* **6**, 313–320.

Bowker, R., Smith, J., Webstern, O. (1985). Odor and corrosion control in sanitary sewerage systems and treatment plant. *EPA Design Manual.* EPA 625/-85/018.

Brandl, J. S., Stover, E. L. (1988). Odorous VOS control and large industrial wastewater treatment plant. *Water Pollution Federation 61th Annual Conference,* 2–6 Oct., Dallas, TX.

Bruchet, A. (1985). Application de la technique de pyrolyse CG-SM à l'étude des matières organiques non volatiles des eaux naturelles ou en cours de traitement. Thèse Université, 22 novembre 1985, Poitiers, France.

Bruchet, A., Alben, K., Shpirt, E. (1986a), (1987). Composition of semi-volatile fractions extracted from coating material leachate at low and high Chlorine doses. *American Water Works Association Annual Conference,* Kansas City, MO.

Bruchet, A., et al. (1986b). Perspective on Methods to concentrate and analyse semi-volatile organics at trace (ng/L) levels in drinking water: simultaneous distillation-extraction: resin adsorption. *American Water Works Association WQTC Conference,* Portland, OR.

Bruchet, A., Anselme, C., Marsigny, O. and Mallevialle, J. (1987). THM formation and organic content: a new analytical approach. *AQUA.*

Burlingame, A. L., et al. (1976a). The molecular nature and extreme complexity of trace organic constituents in Southern California Municipal Wastewater effluents. In *Identification and Analysis of Organic Pollutants in Water,* edited by L. H. Keith. Ann Arbor Science Publishers, Inc., Ann Arbor, MI.

Burlingame, A. L., et al. (1976b). The characterization of trace organic constituents in petroleum refinery wastewater by capillary GC/real-time high resolution mass spectrometry, a preliminary report. In *Identification and Analysis of Organic Pollutants in Water,* edited by L.H. Keith. Ann Arbor Science Publishers, Inc., Ann Arbor, MI.

Buttery, R. G., et al. (1976). Additional volatile components of cabbage, broccoli and cauliflower. *J. Agric. Food Chem.* **24**, 4–289. Cahiers Techniques de la Direction de la Prévention des Pollutions (1984), Les odeurs et les nuisances olfactives, no. 15.

Caspar, G., Burn, M.D. (1934). *Postmortem Bacteriol. J. Infect. Dis.* **54**, 395–403.

Chang, S. L., Woodward, R. L., Kabler, P. W. (1960). Survey of free-living nematodes and amoebas in municipal supplies. *J. Am. Wat. Works Ass.* **52**, 5, 613.

Cheremisinoff, P. N., Young, R. A. (1975). *Industrial Odor Technology Assessment.* Ann Arbor Science Publishers, Ann Arbor, MI, 509 pp.

CITEPA, 1986. Report, Connaissances des émissions de polluants atmosphériques, Journée d'études, Paris, 1986.

Cobb, N. A. (1918). Filter-Bed Nemas-Nematodes of the slow sand filter beds of American cities. *Contrib. Sci. Nemato* **7**, 189.

Cohen, J. M., et al. (1960). Taste threshold concentrations of metals in drinking water. *J. Am. Wat. Works Ass.* **52**, 5, 660.

Collins, R. P., Knakk, L. E., Soboslai, J. W. (1970). Production of geosmin and 2-exo-hydroxy-2-Methylborane by streptomyces odorifer. *Lloydia* **33**, 199.

Cotruvo, J. A. (1985). Organic micropollutants in drinking water: an overview. *Sci. Total Environ.* **47**, 7.

Countaulds Ltd (1975). Procédé d'élimination de composés minérals malodorants dans un gaz résiduaire. Brevet Français No. 46.507/1975.

Crook, B. V., McRbwen, B. A. (1987). Operational benefit of using reduction—oxydation potential in septicity control. *J. Wat. Pollut. Control Fed.* **86**, 1, 20–33.

Delcour, J. A., et al. (1984). Flavor tresholds of polyphenolics in water. *Am. J. Enol. Vitic* **35**, 3, 134.

Demond, A. H. (1985). Leaching of tetrachloroethylene from vinyl-lined pipe. *J. Environ. Eng.* **111**, 1.

Detrie, J. P. (1969). *La Pollution Atmosphérique - Nuisances,* vol. I, Dunod, Paris.

Dorgelo, C. H. (1978). Emission types and odor abatement in a petrochemical industry, *Process Tech.* **33**, 10, 569–575.

Electric Co. Tokyo (1980). Treatment of spent Alkali liquors from deodorization of waste gases. Tokyo Shibana Electric to Ltd, Report No. 79/61, 335, 1979.

Fazzalari, F. A. (1978). Compilation of odor and taste threshold values data. *ASTM Data Series DS 48A.* ASTM, Philadelphia, PA.

Feodoroff, T. (1987). Du muscle à la viande: la part des biotechnologies. *Biofutur* **57**, 16–27.

Flodstad, H. (1985). Penetration of plastic water pipes by gases and solvents. *IWSA, International Conference, Special Subject* No. 13. Monastir, Tunisia.

Frechen, F. B. (1987). Stoffubergange Wasser/luft am beispiel von Gerushsemissionen aus Klaranlagen, GW Abwasser.

Galé, F. (1984). Neutralization of odors emitted by industrial processes and by treatment of industrial and domestic refuse. *Proceeding, Caractérisation et Techniques de Réduction des Nuisances Olfactives dans les Industries de Procédés.* Société Belge de Filtration, Bruxelles, April, 265–288.

Garrison, A. W., Pope, J. D., Allen, F. R. (1976). GC/MS analysis of organic compounds in domestic wastewater. In *Identification and Analysis of Organic Pollutants in Water,* edited by L.H. Keith. Ann Arbor Science Publishers, Inc. Ann Arbor, MI.

Gerber, N. N. (1969). A volatile metabolite of actinomycetes, 2-methylisoborneol. *J. Antibot.* **22**, 508.

Gerber, N. N., Lechevalier, H. A. (1965). Geosmin, an Earthy-Smelling Substance Isolated from actinomycetes. *Appl. Microbiol.* **13**, 935.

Giger, W., Schaffner, C. (1981). Groundwater pollution by volatile organic chemicals. *Studies Environ. Sci.* **17**.

Guiraud, J., Galzi, P. (1981). *L'Analyse Microbiologique dans les Industries Alimentaires.* éd. Usine Nouvelle.

Hargett, G. W., Rybus, M. H. (1982). Elimination of odor from anaerobic Lagoons. *Public Works* **113**, 5, 82–83.

Harkness, N. (1980). Chemistry of septicinity. *Effluent Wat. Treat. J.* **1**, 16–25.

Henley, D. E., Glaze, W. H., Silvey, J. K. G. (1969). Isolation and identification on an odor compound produced by a selected aquatic actinomycete. *Envir. Sci. Technol.* **3**, 268.

Huismann, B. C., de Nie, L. H., Schaefer, J., Maaise, H., de Vriger, Fl. (1979). Composition of waste gas from a sugarbeet pulp drier and their relation to the process conditions. *16ème Assemblée Générale de la Commission Internationale Technique de sucrerie,* Amsterdam, May–June.

Huismann, B. C., de Nie, L. H., Peters, H. F., V. D. Poel, P. W. (1987a). Odour emission and control in the Dutch sugar industry. *18ème Assemblée Générale de la Commission Internationale Technique de Sucrerie,* Ferrara, June.

Huismann, B. C., de Nie, L. H., Peters, H. F., V. D. Poel, P. W. (1987b). Odour emission and control in the Dutch sugar industry. *Zuckerind* **112,** 11, 958–965.

Izaguirre, G., et al. (1981). Geosmin and 2-Methylisoborneol from cyanobacteria in three water supply systems. *Appl. Envir. Microbiol.* **43,** 708.

Jenkins, D., Medsker, L. L., Thomas, J. F. (1967). Odorous compounds in natural waters: some sulfur compounds associated with blue-green algae. *Envir. Sci. Technol.* **1,** 731.

Johnson, L., Josefsson, B., Martorp, P., Ekdand, G. (1981). *Int. J. Environ. Anal. Chem.* **9,** 7, 26.

Kakitani, I. (1979). Treatment of waste gas from recovery boilers in Kraft pulp plants. *Kankyo Shimpojirme* **7,** 30–41.

Kangas, J., Nevalainen, A., Manninen, A., Savolaine, H. (1986). Ammonia hydrogen sulfide and methyl mercaptides in finish municipal sewage plants and pumping stations. *Sci. Total Environ.* **57,** 49–55.

Keith, L. E., Ed. (1976). *Identification and Analysis of Organic Pollutants in Water.* Ann Arbor Science Publishers, Inc., Ann Arbor, MI.

Keith, L. E., Ed. (1981). *Advances in the Identification and Analysis of Organic Pollutants in Water.* Ann Arbor Science Publishers, Inc., Ann Arbor, MI.

Kienov, K. E., Pomeroy, H. D., Kienov, R. R. (1982). Prediction of sulfide build-up in sanitary sewers. *J. Environ. Eng.* **108,** 941–956.

Koe, L. C. C., Brady, D. K. (1985). Assessment of odorous emission from sewage pump stations. *Intern. J. Environ. Studies* **26,** 223–229.

Koe, L. C. C., Brady, D. K. (1986). Olfactometry quantification of sewage odor. *J. Environ. Eng.* **112,** 2, 311–327.

Köhler, H. (1984). Reduction of odour emissions from painting facilities with the aid of biological. *Proceeding, Caractérisation et techniques de Réduction des Nuisances Olfactives dans les Industries de Procédé.* Société Belge de Filtration, Bruxelles, April, 265–288.

Köhler, H., Lachenmayer, U., Hamaus, W. J. (1981). Behandling var geruchsinternsiver Alluft aus der Spanplattenindustrie mit Hilfe du biologischem Alluftanfberintung. *Staub. Reinhalt Luft.* **41,** 3, 102–107.

Koizumi, C., Wader, S., Ohohimer, T. (1987). Factors affecting development of rancing off odor in cooked fish meat during storage at 5 °C. *Nippon Suisan Gakkaishi* **53,** 11, 2003–2009.

Krasner, S. W., et al. (1986). Free chlorine versus monochloramine in controlling off-tastes and odors in drinking water. *Proceedings of the American Water Works Association,* Annual Conference, Denver, CO.

Krasner, S. W., Means, E. G. (1986). Returning recently covered reservoirs to service: health and aesthetic considerations. *J. Am. Water Works Assn.* **78,** 3, 94.

Kreft, P., et al. (1981). Note and comments: leaching of organics from a PVC-Polyethylene-Plexiglas pilot plant. *J. Am. Water Works Assn.* **73,** 10, 558.

Lafaye, P., Gillard, F. (1984). Odoriferous emissions from open by-product tank in coal carbonization processes. *Proceeding, Caractérisation et Techniques de Réduction des Nuisances Olfactives dans les industries de Procédés.* Société Belge de Filtration, Bruxelles, April, 265–288.

Lahaye, P., (1986). Analyse olfactométrique et physicochimique des émissions gazeuses d'installations industrielles de la sidérurgie à chaud. *Clean Air Congress,* 25–29 August, Sydney, Australia.

Lambin, S., Germain, A. (1969). *Précis de Microbiologie I,* éd. Masson, pp. 604–616.

Laplanche, A., Laifa, A., Martin, G. (1986). Étude de l'effet dépolluant d'une étuve de cuisson de peintures équipées de thermoréacteurs. *Poll. Atm.* **1**, 31–35.

Laplanche, A., Lemasle, M., Laifa, A., Martin, G. (1985). Mesure des émissions de styrène au cours de fabrication de pièces en plastiques armées. *Poll. Atm.* **2**, 107–110.

Le Cloirec, P., Lemasle, M., Martin, G. (1988). Les odeurs: Analyses et concentration dans diverses situations. *Poll. Atm.* **119**, 3, 284–288.

Le Cloirec P., Martin, G., Dagois, G. (1989). Déodorisation de gaz de fermentation d'un biostabilisateur d'ordures ménagères: étude pilote. *Tech. Sci. Méth.* **4**, 231–236.

Lemasle, M., Martin, G. (1980). Étude sur les possibilités de désodorisation par voie biologique des unités de peintures. Report, Laboratoire CNGE, E.N.S. Chimie de Rennes, France.

Lery, N., Isnard, E. (1984). La mort et l'enfeu-Laboratoire Médecine Légale et Toxicologie Médicale de Lyon.

Lery, N., et al. (1988). La décomposition des corps en sépulture étanche - II Tentative d'approche méthodologique-Congrès de l'Académie Internationale de Médecine Légale et de Médecine Sociale-Liège-*Acta Medicinae Legalis et Socialis* (in press).

Lin, S.D. (1976). Sources of tastes and odors in water. Parts 1 and 2. Wtr., *Sewage Works,* June, 101, 104 (part 1) and July, 6467 (part 2).

MacKenthun, K. M., Keup, L. E. (1970). Biological problems encountered in water supplies. *J. Am. Water Works Ass.* **62**, 8, 520.

Mallevialle, J., Suffet, I. H. (1987). *Identification and Treatment of Tastes and Odors in Drinking Water.* American Water Works Association Research Foundation, Lyonnaise des Eaux.

Marshall, J. (1985). The diffusion of water and solvents into high density polyethylene. *IWSA Conference,* Monastir, Tunisia.

Marshall, J., Hope, P. S., Ward, M. (1982). Sorption and diffusion of solvents in highly oriented polyethylene. *Polymer Repts.,* vol. 23.

Martin, G., Le Cloirec, P., Laplanche, A., Lemasle, M., Gillet, M. (1986). Étude de situations de pollution odorante dans divers effluents industriels. *7th Clean Air Congress.* Sydney, Australia.

Medsker, L. L., et al. (1969). Odorous compounds in natural waters. 2-Exo-hydroxy-2-Methylbornane, the major odorous compound produced by several actinomycetes. *Envir. Sci. Technol.* **3**, 476.

Meheus, J., Peeters, P., Celens, J. (1984). Theoretical and practical approach by simple laboratory tests: case studies in the water distribution system. *IWSA Conference.* Monastir, Tunisia.

Meyer, B. J., Converse, J. C. (1981). Gas production vs storage time in Swine nursery manure. *Am. Soc. Agro. Eng.* St. Joseph, MI.

Miner, J. R., Hazen, T. E. (1989). Ammonia and amines: component of swine building odor. *Trans. ASAE* **12**, 6, 772–774.

Morton, C., Card, T. (1986). Design of packed towers for odor control. *Proceedings of the AWPRCF Conference,* Philadelphia, PA.

Müller, H., Jüttner, F., De Haar, U. (1982). Cited in Mallevialle and Suffet, 1987.

Nominé, M. (1979). Traitement des odeurs dans l'industrie des sous-produits d'animaux examen de différents procédés. *Rev. Corps Gras* **26**, 12, 493–496.

North, A. A. (1980). *Odour Control, a Concise Guide,* Rapport, Warren Spring Lab., 42 pp.

O'Mahony, M. (1972). Salt taste sensitivity, a signal detection approach. *Perception* **I**, 459.

Paillard, H., Blondeau, F. (1988). Les nuisances olfactives en assainissement: causes et remèdes. *Tech. Sci. Mun.* **2**, 79–88.

Partbridge, P. A., Shala, F. J., Cernansky, N. P., Suffet, I. H. (1983). Characterization and analysis of diesel exhaust odor. *Environ. Sci. Technol.* **21**, 4, 403–408.

Perry, et al. (1984). Cited in Malleviale and Suffet, 1987.

Persson, P. E., Skoy, S., Hasenson, B. (1987). Community odours in the vicinity of a Petrochemical industrial complex. *J.A.P.C.A.* **37**, 1418–1420.

Pottevin, H. (1911). Les cimetières. *Traité d'Hygiène XV,* edited by J.B. Baillière, Paris.

Psarofaghish, G., Adam, H. (1975). Données et solutions au problème des odeurs posé par une station d'épuration implantée en zone urbaine. *EAS* **3**, 75, 194–206.

Raguenes, N. (1988). *Microbiologie Alimentaire,* Vol. 1, Chap 8, Edited by Lavoisier Tec & Doc, Paris.

Reich, K. E., et al. (1985). Diffusion of organics from solvent-bonded plastic pipes used for potable water plumbing. *IWSA Conference,* Monastir, Tunisia.

Reny, R., Peters, C. J., Baker, J. M. (1984). Odour problems associated with Solid waste. *Proceeding, Caractérisation et Techniques de Réduction des Nuisances Olfactives dans les Industries de Procédés.* Société Belge de Filtration, Bruxelles, April, 265–288.

Reissner, R. (1971). Desulfuring combustion gases, Brevet RFA, 2, 000, 059, 15 April 1971.

Rimoux, L., et al. (1988). La décomposition des corps en sépulture étanche - V Biochimie des effluents liquide et gazeux, Congrès de l'Académie Internationale de Médecine Légale et de Médecine Sociale. Liège *Acta Medicinae Legalis et Socialis* (in press).

Roberts, M. C., Johnson, C. E., Miller, R. C. (1971). Removal of H_2S from Waste gases, U.S. Patent No. **2**, 104–250, 1970.

Rosen, A. A., Mashni, C. I., Safferman, R. S. (1970). Recent developments in the chemistry of odour in water: the cause of earthy/musty odor. *Water Trmt. Exam.,* 19–106.

Rosset, R. (1988). *Microbiologie Alimentaire,* Vol 1, Chap 5, edited by Lavoisier Tec & Doc, Paris.

Rozier, J., Carlier, V., Bolnot, F. (1985). Bases microbiologiques de l'hygiène des aliments - École Nationale Vétérinaire - Maisons Alfort - Ed. SEPAIC.

Safferman, R. S., et al. (1967). Earthy-smelling substance from a blue-green alga. *Envir. Sci. Technol.* **1**, 429.

Sigworth, E. A. (1965). Identification and removal of herbicides and pesticides. *J. Am. Water Works Ass.* **57**, 8, 1016.

Silvey, J. K. G., Roach, A. W. (1954). Actinomycetes in the Oklahoma city water suply. *J. Am. Water Works Ass.* **45**, 4, 409.

Simonin, C. (1955). *Médecine Légale Judiciaire.* Maloine, Paris.

Slater, G. P., Block, V. C. (1983). Volatile compounds of the cyanophycease—A review Water. *Sci. Technol.* **15**, 6, 7,7, 181.

Stinson, K. B., Carns, K. E. (1983). Ensuring water quality in a distribution system. *J. Envir. Eng.* **109**, 2/289.

Swarin, S. J., Lipari, F. (1983). *J. Liq. Chromatogr.* **6**, 425–444.

Sweeten, J. M. (1988). Odor measurement and control for the swine industry. *J. Environ. Health* **50**, 5, 282–286.

Thal, M. F., Zettwoog, P., Guillet, P. (1978). Mesure des odeurs. *Nuisances Environ.* **67**, 7 p.

Thurman, E. M. (1985). *Organic Geochemistry of Natural Waters.* Martinus Nijhoff Publishers, The Netherlands.

Tokimoto, Y., Kobayashi, A. (1988). Odor of cooked Japanese Adzuki beans. *Nippon Noycikagahu Kaishi* **62**, 1, 17–22.

Uno Jensen, J. (1984). An example of how a major car assembly plant reduced the odorous emission from their paint bake oven. *Proceeding, Caractérisation et Techniques de Reduction des Nuisances Olfactives dans les Industries de Procédés.* Société Belge de Filtration, Bruxelles, April, 265–288.

Van Der Kooij, D., Visser, W.A., Hijnen, W.A. (1982). Determining the concentration of easily assimilable organic carbon in drinking water. *J. Am. Water Works Ass.* **74**, 10, 540.

Van Gemert, L.J., Nettenbreuer, A.H., Eds. (1977). Compilation of odour threshold values in air and water. National Institute for Water Supply, Voorburg, Netherlands, and Central Institute Nut. Foods Res. TNO, Zeist, Netherlands.

Van Langenhove, H., Roelstraete, K., Schamp, N., Houstmeyors, J. (1985). GC/MS identification of odorous volatiles in wastewater. *Water Res.* **19**, 5, 597–603.

Vicard, J. F. (1985). Nouveau procédé pour le dépoussiérage et la désulfuration combinée par voie humide. *Journ. Agence Qualité Air,* Paris, October.

Vonk, M. W. (1984). The diffusion of water and solvents into high density polyethylene *IWSA Conference* Monastir, Tunisia.

Wajon, J. E., Alexander, R., Kagi, R.I. (1985a). Determination of trace levels of dimethyl polysulphides by capillary gas chromatography. *J. Chrom.* **319**, 187.

Wajon, J. E., Kagi, R. I., Alexander, R. (1985b). The occurence and control of swampy odour in the water supply of Perth, Western Australia. Report submitted to the water authority of Western Australia, by the school of Applied Chemistry, Western Australian Institute of Technology, Bentley, Western Australia, Nov. 1985.

Wajon, J. E., et al. (1985c). Dimethyl trisulfide and objectionable odours in potable water. *Chemosphere* **14**, 1, 85.

Water Research Center (1981). A guide to solving water quality problems in distribution systems. Technical Report medwehhaus, England, June 1981.

Whitfield, F. B., Freeman, D. (1983). Off flavours in crustaceans caught in Australian coastal waters. *Water Sci. Technol.* **15**, 6, 7, 85.

Wilson, C. A. (1973). Odor Control in a kruft pulp mill. *New Zealand Engn.* 10–15.

Wood, S., Williams, S. T., White, W. R. (1983). Microbes as a sources of earthy flavours in potable water. *Rev. Intl. Biodeterioration Bull.* **19**, 3, 4, 83.

Yoo, S. R., Elgas, W. M., Lee, R. (1984). Water quality problems associated with reservoir coating and linings. *Proc. A.W.W.A. Conference,* Dallas, TX.

Zoeteman, B. C. J. (1978). *Sensory Assessment and Chemical Composition of Drinking Water.* Offsetdrukkerij Van der Gang V.V's Gavenhage.

Zoeteman, B. C. J., Kraayeveld, A. J. A., Piet, G. J. (1971). Oil pollution and drinking water odour, H_2O, 16, 37.

Zoeteman, B. C. J., Piet, G. J. (1974). Cause and identification of taste and odour compounds in water. *Sci. Total Environ.* **3**, 103.

Metrology and Sampling

P. Le Cloirec and M. Perrin

9.1 Introduction

The idea of odor is very subjective and its qualification as well as its quantification are not an obvious process. Two methods are currently in use:

1. *Olfactometric analysis.* This analysis brings the olfactive mucus into play and makes global appreciation of the odor possible.
2. *Physical-chemical measurements.*These analyses are classic in chemistry; their goal is to quantify the molecules present in odorous gas or air. However, due to trace levels, odor preconcentration is often necessary.

The purpose of this chapter is to approach the idea of sampling and measuring odors. Both of the above techniques will be discussed.

9.2 Olfactometric Analysis

Odors are related to the presence of specific chemical compounds in the air we breathe: the idea of measuring odors by chemical or physical-chemical analytical methods comes from this obvious fact. In many cases, qualitative and quantitative knowledge of atmospheric composition are not sufficient to understand the odorous properties of molecules.

The reason for this is that the correlation between the physical-chemical properties of molecules and their odorous properties has not been established for the complex mixtures encountered in industry (Chapter 5). Moreover, most odorous products exist at such low concentrations that frequently even the most sensitive analyzers cannot detect them.

Olfactometric analysis is therefore based on the only detectors available, that is to say, the olfactive mucous membranes. It is clear that this poses practical problems. Indeed, the olfactive sensation that results from the interaction of a given number of odorous molecules with the mucous membranes varies, not only from one individual to another, but also for the same individual, depending on that individual's physiological state and the sniffing conditions. However, thanks to works accomplished by researchers of various origins, the technique for measuring odor levels, or olfactometric analysis, is founded on a serious scientific basis and can be used profitably.

Olfactometric analysis consists of measuring the following:

1. The concentration of an odorous mixture, expressed in threshold units
2. The odorous intensity of an atmosphere, generally expressed in relation to the level on a range given by a reference scale

These measurements are carried out by processing the verbal responses of a jury of experts consisting of a representative sample of the standard population. Variable dilutions of the odorous atmosphere to be studied are presented to this jury. The accuracy, consistency and reproducibility of the results essentially depend on the jury's selection and training as well as the adopted working procedures, especially those applied in the field.

9.2.1 Method for Measuring the Concentration of an Odorous Mixture in Threshold Units

9.2.1.1 Definition

A threshold concentration can be defined for any odorous pure compound or compound mixture. For this concentration, 50% of the responses of the individuals in the jury express the perception of the odor, 50% express the nonperception. By definition, the concentration of the odorous compound is then equal to the perception threshold.

Any other concentration is expressed with respect to this threshold concentration, which is used as the unit of concentration measurement.

9.2.1.2 Measurement Principle (Norm AFNOR, 1986)

After having been diluted by an appropriate nonodorous gas, the odorous mixture is presented to each individual in the jury, who indicate individually whether or not they can detect the odor of the mixture. On the basis of successive tests, an estimate of the dilution rate, for which the odor perception probability is equal to 50%, is established for each of the subjects.

9.2.1.3 Sampling

Sampling of odorous atmospheres is described in the French Norm AFNOR Pr X43-104, which is currently being published. This is an important phase in the odor measurement process; it improves the quality and reliability of measurement results.

The choice of a sampling method depends on the type of olfactometric analysis to be done; two types of olfactometric analysis can be considered:

1. On-line olfactometric analysis
2. Off-line olfactometric analysis

The choice will depend on the case to be studied. A specific methodology must be followed during sampling, sample conditioning, and transfer toward the olfactometer or container in such a way that the sample is representative of the total flux of the gas to be analyzed.

9.2.1.3.1 On-Line Olfactometric Analysis Sampling

This sampling technique is applied in the case of odorous atmospheres originating from canalized or canalizable sources (chimney emissions, storage tank emissions, biofilters, mixed liquid volumes such as wastewater pumping stations). Sampling for on-line olfactometric analysis is only used for emissions whose concentration is constant for the duration of sampling.

Equipment for on-line olfactometric analysis consists of a sampling probe, a pump or air pump, an outlet pipe from the olfactometer to the exterior, and an inlet pipe. After being sampled with a probe, the odorous atmosphere is sent through a pipe into the olfactometer at atmospheric pressure.

9.2.1.3.2 Off-line Olfactometric Analysis Sampling

This sampling technique can be applied to all odorous sources, whether canalized, canalizable, diffuse, or, more generally, to all sources for which the odor varies as a function of time. Diffuse sources are liquid, semiliquid, or solid surfaces such as ponds or settling tanks.

Necessary equipment for off-line olfactometric analysis sampling essentially consists of a sampling case containing a sampling bag, a pump, and a negative pressure gauge. It may also include a sample transfer system.

The odorous atmosphere to be analyzed is sampled in an adapted container, most often a plastic bag whose volume must be sufficient for olfactive analysis. Materials used for the bags must have the following properties:

1. No odor of their own
2. Chemical inertia
3. Negligible absorption of odorous substances
4. Low gas permeability
5. Sufficient resistance with respect to mechanical constraints

The holding time of an odorous atmosphere between sampling and measurement must be as short as possible and must not exceed 24 h.

9.2.1.3.3 Comparison of Sampling Techniques for On- or Off-line Olfactometric Analyses

The advantage of on-line sampling techniques is to reduce the risks of composition modification of the gaseous sample by chemical reactions or absorption due to the short time lapse between sampling and analysis. Their disadvantage is the necessary use of aerated rooms in order to isolate the experts from surrounding air that is always more or less odorous, means that they are difficult and expensive to put into practice.

Off-line sampling techniques are obligatory in all cases when odorous emissions are not constant. These techniques improve measurement precision by placing the experts in the best possible environmental conditions.

9.2.1.4 Equipment

Perception threshold odor measurement of a gaseous effluent essentially requires an olfactometer. The olfactometer is a device that makes it possible to control the dilution of an odorous mixture by an odorless gas and convey it to a subject. It must comply with a certain number of requirements with respect to the subject and with the odorous mixture. Only dynamic olfactometers comply with those requirements as the dilution rate must vary rapidly from 10 to 10,000. The olfactometer-subject coupling device should allow sniffing as well as prevent absorption and air drafts. It is built with materials that are hardly adsorbent, and a three-channel system is highly recommended because of the so-called "forced choice" procedure described in Section 9.2.1.6. The outlet flowrate should be around 2 m³/h.

9.2.1.5 Jury Selection

The larger the number of subjects composing a jury, the more reliable and reproducible the measurement is likely to be. Thus, a compromise between expense and validity of the results must be found. Depending upon the desired results, it is recommended to have at least the following:

1. 16 subjects when the desired result is a precise idea of the threshold perception value of a representative population group
2. 8 subjects in most cases
3. 4 subjects in cases of comparative measurements

The subjects must be qualified in function with their individual perception threshold for five pure products that represent the main classes of odorous compounds. For each of these products, the threshold of each subject must be situated between 0.1 and 10 times the average threshold given as a reference. This qualification is verified from time to time.

9.2.1.6 *Operating Procedure*

Perception threshold concentration is estimated by random presentation of a pre-established series of dilutions in geometrical progression with a ratio of $\sqrt{2}$ and calculated from a dilution close to the perception threshold. These dilutions are presented through one of the three channels of the olfactometer, the two others being filled with odorless air. After sniffing, the subject must indicate which of the three channels is odorous. The response is forced choice, that is to say that doubtful responses are not acceptable: in case of hesitation, the subject must guess. Graphic statistical processing of the totality of responses given to the series of dilutions presented makes it possible to determine the individual dilution factor to the perception threshold.

9.2.2 Methods for Measuring the Odorous Intensity of an Atmosphere

9.2.2.1 *Definition*

The odorous intensity of an atmosphere is the sensational intensity for a concentration of stimulus higher than that corresponding to the perception threshold. Odorous intensity is a function of the concentration of the odorous mixture.

9.2.2.2 *Measurement Principle*

Odorous intensity is measured in arbitrary units through comparison with psychophysical measurements performed by a jury of experts whose responses are the subject of statistical treatment.

9.2.2.3 *Sampling*

The odorous intensity of an atmosphere can be measured directly in the environment under consideration. When olfactive analysis of the samples is performed in the laboratory, the sampling procedure described in Section 9.2.1.3 must be adopted.

9.2.2.4 *Jury Selection*

In preliminary tests, the experts must be capable of distinguishing samples of odorous pure compounds of different intensities. Their selection will then be completed by the tests described in Section 9.2.1.5.

9.2.2.5 *Measurement Methods*

9.2.2.5.1 Arbitrary Unit Implementation

This method consists of asking the experts to give a numbered rating to the perceived intensity value, either in relation to a reference with arbitrarily fixed intensity values or without reference. The subject evaluates each stimulus separately using arbitrary representations from which an individual reference system develops little by little.

9.2.2.5.2 Comparison Method

This method makes it possible to avoid the use of arbitrary units that are not always practical. It consists of using an intensity reference constituted with the help of one or several samples containing one or several known concentration(s) of carefully chosen pure compounds. The measurement consists of comparing the stimulus intensity with the intensity of two consecutive samples from the reference scale.

9.2.3 Olfactometric Analysis Applied to Industry

A certain number of industrial branches are responsible for pollutant emissions in working areas or the environment which, even if they do not have toxic or irritant properties, are no less disagreeable because of their malodorous character.

Thus, the problem of measuring odors is an additional one for engineers responsible for finding solutions to these problems.

9.2.3.1 Measurements at Emission and in the Environment

Odorous product emissions can be characterized by the measurement of two parameters at the source: the concentration expressed in threshold units and/or the intensity of the mixture.

Measurements in the environment are intended to allow a quantitative evaluation of the level of odorous pollution created in an area and to which the neighboring populations are exposed. In any case, these measurements will not allow an evaluation of the nuisance experienced by these populations, the concept of nuisance requiring psychosociological parameters. The only way to quantify the nuisance is to appeal directly to the concerned populations by way of a survey.

9.2.3.2 Application Examples

Table 9.1 shows which olfactive measurements should be applied to problems most frequently encountered: concentration measurement and/or measurement of odorous intensity.

9.3 Chemical Analysis

An alternative to olfactometric analysis for odor quantification is chemical analysis. This analysis can be approached by either a qualification and quan-

Table 9.1.   Olfactometric Measurements to be Applied for Different
Industrial Problems

	Olfactometric Analysis to be Applied	
Problems to be Treated	Odor Concentration Measurement	Odorous Intensity Measurement
Identification and classification of different odorous sources according to their pollution level	+	+
Purification equipment optimization	+	+
Development of less odorous manufacturing processes	+	+
Purification equipment commissioning	+	+
Waste stream compliance monitoring with respect to legislation	+	
A priori determination of the consequences of a waste stream on the environment	+	
Identification of the areas affected by odorous gas concentrations higher or equal to threshold concentration	+	

tification by a family of products or determination by specific products. These procedures are often related to the sampling method.

The main sampling methods and their laboratory analysis will be described in the following paragraphs.

9.3.1 Atmosphere Sampling

Various methods of sampling the effluent to be analyzed can be used, from simple vial or pocket filling to the absorption of specific materials or selective absorption by a product family. These methods will be presented and discussed according to their applications.

9.3.1.1 Container Sampling

When direct analysis of the gas is possible without prior concentration, sampling is performed in a vial or in a pocket. Vials are made of glass and they are equiped with a Teflon stopper. Pockets are made of rubber or plastic covered with specific polymers such as Teflon, Mylar, or Tedlar (Guérin, 1981). The coatings avoid the possible absorption of the compounds to be analyzed on the container walls. Schnetzle et al. (1975) identified several materials that minimize sample loss. Thus, absorption is low on Tedlar, Teflon, and Mylar. Figure 9.1 shows the absorption of ethylbenzene on different coating films for sampling bags. Butler et al. (1988) noted the possible permeability of nylon coated bags and recommended the use of Tedlar. However, because of possible interactions between the chemical species themselves or with the coatings, the analysis must be performed within 2 h after sampling (French Ministry of Environment recommendation, 1979). Container filling can be done either by positive or negative pressure using a pump. In the case of glass vial use, a vacuum is created prior to sampling. A valve and a septum make it possible to withdraw gas using a syringe prior to analysis.

9.3.1.2 Adsorption Sampling

Because of low concentrations occuring in the gas to be analyzed, it is often necessary to perform a concentration prior to analysis. This concentration is implemented by the use of different types of adsorbing materials. Chapter 11 of this book gives the general principles of adsorption.

Generally, the trapping technique consists of passing a volume of gas through a tube filled with an adsorbant. The treated volumes are variable depending, of course, on the pollutant concentration. Belin (1984) passed 33 L of urban air on 250 mg of an adsorbant within an hour. However, vol-

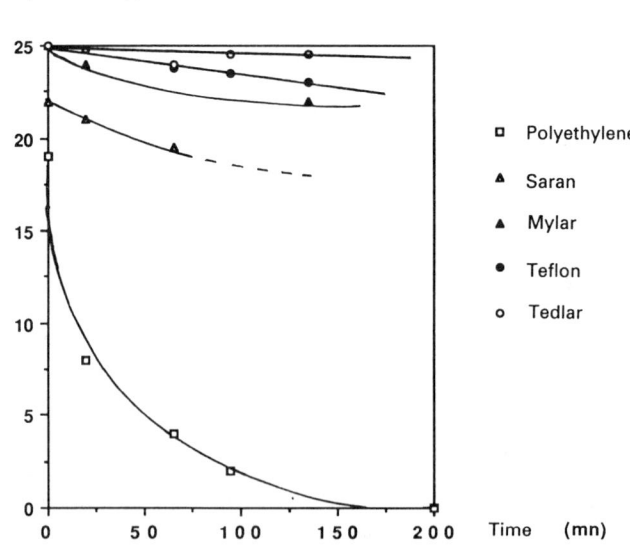

Figure 9.1. Ethylbenzene absorption on various sampling bag coatings (Schnetzle et al., 1975).

umes ranging from 10 to 3,000 L are reported in the case of the absorption of odorous molecules issued from a wastewater treatment plant (Zeman et al., 1984). Adsorbant masses depend upon the nature of the adsorbing material. It generally ranges from 50 to 1,000 mg either in monolayers or in adsorbant mixtures.

A wide variety of adsorbants are currently in use. They can be prepared especially for this kind of analysis using a chemical washing or a solvent treatment. However, a great variety of ready-to-use tubes are presently available on the market (Supelco, 1988b; Arelco, 1988). Among adsorbants, activated carbon is the most widely used; it is found under several trademarks such as Carbotrap, which is constituted of graphite, hence developing a specific surface area of around 100 m² g⁻¹(Cheremisinoff and Ellerbusch, 1978). Other kinds of adsorbants are also found such as silica gel or activated alumina (Guérin, 1981). Gas chromatography column fillings are also used for the absorption of odorous molecules: Tenax, Porapack, Chromosorb, and macroreticulated resins, such as XADs. Table 9.2 shows some adsorbing materials with their use for odorous products or families of odorous products. This table shows that the adsorbants are not selective with respect to families of compounds. Therefore, the quantification of the products requires separation prior to analysis. Additionally, absorption competition may occur thereby interfering with the quantification of odorous compounds (Guérin, 1981). It is possible to use tubes made of several layers either of the same material with different sizes or with two different materials (Supelco, 1988a). This enhances adsorption efficiency.

Desorption is an important process for the analysis of trapped molecules. A rather old method (Peters and Weil, 1930, Peters and Laghman, 1937)

Table 9.2. Several Adsorbants Used to Trap Odorous Molecules

Products or Families of Compounds	Adsorbants	References
Mineral acids	Na₂CO₃ 5% on Chromosorb	NIOSH, 2nd ed., Vol. 7
Organic acids	Carbotrap or XAD or Tenax	Supelco, 1988c
Organic compounds	Activated carbon	NIOSH, 2nd ed., Vols. 1–7
Mercaptans and Alkylpolysulfides	Activated carbon or Tenax	Zeman et al., 1984
Ammoniac	Na₂CO₃ 5% on Chromosorb	NIOSH, 3nd ed., Vol. 1
Amines	Activated silica gel	NIOSH, 2nd ed., Vol. 1
Alcohols	Activated silica gel	NIOSH, 2nd ed., Vols. 1–7
Formaldehyde and Acroleine	2-hydroxymethyl piperidine on Supelpack 20 N	OSHA Manual
Exhaust gas	Silica gel	Partbridge et al., 1987

consists of a fractionated desorption. The material is heated to successive temperature steps so that the molecules desorb one after the other. A good separation is often difficult to achieve when the sampled gas is a complex mixture. Today it is preferred to heat the material to the point where total desorption is achieved. The desorbed molecules are then separated by gas chromatography, for instance. The equipment that is presently available on the market allows a rather fast thermal shock: the temperature goes from 30 to 400 °C within 26 ± 2 seconds (Supelco, 1988b). An alternative to thermal desorption is to extract the adsorbed molecules using a solvent. This technique was successfully used by Zeman and co-workers (1981, 1984) for the trapping and the desorption of sulfurous compounds. The extraction is performed using 2–3 mL of dichloromethane with reflux during 6 h. The extract is then concentrated in a Kuderna Danish apparatus prior to chromatographic analysis. Other solvents can be used as well, such as ether or ethanol (Guérin, 1981). In the case of complex mixtures, Zeman and Koch (1981) suggested absorption on silica gel after a prior desorption of activated carbon using dichloromethane and followed by a selective elution using hexane to remove hydrocabon compounds, such as alcohols. Then a final elution using ether is used prior to analysis of osmogenic compounds. This method makes it possible to remove a great number of the molecules interfering with products that are looked for and analyzed by chromatography. Another method uses the two preceeding approaches. It consists of a thermal desorption followed by bubbling with a carrier gas in an appropriate solvent (Sulpelco, 1988a). The liquid sample can then be concentrated and analyzed.

9.3.1.3 Absorption Sampling

When the odorous pollution is complex as in most actual situations, it is often interesting to sort the compounds to be analyzed by families of products. Furthermore, a previous concentration is often necessary in the case of diluted pollution (Cheremisinoff, 1975). The absorption principle consists of a selective fixation of the compound to be analyzed as a solution or as a precipitate. More details about absorption are given in Chapter 10.

Absorbers should allow rapid and complete absorption. Several devices are available, for example, Lamn's, Schaw's, Cauen's absorbers, pellet absorbers, etc. (Guérin, 1981). Simple scrubbing flasks, like Durand's (Fig. 9.2) can be used as well. Bostroem (1965) showed that the efficiency of different absorbers is equivalent if the liquid height is 3 cm and if the gaseous flowrate is lower than 360 L/h.

Other techniques based on absorbtion phenomena are used, such as leaching, permeation, or filtration on impregnated support. This last possibility is used in Draeger's method. These systems will be developed further in Section 9.3.2.3

The absorbing solution is chosen according to the products to be trapped or to the analysis technique (Table 9.3). For instance, in the case of a gas

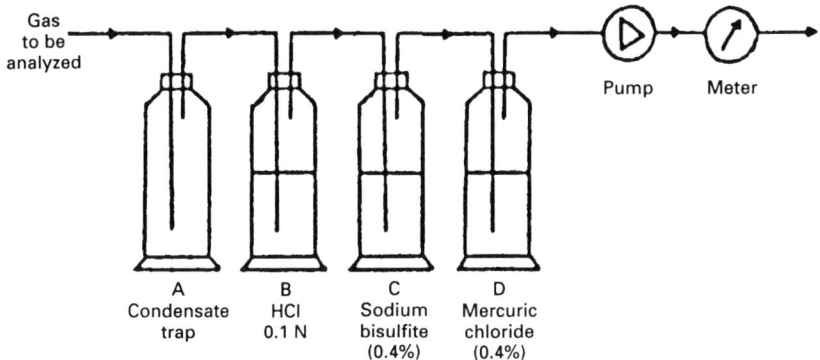

Figure 9.2. Trapping and concentration apparatus for odorous molecules (Le Cloirec et al., 1988).

containing only H_2S, an iodine solution or a bromine solution can be used (iodometric or conductimetric analysis, respectively).

In the case of a real gaseous effluent, a selective sampling system (Fig. 9.2) is used for trapping the products by families (CNGE Report, 1979; Le Cloirec et al., 1988). It is made of a series of scrubbing flasks preceeded by a condensate trap. Hydrochloric acid (0.1 N) makes it possible to trap ammoniac and amines, whereas sodium bisulfite (0.4%) traps aldehydes and ketones. On the other hand, mercuric chloride (0.4%) precipitates H_2S and mercaptans as mercuric sulfide. These compounds are then globally quantified by gravimetric analysis. However, this methodology by trapping and

Table 9.3. Several Scrubbing Solutions for the Absorption of Odorous Compounds Families

Products or Families of Products	Absorbants	References
Acids (SO_2, ..)	H_2O_2	Norm Afnor, 1977
Ammoniac	HCl 0.1 N or boric acid 0.5%	Le Cloirec et al., 1988 Fukuyama et al., 1986
Amines	HCl 0.1 N	Thal et al., 1981 Nominé, 1979 Le Cloirec et al., 1988
Aldehydes and ketones	Sodium bisulfite 0.4%	Thal et al., 1981 Le Cloirec et al., 1988
H_2S and mercaptans	Hg Cl_2 0.4% aqueous solution	Le Cloirec et al., 1988 Thal et al., 1981

concentration is not specific enough for other more or less odorous products, such as solvents. The solvents will end up in the aqueous solutions of the traps because of their solubility. The volume of solutions range from 100 to 200 mL. The gaseous flowrates are approximately 100 L/h. The scrubbing time should be adjusted as a function of the guest concentration so as to obtain an adequate analytical precision.

For instance, for quartering the sampling time ranges from 1 to 2 h, whereas for a wastewater treatment plant, the sampling time ranges from 5 to 10 h depending on the location of the sampling (confined or free atmosphere). The sampling is performed during a given time, hence giving an average value on the sampling period. They take into account possible concentration surges during the sampling time

Fukuyama et al. (1986) used boric acid at 0.5% as a scrubbing solution for the study of wastewater treatment plant gases. In this work, the gaseous flowrate was set to 90 L/h.

9.3.2 Analysis

After sampling, trapping, or concentration by compounds or product families, qualitative and quantitative analysis of the odorous molecules should be performed. A summary of the usable analytical techniques according to the compounds or the families of compounds is presented in Table 9.4. These techniques are described in the following paragraphs.

Table 9.4. Summary of the Principle Analytical Techniques Used Depending on the Compound. [GC/MS: Gas Chromatography (GC) Coupled with Mass Spectrometry (MS).]

Products or Families of Products	Analysis	Observation
H_2S	Iodometry Gravimetry Colorimetry GC	Compound alone
Mercaptans	Gravimetry GC/MS	Global quantification by flame emission
SO_2	Volumetry GC/MS	Acidity Flame emission
NH_3	Volumetry Colorimetry	Possible interferences
Amines	Volumetry GC/MS	Global quantification by flame emission
Aldehydes and ketones	GC/MS	Flame emission
Alcohols	GC/MS	Possible interferences with amines and aldehydes

9.3.2.1 Gravimetric Analysis

This method is used more specially to determine mercaptans and H_2S after scrubbing and precipitation by mercuric chloride (Section 9.3.1.3). This set of products is quantified as H_2S (Thal et al., 1981; Le Cloirec et al., 1988). Also, SO_2 and SO_3 are determined by nephelometry as $BaSO_4$ precipitated (Guérin, 1981).

9.3.2.2 Volumetric Analysis

In the case of an effluent containing H_2S or SO_2 alone, the volume of gas necessary to discolor an iodine solution is measured (Guérin, 1981). In the specific case of SO_2 an acid gaseous pollution index is determined by the transformation of SO_2 into H_2SO_4 by hydrogen peroxide. The acidity is then measured by a solution of sodium borate (Norm AFNOR, 1977). For nitrogenous compounds (amines and ammoniac) HCl scrubbing solutions can be measured by the Kjeldhal method (Rodier, 1975). It should be pointed out that the amine concentrations should be rather high and that the quantification is global.

9.3.2.3 Colorimetric or Infrared (IR) Absorption Analysis

Colorimetric analysis is meant for the determination of aqueous compounds resulting from NH_3 trapping by HCl (Norm AFNOR, 1975). For H_2S the time required for color development of a paper impregnated with lead acetate is measured (Guérin, 1981). Perret (1986) noted the possibility of continuous determination of NH_3 or SO_2 by IR absorption.

The specific colorimetric analysis is possible by the use of Dräger's tubes (Dräger, 1989). The gas is directly pumped into a tube containing a product or a phase susceptible to develop color by a reaction with a specific product. The color height is measured in a graduated tube through which a given volume of gas containing an odorous product is pumped. A large number of tubes allowing measurement of NH_3, H_2S, mercaptans, solvents, etc. is available. This system is very practical for field measurements and gives a rather good estimate of odorous compound concentrations. However, interferences are sometimes possible in the case of complex mixtures.

9.3.2.4 Gas Chromatography

Because of their volatility, odorous compounds can be qualified and quantified quite easily by gas chromatography. Le Cloirec et al. (1988) proposed chromatographic analytical conditions for three families of compounds generally responsible for odors: sulfurous compounds, aldehydes and ketones, and aminated compounds. Table 9.5 shows these analytical conditions.

Table 9.5.   Gas Chromatographic Analysis Conditions of Odorous Compound Families

Compounds	H_2S and Mercaptans	NH_3 and Amines	Aldehydes and Ketones
Pretreatment Concentration		Absorption in HCl	Absorption in Sodium Bisulfite
Column	Poly MPE 12% and H_3PO_4 5% impregnated Chromosorb T	Penwalt	Carbovax 20 M 4% 0.8% KOH on Carbovax B
Precolumn		ascarite	ascarite
Oven temperature (°C)	30	80–120	90–120
Temperature rate (°C/mn)		5	5
Injector temperature (°C)	40	200	200
Detector temperature (°C)	40	200	200
Carrier gas	N_2	N_2	N_2
Pressure (bars)	0.75	1	1
Injected volume (μl)	100	5	5
Detector	Flame emission	Flame ionization	Flame ionization

It should be noted that this technique, which was previously designed for quartering gases (CNGE Report, 1979), has been applied to a large number of both food and chemical industries. For more details, Chapter 8, which exposes odorous pollution sources, can be referred to. Absorbed compounds, such as alcohol, can be detected by this technique. Interference is possible with amines. An example of a chromatogram is given by Figure 9.3. It represents the real case of vault gases (Chapter 8).

When the mixture is very complex, capillary columns followed by a flame ionization detector is often useful in determining aminated compounds, solvents (alcohols), ketones, and aldehydes. Flame emission is used for the detection of sulfurous compounds (Huang et al., 1979; Le Sauze et al., 1989). A specific detector is also used for the determination of nitrogenous compounds (amines). A large number of capillary columns are presently available on the market for the determination of solvents and odorous products. Some columns even accept the direct injection of aqueous solutions. As an example, a few phases of capillary columns used by Zeman and co-workers (1981, 1984) for the analysis of wastewater treatment plant gases are given. Sulfurous products are analyzed on an OV 101 column whereas aminated compounds are analyzed on an UNCOM HB 5100 WGA column. The resulting chromatograms are given in Figures 9.4 and 9.5 where compounds based on pyridine and thiophene are featured. Partbridge et al. (1987) used a capillary column SE 30 in the case of diesel exhaust gas. The sample was directly injected without previous concentration.

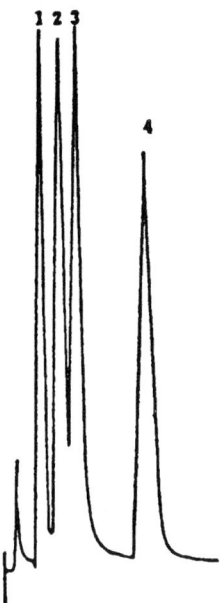

Figure 9.3. Chromatographic analysis of amines. Case of vault gases (Le Cloirec, 1988). **1**: Trimethylamine; **2**: Isopropylamine; **3**: Ethanol; **4**: Methylethylketone.

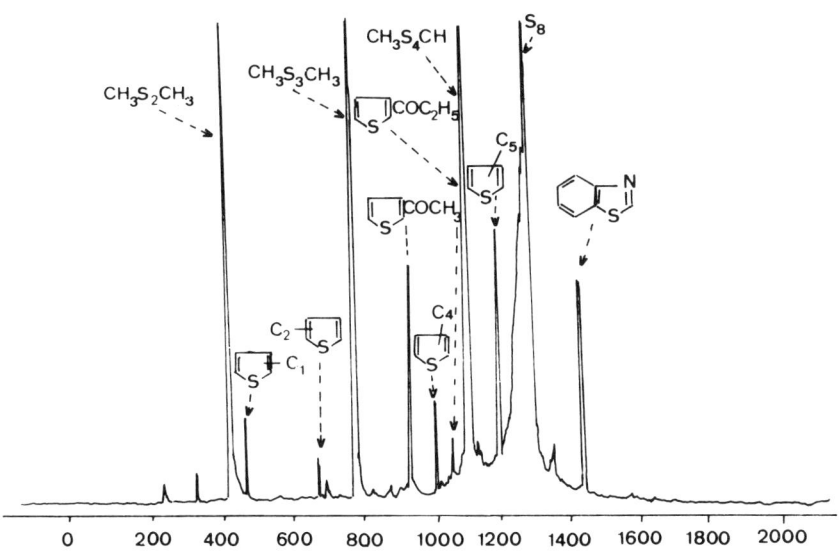

Figure 9.4. Chromatographic profile of sulfurized odorous compounds (Zeman and Koch, 1981).

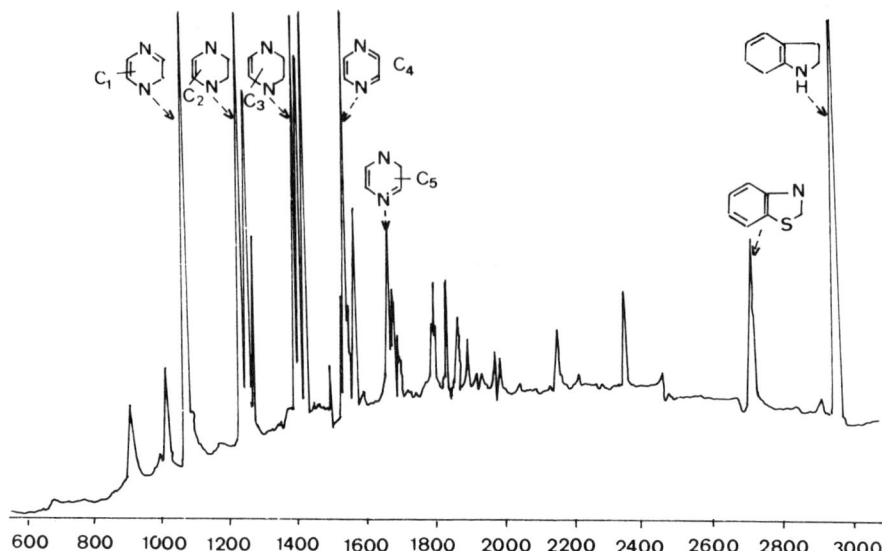

Figure 9.5. Chromatographic profile of nitrogenous odorous compounds (Zeman and Koch, 1981).

Because of the complexity of the resulting chromatograms, mass spectrometry (electron ionization) makes it possible to qualify compounds separated by a column. Similarly, Zeman and Koch (1983) implemented the coupling of chromatography with mass spectrometry for scatole and thiophene analysis. In this case, they used chemical ionization by NH_3. Tokimoto and Kabayashi (1988) qualified and quantified a large number of compounds (fatty acids, lactones, alcohols, terpenes) for a food processing plant using GC/MS.

9.4 Conclusion

This chapter describes two main families of analytical techniques for odor measurements: olfactometric analysis and physical-chemical analysis. These two techniques are complementary and must be applied in a global approach to study an odorous source. Such an approach makes identification and quantification possible as well as the implementation of curative treatments. The physical-chemical analysis precisely determines the yield of the treatment plant thereby allowing the comparison of the treatment technologies. On the other hand, olfactometric analysis is more of an environmental approach, that is to say, a nuisance determination and removal technique.

Bibliography

Arelco (1988). *Catalogue hygiène toxicologie.*

Belin, C. (1984). Measurements of trace organic gaseous compounds in the atmosphere Proceeding, Congrès Société Belge de Filtration, Bruxelles, April.

Bostroem, C. E. (1965). *J. Air Pollut.* **9**, 333–341.

Butler, F. E., Coppedge, E. A., Suggs, J. C., Knoll, J. E., Mildgett, M. R., Sykes, A. L., Hartman, M. W., Steger, J. L. (1988). Development of a method for determination of methylene chloride emissions at stationary sources. *J. Air Pollut. Control Ass.* **38**, 3, 272–277.

CNGE Report (1979). Protocole pour le prélèvement et l'analyse chimique de gaz malodorants. École Nationale Supérieure de Chimie de Rennes, France.

Dräger (1989). *La Maîtrise des Gas Polluants à l'Aide des Tubes Dräger.* Société Draeger-Brandt S.A., Strasbourg.

French Ministry of Environment and Living Standards (1979). Service de l'Environnement industriel, Protocole pour le prélèvement et l'analyse chimique des gaz malodorants.

Fukuyama, J., Irone, S., Ose, Y. (1986). Devalorisation of Exhaust gas from wastewater and night soil treatment plant by activated sludge. *Tox. Environ. Chim.* **12**, 87–109.

Guérin, H. (1981). *Traité de manipulation et d'analyse des gaz,* 2ème éd. Masson, Paris.

Huang, J. Y. C., Wilson, G. E., Schroepfer, T. W. (1979). Evaluation of activated carbon adsorption for sewage odor control. *J. Water Pollut. Control Fed.* **51**, 1054–1062.

Kiselev, A. V., Yashim, Y I (1969). *Gas Adsorption Chromatography.* Plenum Press, New York.

Le Cloirec, P., Lemasle, M., Martin, G. (1988). Mesure des odeurs de divers effluents, un protocole d'analyses chimiques, des concentrations dans divers effluents. *Pollut. Atm.* **3**, 284–288.

Le Sauze, N., Laplanche, A., Martin, G., Paillard, H. (1989). *Wasser,* Berlin.

NIOSH (National Institute of Occupational Safety and Health) cited in Supelco, 1988c.

Nominé, M. (1979). Traitement des odeurs dans l'industrie des sous-produits animaux: examen des différents procédés. *Rev. Fr. Corps Gras* 493–496.

Norme AFNOR (1975). Détermination de l'azote ammoniacal. NF T 90 015.

Norme AFNOR (1977). Détermination d'un indice de pollution acide exprimé en SO_2. NF X43-016.

Norme AFNOR (1986). Qualité de l'air. Méthode de mesurage de l'odeur d'un effluent gazeux. Détermination du facteur de dilution au seuil de perception. NF X43-101.

Norme AFNOR Pr X43-104. Qualité de l'air. Atmosphères odorantes. Méthodes de prélèvement. In press.

OSHA (Occupational Safety and Health Administration) cited in Supelco, 1988c.

Partbridge, A., Shala, F. Y., Carnansky, N. P., Suffet, I. H. (1987). Characterization and analysis of diesel exhaust odor. *Environ. Sci. Technol.* **21**, 403–408.

Perret, R. (1986). Techniques de mesure des polluants gazeux à l'émission. *Poll. Atm.* **1**, 43–53.

Peters, K., Laghman, W. (1937). *Z. Phys. Chimie A* **180**, 51–57.

Peters, K., Weil, K. (1930). *Z. Phys. Chimie A* **148**, 1–26.

Rodier, J. (1975). *L'Analyse de l' Eau*, Dunod, Paris.

Schnetzle, Prater, Ruddel (1975). *J. Air Pollut. Control Ass.* **24**, 9, 925–932.

Supelco (1988a). *Reporter* **7**(3), 1–3.

Supelco (1988b). *Catalogue General Chromatography Supplies*.

Supelco (1988c). *Reporter,* **7**(2), 1–3.

Thal, M., David, M., Bonscau (1981). Mesures d'odeurs sur des équarrissages. *Poll. Atm.* **91**, 223–227.

Tokimoto, Y., Kobayashi, A. (1988) Odor of cooked Japanese Adzuki Beans. *Nippon Nugeikugaku Kaishi* **62**(1), 17–22.

Zeman, A., Koch, K. (1981). Determination of odorous volatiles in air using chromatographic profiles. *J. Chromat.* **216**, 199–207.

Zeman, A., Koch, K. (1983). Mass spectrometry analysis of malodorous air pollutants from sewage plants. *Int. J. Mass. Spectro. Ion. Phys.* **48**, 291–294.

Zeman, A., Teichmann, H., Koch, K. (1984). Detektion im ppb Bereich. Kapillargas chromtographie (CGC) mit osmogenenspezifishn gunf. *Wasser Abwasser* **125**, 563–569.

Zettwood, P. (1978). Olfactométrie dan l'industrie. Application à la mesure des odeurs à l'émission et dans l'environnement. *Techniques de l' Ingénieur* 1–8.

10

Treatment of Odors by Scrubbing and Oxidation

A. Laplanche and G. Besson

10.1 Introduction

Among the methods currently used to remove organic products responsible for odors in a gaseous flux is to scrub with an aqueous solution. This is an operation that consists of transferring the compound or compounds to be removed from a gaseous phase to a liquid phase. This technique, based on gas-liquid mass transfer, is sometimes accompanied by a chemical reaction. If the transferred compound is not modified, only physical absorption occurs.

In the liquid phase, if one changes the pH of the product to be removed to improve its "apparent solubility" by favoring its dissociation, the process is a mass transfer accompanied by an instantaneous chemical reaction: the dissolution of the organic compound into soluble ionic forms. This operation is called acid-alkaline scrubbing.

Finally, the use of an oxidant is susceptible to improve the efficiency of aqueous scrubbing. Pollutant destruction by oxidation cannot only improve the transfer, but continuously regenerate the scrubbing solution at the same time. Oxidation is done simultaneously or following acid-alkaline scrubbing. The operation is thus either oxidized scrubbing or acid-alkaline scrubbing followed by oxidation.

When the compounds to be removed are soluble or dissociable in water, the scrubbing liquid is a pH controlled or noncontrolled aqueous solution. This is the usual way compounds such as ammoniac, amines, hydrogen sulfide, mercaptans, acids, certain alcohols, and phenols are removed. When it is necessary to treat products that have a low water solubility or those that

are not water-soluble (alphatic and aromatic hydrocarbons, terpenes, aldehydes, ketones, esters, carbon sulfide, . . .), solvents are used (oils, petroleum derivatives, heavy fuels). The solvent is chosen as a function of the
polarity of the compound to be absorbed. After use, these solvent mixtures
can be recycled in the manufacturing process, distilled for reuse or inciner-
ated to produce energy or steam.

When the gas to be purified has a complex composition, a chain of several
scrubbing units working in series under different conditions would constitute
a complete purification process. Several diagrams of the principles of these
techniques are given in Figure 10.1.

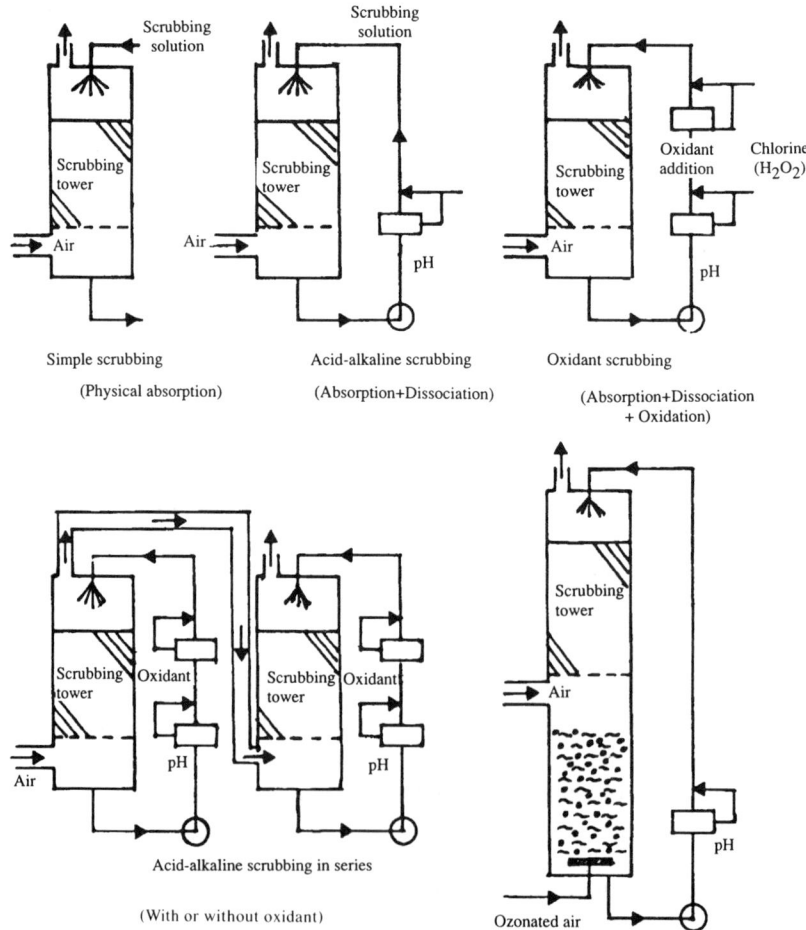

Figure 10.1.   Odor treatment flowcharts of scrubbing and oxidation.

In the following discussion, different points will be addressed successively that will make it possible to approach the design of an installation in simple cases:

- *Theoretical aspects of mass transfer* with and without chemical reaction
- *Hydrodynamics elements* and scrubbing tower technology
- *Physical-chemical properties* and reactivity of the principal oxidants (chlorine, ozone, hydrogen peroxide) with several major odorous compounds.

Several examples of use and industrial plant implementation will illustrate the preceding points and finish the chapter.

10.2 Theoretical Aspects of Mass Transfer

During any scrubbing treatment of gas contaminated by odorous products, there is mass transfer from the gaseous phase to the liquid phase. The efficiency of an operation depends on the quantity of the product transferred and, thus, on equations which govern gas-liquid mass transfer.

10.2.1 Definition of Mass Transfer: Fick's Law

When there is a concentration gradient in a fluid consisting of at least two components, there is a tendancy for each constituent to go in a direction that tends to reduce the concentration gradient. This phenomenon is called "mass transfer."

In one phase, mass transfer by diffusion in a given direction z of an odorous substance S is given by Fick's Law. If C is the concentration of the substance, one has the following (see Section 10.7 for notation):

$$J = - D\frac{dC}{dz} = - C_T D \frac{dy}{dz}$$

where J is the molar flux density in direction z related to a plane where there is no mass transport (mole/s m^2), D is the diffusivity of S in the mixture (m^2/s), dC/dz is the concentration gradient in direction z (mole/s m^2), C is the concentration of substance S, C_T is the total molar concentration, and y is the molar fraction of substance S in the considered phase. If transport is taken into account, the total transferred quantity becomes

$$N = J + U C$$

U being the average velocity of the fluid in the direction of the plane where there is no mass transport.

Remark: Odorous products are always in low concentrations in comparison to the total concentration ($C \ll C_T$). For this reason, even if one considers diffusion through an immobile gas, the same analytical expression of the transferred flux is obtained.

In the case of a liquid-gas transfer, there is mass transfer through an interface; a concentration gradient can exist within each of these phases. The phenomenon is thus governed by the thermodynamic equilibrium between the gaseous phase and the liquid phase and by the simultaneous mass transfer within each phase. Conditions existing in the proximity of the interface are very difficult to observe or explore experimentally.

Based on the fact that mass transfer resistance is found entirely within a small distance of the interface, several theories about the quantification of mass transfer have been written.

10.2.2 The Different Theories

The oldest theory is the *two film theory* suggested by Witman in 1923. It suggests that, for each phase, resistance to mass transfer is located in a thin stagnant film that is separate from the interface. The transfer through the films is treated as a molecular diffusion at a steady state. The fluid flow is turbulent over the stagnant film surface.

The *penetration theory* (Higbie, 1935) suggests that the transfer is due to the fluid renewal by turbulence at the interface. For a given period, the problem is treated as a mass transfer in a transient process.

The *interface renewal theory* (Danckwerts, 1951) modifies the preceding approach by considering that the material renewed at the suface remains there during different periods of time.

In 1958, Toor and Marchello put forward a more general expression, the *film penetration theory,* which shows that the preceding theories are boundary conditions of their theory.

However, the two film theory is still widely used even if the original concepts do not seem as close to reality as those of more recent theories. Indeed, the mathematical approach is much simpler and, most of the time, the results are nearly identical to those obtained with more complex approaches. This theory will thus be developed for scrubbing towers.

10.2.3 The Two Film Theory

In two film theory application, steady state is assumed and mass transfer is treated in each phase as a molecular diffusion. In this theory, the boundary layer is quiescent, in which case $N = J$.

When considering an equilibrium relationship between the liquid phase and the gaseous phase, the following terms can be defined: p is the partial pressure of S in the gaseous phase; C_E is the theoretical concentration of S, which, in the liquid phase, will be in equilibrium with p; C is the concentration of S in the liquid phase; p_E is the theoretical partial pressure of S, which, in the gaseous phase, will be in equilibrium with C; and p^* and C^* are the partial pressure and concentration of S at the interface in each of the two phases and in equilibrium.

If the gaseous and liquid film thicknesses of each side of the interface are called δ_G and δ_L (Figure 10.2), the integration of Fick's law allows one to write, per unit transfer surface area,

$$N = \frac{D_G}{\delta_G RT}(p - p^*) = \frac{k_G}{RT}(p - p^*)$$

$$N = \frac{D_L}{\delta_L}(C^* - C) = k_L(C^* - C)$$

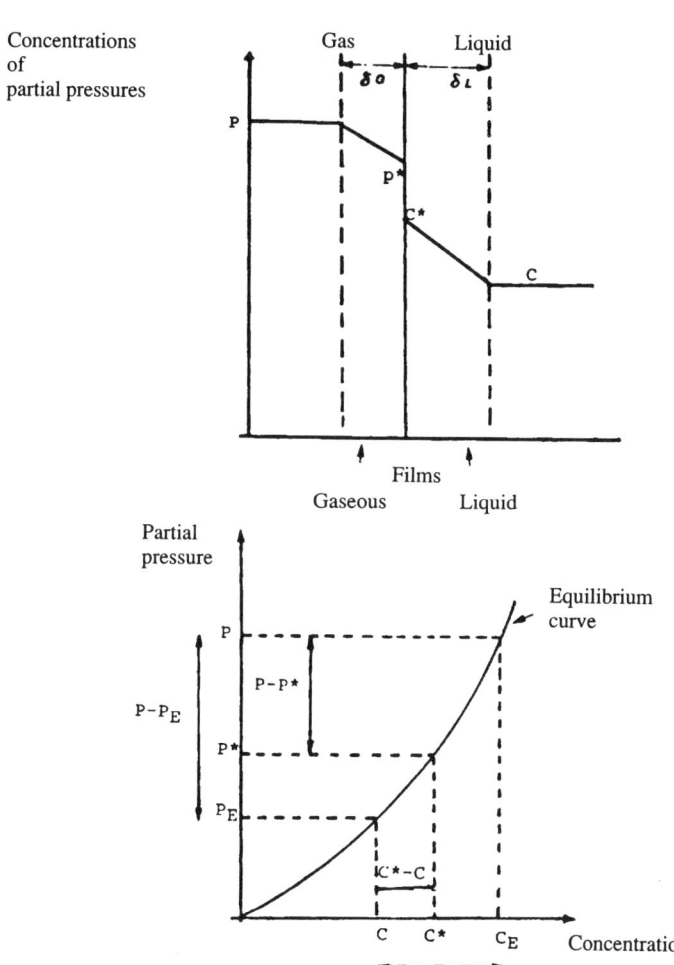

Figure 10.2. Diagram of films and driving forces on each side of the interface.

where D_G and D_L are the respective diffusion coefficients of the pollutant in the gaseous and liquid phases and k_G and k_L are the terms called compartmental or film transfer coefficients.

In the case of diluted aqueous solutions, equilibrium equation between the gaseous phase and the liquid phase is often very close to Henry's ideal equation. The equilibrium curve is thus a straight line:

$$p = HC_E \text{ or } p_E = HC \ (H = \text{atm m}^3/\text{mol})$$

where H is the Henry's constant of the compound to be tranferred. This relationship is written in other forms:

$$-p = HC_Tx_E \text{ or } H'x_E \text{ with } H' = C_TH \ (H' = \text{atm/molar fraction})$$

$$-yP = HC_Tx_E \text{ or } y = H''x_E \text{ with } H'' = C_TH/P \ (H'' \text{ is dimensionless})$$

where P = total pressure. In practice, the unit and thus the absolute value of Henry's constant will depend upon the parameters chosen to represent the equilibrium.

For the sake of easy measurements, it is often more practical to group the two films in the reasoning. Combining the different equations, overall transfer coefficients K_G and K_L are defined so that the tranferred flux per unit area is written:

$$N = \frac{K_G}{RT}(p - p_E)$$

$$N = K_L(C_E - C)$$

The coefficients are linked through the following relationships:

$$1/K_G = 1/k_G + H/k_L \text{ and } 1/K_L = 1/k_L + 1/Hk_G$$

In the case where the compound to be transferred diffuses easily in the gaseous phase but is hardly soluble in the liquid phase, all the mass transfer resistance is concentrated in the liquid film. One can then write

$$p^* = p, \ C^* = C_E, \text{ and } 1/K_L = 1/k_L$$

If a is the interfacial area per unit volume (m^2/m^3), for an elementary volume of column dV, the amount of tranferred pollutant dN' will be

$$dN' = k_Ga \ dV(p - p^*) = K_Ga \ dV(p - p_E)$$

$$dN' = k_La \ dV(C^* - C) = K_La \ dV(C_E - C)$$

The integration of these equations over the totality of the volume V of a reactor should take into account the concentration or the partial pressure variations over the length of the scrubbing tower.

10.2.4 Mass Transfer with Chemical Reaction

The transfered pollutant may react with components of the liquid phase. This reaction modifies the concentration profile near the interface and thus changes the transfer conditions.

At steady state, in a given direction and for a given point in the column, one can write, for the pollutant, $dN/dt = r\, d\, v$, where r represents the local reaction rate. If only one reaction of the pollutant S with the reactant R is taken into account and follows the stoichiometric relationship,

$$S + \upsilon R \rightarrow \text{Products}$$

the overall transformation yield can be written as

$$r = k\, C^m\, C_R^n\, \beta$$

where β is the liquid volumic fraction in the reactor and k is the reaction rate constant.

Calculations have been developed for pseudo first-order reactions and for second-order reactions (with $m = 1$, $n = 1$). When performing a mass balance on an elementary film section, the integration of the resulting differential equation makes it possible to write the pollutant transfer with an enhancement factor E:

$$dN' = k_L aE(C^* - C)dV$$

The value of E is accessible through three dimensionless numbers:

1. The Hatta number:

$$Ha^2 = kC_R D_L / k_L^2$$

2. The "concentration-diffusion" Z number:

$$Z = \frac{D_R C_R}{\upsilon D_L C^*}$$

   where D_R is the reagent liquid diffusivity
3. The "reaction-transfer" R number:

$$R = \frac{kC_R \beta}{k_L a}$$

The values of the dimensionless numbers are given in Table 10.1, and Figure 10.3 reproduces Van Krevelen's and Hoftijzer's diagram (1948), referred to in Table 10.1.

Table 10.1. Different Cases of Mass Transfer with Chemical Reaction

Chemical Reaction	Dimensionless Number	Concentration Profiles	C Value	E Value	Transfer Equation	Parameter to Be Favored
Very slow	$Ha < 0,3$ $R < 0,1$		$C \# C^*$	$E = 1$	$dN = k_La(C^* - C)\,dv$ or $dN = kC^*C_R.\,dv$	Liquid volume
Slow	$Ha < 0,3$ $0,1 < R < 10$		$0 < R < C^*$	$E = 1$	$dN = k_La(C^* - C)\,dv$	Liquid volume
Slow	$Ha < 0,3$ $R > 10$		$C \# 0$	$E = 1$	$dN = k_La\,C^*\,dv$	Interface surface area
Moderately rapid	$0,3 < Ha < 5$		$0 < C < C^*$	$E = f(Ha,Z)$ Van Krevelen's diagram	$dN = k_LaE(C^* - C)\,dv$	Liquid volume and interface surface area

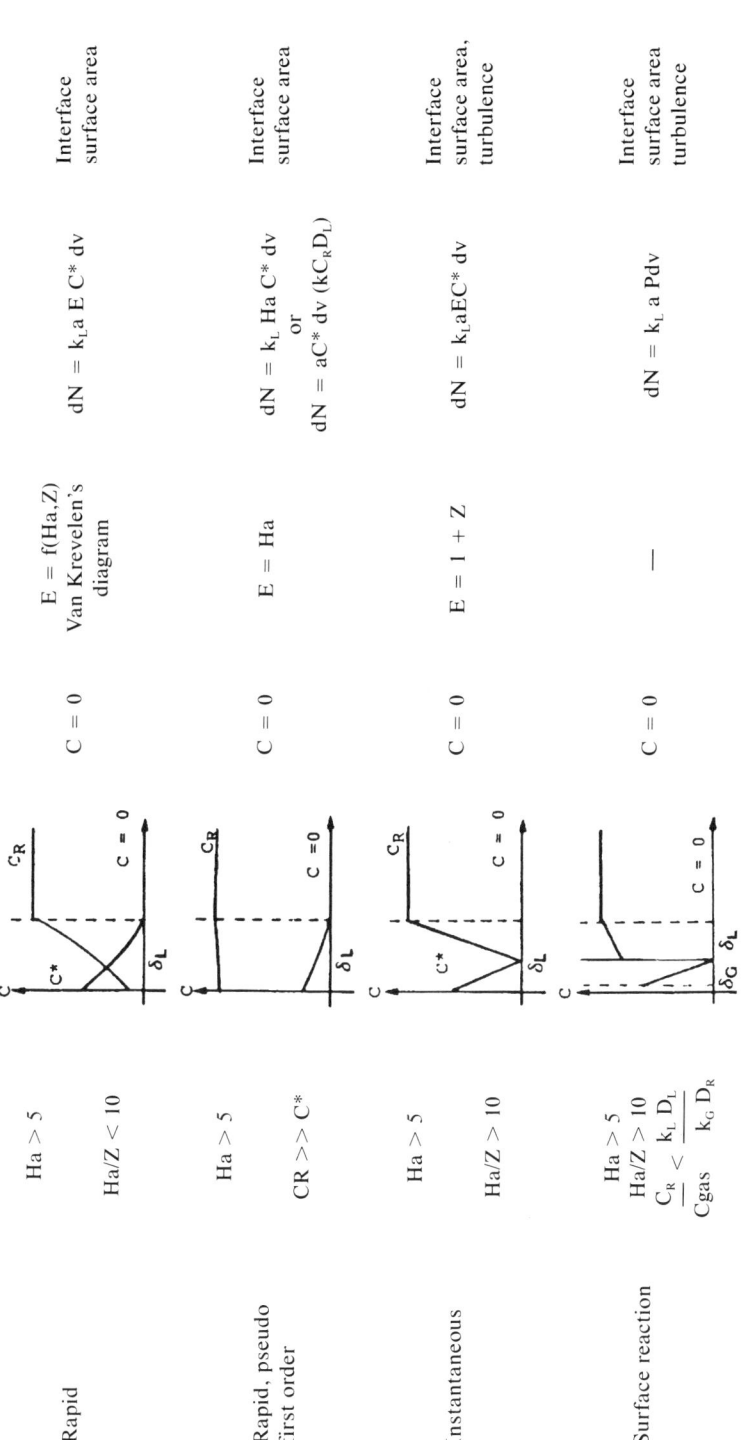

Rapid	$Ha > 5$ $Ha/Z < 10$	$C = 0$	$E = f(Ha,Z)$ Van Krevelen's diagram	$dN = k_L a\, E\, C^*\, dv$	Interface surface area
Rapid, pseudo first order	$Ha > 5$ $C_R \gg C^*$	$C = 0$	$E = Ha$	$dN = k_L\, Ha\, C^*\, dv$ or $dN = aC^*\, dv\, (k C_R D_L)$	Interface surface area
Instantaneous	$Ha > 5$ $Ha/Z > 10$	$C = 0$	$E = 1 + Z$	$dN = k_L a E C^*\, dv$	Interface surface area, turbulence
Surface reaction	$Ha > 5$ $Ha/Z > 10$ $\dfrac{C_R}{Cgas} < \dfrac{k_L\, D_L}{k_G\, D_R}$	$C = 0$	—	$dN = k_L\, a\, Pdv$	Interface surface area, turbulence

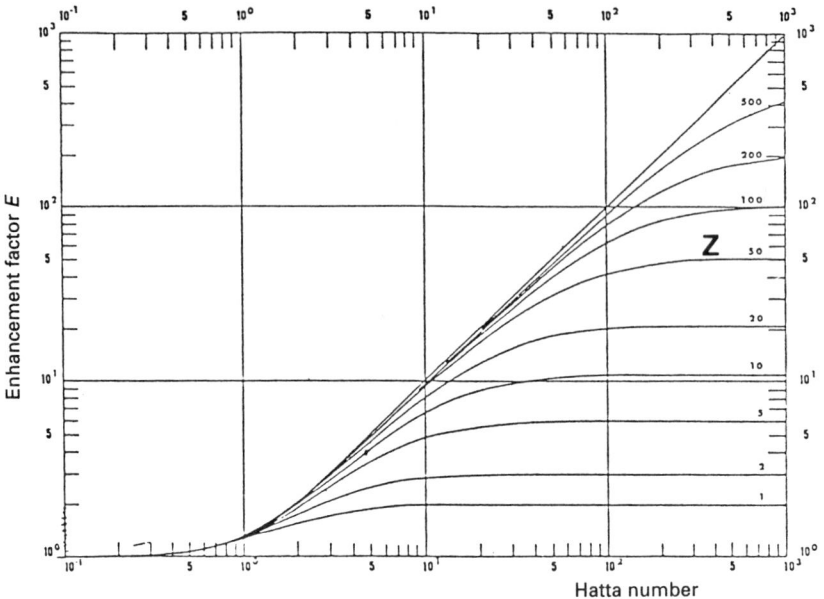

Figure 10.3. Van Krevelen's and Hoftijzer's diagram (1948).

Please note the following.

1. Ionic reactions (acid-basic scrubbing) are generally instantaneous. The value of k is very high, and the reaction occurs entirely within the liquid film. In the case of a scrubbing operation performed with an oxidizing solution, the influence of the oxidant depends upon its concentration and the oxidant-pollutant reaction rate (Table 10.1). When performing an acid-basic scrubbing in the presence of an oxidant, reagents R_1 and R_2 react with rate constants k_1 and k_2. It is then possible to define Hatta's number as

$$Ha^2 = \frac{(k_1 C_{R1} + k_2 C_{R2}) D_L}{k_L^2}$$

and the two Z numbers as

$$Z_1 = \frac{D_{R1} C_{R1}}{v_1 D_L C^*} \text{ and } Z_2 = \frac{D_{R2} C_{R2}}{v_2 D_L C^*}$$

If both reactions are instantaneous, the selectivity depends on the ratio

$$D_{R1} C_{R1}/D_{R2} C_{R2}$$

If the acid-basic reaction is instantaneous and the other reaction is rapid, the oxidation reaction will be favored. If both reactions are rapid, mass transfer will not influence the selectivity of the operation.

2. When the reaction is not first order ($r = kC^m C_R^n$), Hikita and Asai (1964) showed that the reaction regime always depends on the Hatta number defined as

$$Ha^2 = \frac{1}{k_L^2} \frac{2}{m+1} k \, D_L \, C^{*(m-1)} \, C_R^n$$

and it is possible to use the formulas given in Table 10.1 with errors lower than 8%.

3. When simultaneous absorption of several gases occurs, the situation becomes very complicated, and usable diagrams such as Van Krevelen's and Hoftijzer's (1948) do not exist.

A synthesis on this topic was published by Ramachandran and Sharma (1980).

10.2.4.1 Example of Dimensionless Number Calculation and of the Use of Van Krevelen's and Hoftijzer's Diagram (1948)

The absorption of an odorous compound is accelerated by an oxidation. Calculate the transfer acceleration coefficient E for the following conditions:

- Oxidant solution concentration: 5×10^{-3} mol L^{-1}
- Reaction rate coefficient: 10^5, 10^7, 10^9 M s^{-1}
- Reaction stoichiometric coefficient: 1
- Oxidant diffusivity: 3×10^{-9} m^{-2} s^{-1}
- Odorous compound diffusivity: 1.5×10^{-9} m^{-2} s^{-1}
- Odorous compound interface concentration: 10^{-3} mol L^{-1}
- Mass transfer coefficient without reaction: 2×10^{-4} m s^{-1}
- Interfacial area per unit volume: $a = 200$ m^2 m^{-3}

Results:

	k	10^5	10^7	10^9
Ha		4.33	43.3	433
Z		10	10	10
E		4	10	11

In the first case, $Ha/Z < 10$; the overall regime is moderately rapid and the value of E is obtained from the Van Krevelen diagram at the intersection of $Ha = 4.33$ and the $Z = 10$ curve ($E = 4$). In the second case, the reaction is rapid and the $E = 10$ value is obtained in the same manner. In the third case, $Ha/Z > 10$, the reaction is instantaneous and $E = 1 + Z = 11$.

10.3 Absorption Tower Technology and Implementation

Odor treatment by transfer of odorous compounds from a gaseous phase to a liquid phase (with or without chemical reaction) necessitates the use of gas-liquid contactors.

Table 10.2, completed with Trambouze's work on reactors (Trambouze et al,. 1984), summarizes the principal existing reactors with their main characteristics.

In odor treatment by scrubbing with or without chemical reaction, the involved flowrates are quite important most of the time. Furthermore, the pollutants are often hardly soluble and the involved chemical reactions are more or less fast. It is important to get a pressure drop as low as possible on the gaseous stream while having a high transfer yield.

The packed tower is most often used in order to comply with all these requirements and the following development is more specific to this type of contactor. However, one will prefer to use gas or liquid driven Venturis or spray towers to complete instantaneous reactions involving more or less dusty gases.

10.3.1 Packed Tower Technology

10.3.1.1 Column Description

A packed tower is generally constituted of a cylindrical body containing

1. A bottom part used as a recycle capacity with a volume related to the pump flowrate (generally corresponding to about a 1.5 min. residence time)
2. A gas inlet
3. A packing support

Table 10.2. Gas-Liquid Contactors and Parameters Order of Magnitude

Contactor Type	Liquid Hold Up(%)	Kg (m/s)	Kl (m/s)	Interface Surface Area (m²/m³)
Bubble tower	> 70	1 to 5 10^{-2}	1 to 5 10^{-4}	100/500
CSTR	> 70	1 to 5 10^{-2}	1 to 6 10^{-4}	200/2000
Packed tower	5/15	1 to 5 10^{-2}	0.5 to 5 10^{-4}	50/350
Tray tower	60/80	1 to 5 10^{-2}	0.5 to 1 10^{-4}	25/100
Spray tower	< 30	0.5 to 2 10^{-3}	0.01 to 2 10^{-4}	10/100
Liquid driven Venturi scrubber	< 5	0.2 to 3 10^{-3}	0.7 to 8 10^{-4}	400/5000
Gas driven Venturi scrubber	< 5	0.1 to 2 10^{-3}	0.3 to 7 10^{-4}	900/20000

4. A packing material, divided into one or several zones
5. A liquid feed distributor with redistribution if needed
6. A mist eliminator capable of stopping the liquid droplets carried over in the gas stream

10.3.1.2 Packing Support

The packing support has a double role:

1. It must hold the weight of the packing and of the liquid held on it (see Table 10.2)
2. It must play the role of a fluid distributor at the bottom of the column

For this latter role, it is important that the grid have a homogenous relatively high void fraction (at least 75%).

10.3.1.3 Packing Material

Packing material is the essential element of packed towers. Its choice is driven by several criteria: efficiency, cost, implementation, and column hydrodynamics. The shape and the main characteristics of the packing materials are given in Table 10.3. In a tower, materials can be arranged as regular or stacked packing but, in general, the packing is random.

Packing material can be classified into two families.

1. The first types (Raschig and Pall rings, saddles) are characterized by their geometric surface, which is often different from their active surface because of the occurence of channeling (their shape favors a centrifuge effect leading the liquid towards the walls).
2. The second types (Tellerettes, Chem Pack) are not characterized by their surface but rather by the number of contact points with neighboring material in the packing. An ever renewed drop is created at the contacting points, thereby creating the active surface.

Channeling does not exist in this latter type of packing material.

Each packing material is characterized by a packing factor F or packing factor that, for early rings, used to depend on their geometrical shape ($\mathcal{A}/\varepsilon^3$, with $\mathcal{A}$ = surface area and ε = void factor). Today, this factor is determined experimentally. It is used in the definition of the column hydrodynamics. The choice of the diameter d of the packing is more or less related to the diameter of the column Dc.

It is generally assumed that the following ratio must be respected:
$12 < Dc/d < 60$

10.3.1.4 Liquid Feed Distribution

For proper column operation, the liquid must form a continuous film on the packing surface.

Table 10.3. Main Characteristics of Selected Packing Materials

Packing Type	Diameter (mm or ref)	Specific Surface Area (m²/m³)	Void Fraction vol/vol	Packing Factor (m²/m³)
Ceramic Raschig	19	240	0.72	840
rings	25	200	0.73	525
	38	130	0.68	310
	51	95	0.75	210
	76	70	0.75	120
Metal Raschig	19	290	0.93	510
rings	25	210	0.93	380
	38	140	0.94	270
	51	100	0.94	190
	76	70	0.94	105
Plastic Raschig	25	193		456
rings	38	122		278
	51	102		190
Ceramic Pall	25	220	0.73	278
rings	38	165	0.76	158
	51	120	0.77	112
	76	105	0.68	
Metal Pall rings	25	240	0.94	155
	38	145	0.95	92
	51	105	0.95	66
	76	78	0.95	
Plastic Pall rings	16	322	0.67	318
	25	190	0.90	210
	38	122	0.91	118
	51	98	0.92	82
	85	66	0.92	55
Ceramic Berl	19	300	0.67	560
saddles	25	250	0.70	360
	38	150	0.75	210
	51	110	0.77	150
Ceramic Intalox	19	300	0.73	475
saddles	25	250	0.75	320
	38	160	0.74	170
	51	110	0.75	130
	76	92	0.77	77
Plastic Super	25			131
Intalox	50			92
Plastic Tellerettes	25	300	0.75	110
	50	110	0.90	50

This characteristic of the liquid film mainly depends on the quality of the liquid feed distributor for random packing material.

Distributors that are most frequently used are:

- Tube trays with a minimum of 50 distribution points per m² of column section
- Or preferentially solid cone sprayers.

10.3.1.5 Entrainment Eliminators

Since the gas is most often circulated countercurrent with respect to the liquid, it can carry fine liquid droplets. For instance, air circulating at 1 m/s in a column can contain less than 0.3-mm-diam droplets. The liquid distributor must then be coupled to an entrainment eliminator liable to remove transiting droplets with diameters ranging from 50 to 500 μm. Knitted mesh layers or venetian blind arrangements are often used for this purpose.

The operating mode of these entrainment eliminators is related to the impact of fluids on a solid obstacle (wire, wall). The impact causes the light gas to deviate and continue its flow, whereas the heavier liquid, having a much higher kinetic energy (1,000-fold), is stopped by the obstacle. The efficiency of the element is related to the velocity, which is an important factor in the kinetic energy transported by the droplets. This velocity is limited by the possible reentrainment of liquid by the transiting gas.

This reentrainment depends on the transit velocity of the gas, its density, and the type of equipment. For the air-water system, the reentrainment velocity varies between 3 and 5 m/s according to the type of entrainment eliminator. The highest reentrainment velocities are obtained with eliminators with a centrifugal effect or with closely packed zigzag elements. For mesh layers and classical venetian blind elements, 2.5–2.8 m/s velocities are generally used for avoiding reentrainment.

10.3.2 Packing Tower Hydrodynamics

The study of the packing tower hydrodynamics makes it possible to solve the following problems:

What is the optimum gas velocity? (column diameter)

How wet is the packing, in other words, what is the percentage of the packing surface that is covered with a liquid film, thus in nominal operating conditions?

What is the gas pressure drop per unit of tower height?

What is the liquid hold up in a packed tower?

10.3.2.1 Overview of the Pressure Drop in a Packed Tower

The pressure drop evolution in a packed tower operating in the countercurrent mode is given by Figure 10.4. This figure roughly indicates three areas:

1. An area where the gas velocity is low: the liquid film at the surface of the packing material is thin and it is not distorted by the gas flow. The logarithmic variation of the pressure drop is proportional to the mass velocity according to a slope value of about 2.
2. When the liquid mass velocity becomes higher, the loading point is reached where the slope exceeds 2 and where the gas creates turbulence at the liquid surface. This zone is preferable as the pressure drop remains acceptable while the mass transfer from one phase to the other is favored.
3. If the velocity still increases, the liquid flow is soon blocked and the liquid accumulates in the tower. This is the flooding point characterized by a very steep increase in the pressure drop.

Practically, it will be quite interesting to operate in the loading zone, which corresponds to gas flowrates ranging from 50 to 80% of the flooding gas flowrate. As it is much easier to forsee the flooding point than to calculate the loading point, the operating velocity will be calculated with respect to the flooding velocity.

10.3.2.2 Flooding Velocity Determination

Numerous experimental studies have been carried out on the air-water system. In this case, the most classical determination relies on Sherwood et

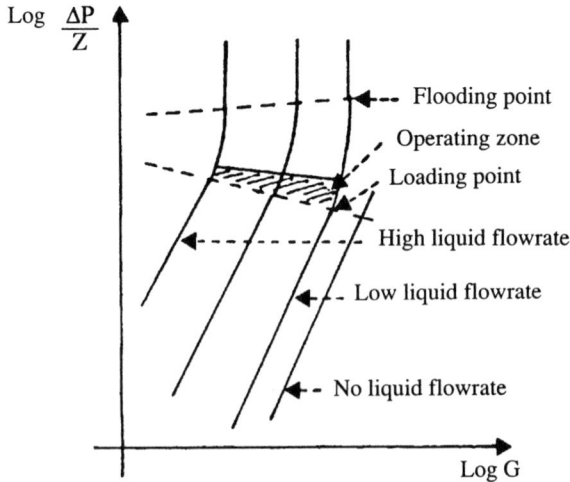

Figure 10.4. Pressure drop in a packed tower.

al.'s (1938) graphical method, which is based on the knowledge of the packing factor F (Figure 10.5). The B term is calculated as

$$B = \frac{L}{G} \sqrt{\frac{\rho_G}{\rho_L}} = \frac{U_{SL}}{U_{SG}} \sqrt{\frac{\rho_L}{\rho_G}}$$

Then a vertical is drawn to the intercept with the corresponding flooding curve. The ordinate of the intercept gives an A value where

$$A = \frac{F}{g} \frac{\rho_{H_2O}}{\rho_L} \frac{\rho_G}{\rho_L} \left(\frac{\mu_L}{\mu_{H_2O}}\right)^{0.2} U_E^2$$

Knowing the value of A, one can deduce the value of U_E, then the value of U_T and the value of the diameter D of the tower. In these expressions, U_{SL} and U_{SG} are the superficial velocities of the liquid and the gas, respectively, in m/s and flowrate per unit empty column section, L and G are the liquid and gas massic flowrate (kg/s), ρ_L, ρ_G, and ρ_{H_2O} are the specific density of liquid, gas, and water, respectively (kg / m³), μ_L is the kinematic viscosity of the liquid (Pl), μ_{H_2O} is the kinematic viscosity of water ($= 10^{-3}$ Pl at 20 °C), g is the gravitational constant (9.81 m/s²), F is the packing factor (see

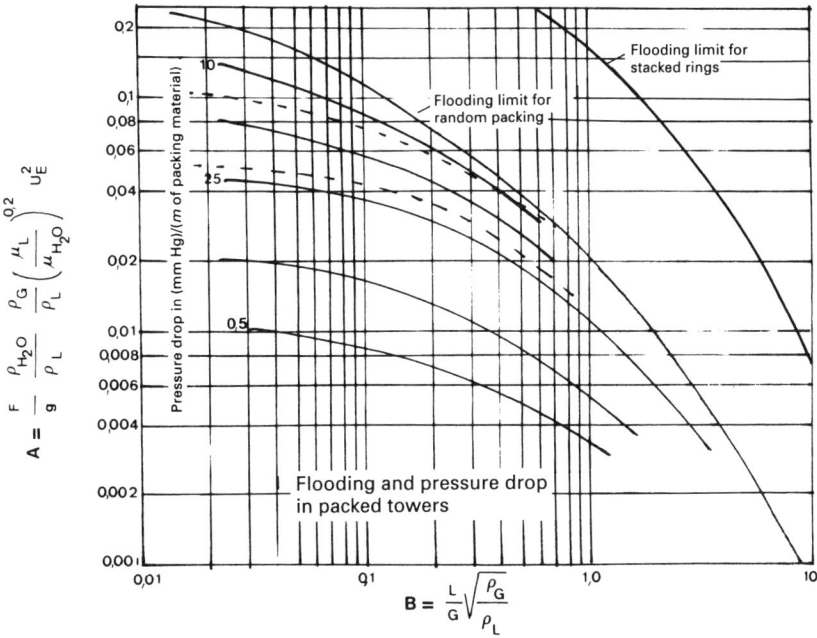

Figure 10.5. Sherwood's graphical relationship (1938).

Table 10.3) (m^2/m^3), U_E is the flooding velocity (m/s), and U_T is the operating velocity (m/s) ($U_T = 0.5$–$0.8U_E$.

Instead of using the graph, it is sometimes more convenient, for computerized models, to use the following mathematical relationships between A and B at the flooding point:

$$\text{Log } A = 0.111,7 - 4.012B^{0.25} \text{ (Trambouze et al., 1984)}$$

$$\text{Log } A = -0.380 - 3.61B^{0.286} \text{ with } B < 10 \text{ (Copigneaux, 1986)}$$

10.3.2.3 Pressure Drop

It is very easy to use the graph given in Figure 10.5 to obtain a pressure drop. A is recalculated with the operating velocity and the intercept of A and B makes it possible to read the pressure drop directly. It is also possible to use the relationships defined by Trambouze et al. (1984):

$$\Delta P/Z \, (pa/m) = 98 \, (A_T/A_E) \, (C_1 + C_2 \, A_T/A_E)$$

with

$$C_1 = 21.79 - 36.19 \, B_T^{0.25} + 16.60 \, B_T^{0.5}$$

$$C_2 = 7.06 + 10.30 \, B_T^{0.25} - 10.36 \, B_T^{0.5}$$

The subscripts T and E refer to operating and flooding conditions, respectively.

Numerous graphs allowing an estimate of the pressure drop are given in *Perry's Chemical Engineers' Handbook* (1984), in Chapter 18, which is devoted to packed towers.

10.3.2.4 Wetting Rate

The liquid flowrate must be high enough to insure a continuous film all over the packing material. Failing that, the packing loses its efficiency.

Morris and Jackson (1953) introduced the concept of a minimum wetting rate, defined as $L/A \, \rho_L > 25 \times 10^{-6} \, m^2 \, s^{-1}$. For a value higher than $2 \times 10^{-4} \, m^2 \, s^{-1}$, the flow is no longer in the form of a continuous film and the liquid floods the packing thereby decreasing the efficient interface surface area.

These equations can be used for aqueous solutions whose physical-chemical properties are not too remote from those of pure water. For scrubbing with solvents or with petroleum cuttings it is important to experimentally determine flooding velocities and pressure drops, as the application of the equations may lead to erroneous results.

10.3.3 Transfer Units and Estimation of Required Column Height

10.3.3.1 Scrubbing Balance: The Operating Line Concept

Consider a scrubber to treat an incoming gas flowrate G_m (mole/m^2 s) with a pollutant partial pressure $p_e = y_e P_e$, which must leave the scrubber with a pollutant partial pressure $p_s = y_s P_s$. Scrubbing is performed by the use of a solution whose flowrate is L_m (mole/m^2 s). This solution contains the molar fraction x_e of pollutant as it enters and will leave with the molar fraction x_s depending only on the mass balances if there is no chemical reaction. p_e and p_s are the pollutant partial pressures entering and leaving the tower, P_e and P_s are the total pressures entering and leaving the tower, y_e and y_s are the molar fractions of the pollutant in the gaseous phase, $y_e{}^*$ and $y_s{}^*$ are the molar fractions of pollutant, which will be in equilibrium in the gaseous phase with the molar fractions x_e and x_s in the liquid, x_e and x_s are the molar fractions of the pollutant entering and leaving the tower in the liquid phase, and G_m and L_m are the molar flowrates of inert gas or scrubbing liquid free from odorous compounds per unit of tower section. In the case of odor treatment, the quantity of pollutant is low with respect to total flowrates. Therefore, it can be assumed, without too large a margin of error, that G_m and L_m represent the total flowrates.

Scrubbing efficiency is defined with respect to a theoretical maximum efficiency that corresponds to the equilibrium state:

$$y_s{}^* = \frac{H'}{P_s} x_s \text{ or } y_e{}^* = \frac{H'}{P_e} x_e$$

In a cocurrent process, the efficiency is Eff $= (y_e - y_s)/(y_e - y_s{}^*)$, whereas for a countercurrent process Eff $= (y_e - y_s)/(y_e - y_e{}^*)$. In theory, the removal can be much more important in a countercurrent process than in a cocurrent process.

Write the mass balances for both the entire tower and any one point on it, considering that the amount of pollutant is low and thus that the flowrates are equal all along the reactor. Cocurrentwise and per unit surface area, one gets

$$G_m (y_e - y_s) = -L_m (x_e - x_s)$$

for the totality of the tower and

$$G_m (y_e - y) = -L_m (x_e - x)$$

at any one point. In (x,y) coordinates, one gets the equation of a straight line passing by (x_e,y_e) and (x_s,y_s), with a slope of $-L_m/G_m$.

Countercurrentwise and per unit surface area, one gets

$$G_m (y_e - y_s) = L_m (x_s - x_e)$$

for the totality of the tower and

$$G_m (y_e - y) = L_m (x_s - x)$$

at any one point. In (x,y) coordinates, one gets the equation of a straight line passing by (x_s, y_e) and (x_e, y_s), with a slope of L_m/G_m. The straight lines representing the evolution of the system during the scrubbing operation are called operating lines (Figure 10.6). Generally, scrubbers are operated countercurrentwise.

10.3.3.2 Transfer Units Definition

According to the mass transfer theory, the following equations can be written for a cross section of a column A with a dz height:

$$dN' = k_G a (p - p^*) A dz$$

$$dN' k_L a (C^* - C) A dz$$

Considering that the pollutant concentrations are low enough, the G_m and L_m are little different from the total molar flowrates and x and y (total molar concentrations in the gas and the liquid streams) are close to the molar concentrations in the gas and the liquid, free from odorous compounds. Therefore,

$$C^* = C_T x^* \text{ and } C = C_T x$$

Hence, for the gas stream, it can be written that the concentration variation equals the transferred amount:

$$AL_m dx = k_L a (C^* - C) A dz E$$

Separating the variables, one gets

$$\int_0^z dz = \int_{x_e}^{x_s} \frac{L_m}{k_L a C_T E} \frac{dx}{(x^* - x)}$$

If it is assumed that $E = 1$ and that L_m and k_L are constant over the height of the column, one gets

$$Z = \frac{L_m}{k_L a C_T} \int_{x_e}^{x_s} \frac{dx}{(x^* - x)} \text{ or } Z = H_L \times N_L$$

where H_L is the height of the transfer unit equal to $L_m/k_L a C_T$ and N_L is the number of transfer units equal to $\int_{x_e}^{x_s} \frac{dx}{(x^* - x)}$.

The same type of equations can be written when applied to either the gaseous phase or to the overall coefficients:

$$Z = \frac{G_m}{k_G a P} \int_{y_s}^{y_e} \frac{dy}{(y - y^*)}$$

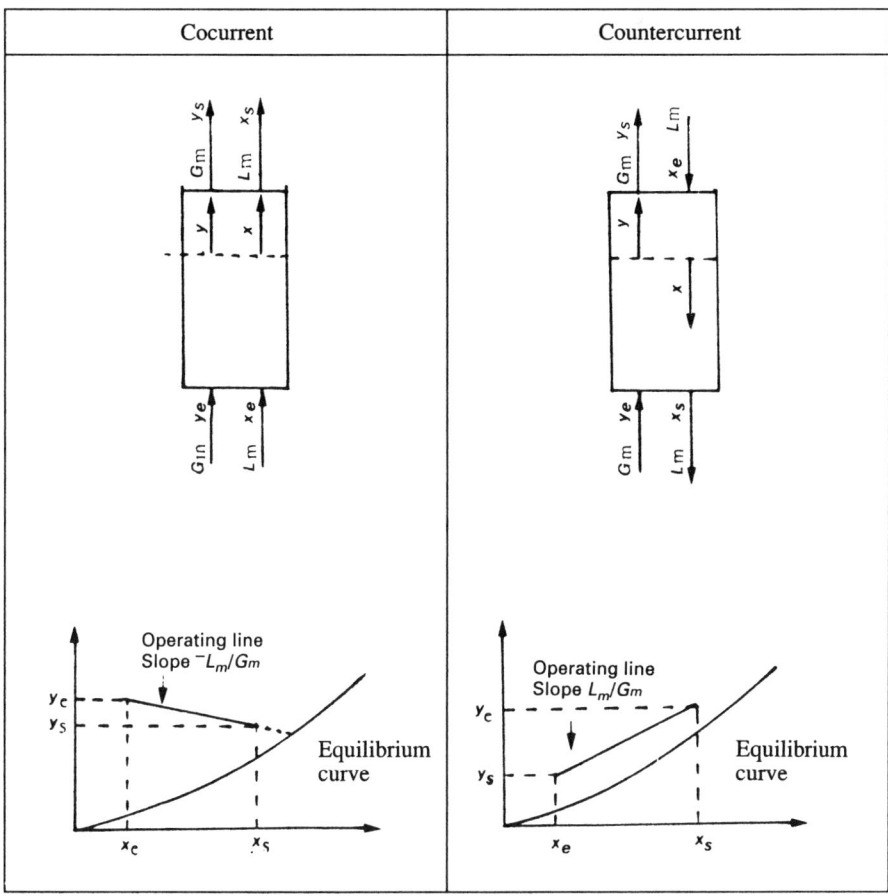

Figure 10.6.   Scrubbers and operating lines.

$$Z = \frac{G_m}{K_G \, a \, P} \int_{y_s}^{y_e} \frac{dy}{(y - y_E)}$$

$$Z = \frac{L_m}{K_G \, a \, C_T} \int_{x_e}^{x_s} \frac{dx}{(x_E - x)}$$

By writing that the total height of the column Z is the product of the height of a transfer unit times the number of transfer units, one gets

$$\left. \begin{array}{l} Z = H_L \times N_L \\ Z = H_G \times N_G \end{array} \right\}$$ taking into account mass transfer resistance in each individual film

$$\left. \begin{array}{l} Z = H_{OL} \times N_{OL} \\ Z = H_{OG} \times N_{OG} \end{array} \right\}$$ taking into account mass transfer in both films

Knowledge of the flowrates, the concentrations and one coefficient makes it possible to evaluate one of the following terms: H_L, H_G, H_{OL}, H_{OG}.

The following relationships can be written between the different transfer unit heights, like between the different transfer coefficients:

$$H_{OG} = H_G + \frac{mG'}{L'} H_L$$

$$H_{OL} = H_L + \frac{L'}{mG'} H_G$$

where m is the slope of the equilibrium curve and G' and L' are the mass flowrates per unit of surface area (kg/m² s). If the compound follows Henry's law, then one gets $m = H''$. H_G and H_L graphs and tables are given in the numerous books and articles dealing with absorption (Coulson and Richardson, 1980; Morris and Jackson, 1953; Perry and Green, 1984; Eckert, 1961).

10.3.3.2.1 Evaluation of the Number of Transfer Units

The number of transfer units can be calculated through different ways:

1. For a rapid evaluation, one can write

   $$H_{OL} = (x_s - x_e)/(x_E - x)_{average}$$

   The value of $(x_E - x)_{average}$ is evaluated through a logarithmic average of the inlet and outlet conditions:

   $$(x_E - x)_{average} = \frac{(x_{sE} - x_s) - (x_{eE} - x_e)}{\ln\left[(x_{sE} - x_s)/(x_{eE} - x_e)\right]}$$

2. For a more accurate evaluation, and if the equilibrium curve is known, the integral value can be calculated pointwise.
3. In the case of an acceleration coefficient greater than 1, and since this coefficient is concentration dependent, it is necessary to evaluate it at any point of the column. Therefore, the design can only be performed by operating on finite column elements dz to which finite values of the variables are assigned.

10.3.3.2.2 Special Case of Diluted Solutions where Henry's Law Applies

Performing a molar balance between the top of the column and any point of it, one gets:

$$G_m (y - y_s) = L_m (x - x_e)$$

If $x_e = 0$, the equation is simplified to $x = (G_m/L_m)(y - y_s)$, and applying Henry's law, $y_E = H''x = (H'' G_m/L_m)(y - y_s)$.

Applying the definition of

$$N_{OG} = \int_{y_s}^{y_e} \frac{dy}{(y - y_E)}$$

and substituting y_E for its value, one gets

$$N_{OG} = \int_{y_s}^{y_e} \frac{dy}{y\left(- \dfrac{H''G_m}{L_m}\right) + \dfrac{H''G_m}{L_m} y_s}$$

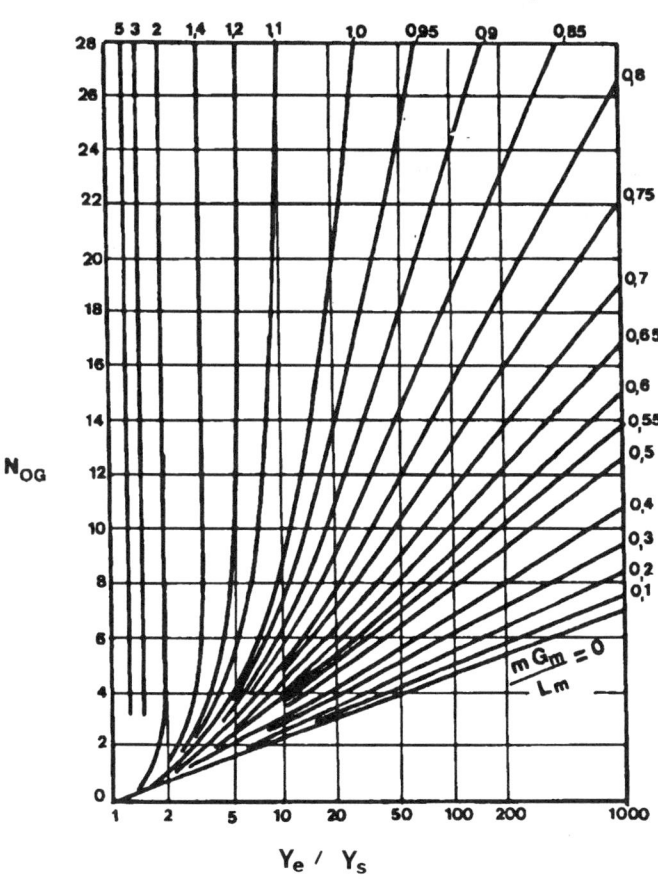

Figure 10.7. Number of transfer units in the special case of diluted solutions following Henry's law.

Integrating, we have

$$N_{OG} = \frac{1}{1 - H''G_m/L_m} \ln\left[\left(1 - \frac{H''G_m}{L_m}\right)\frac{y_e}{y_s} + \frac{H''G_m}{L_m}\right]$$

From this equation, Colburn (1939) drew the graph shown by Figure 10.7 that makes it possible to get in a limited number of cases, the number of transfer units as a function of the parameters y_e/y_s and $H''G_m/L_m$.

10.4 Physical-Chemical Characteristics and Reactivity of Odorous Compounds

10.4.1 Equilibrium Between Phases

Mass transfer is ruled by the concentration gradient that exists between the limit value of the pollutant concentration in the scrubbing solution and the real concentration

$$N = K_L (C_E - C) \text{ (by unit transfer area)}$$

C_E being the limit value in the liquid phase for a given partial pressure in the gaseous phase. The most general way to express the solubility is to write the relationship of equilibrium liquid/vapor: $y = Kx$ (see Chapter 4).

In the case of scrubbing with complex composition solvents, there are few theoretical ways to determine equilibrium curves and it would be necessary to carry out experimental work.

In the case of scrubbing with an aqueous solvent, the quantity of odorous compound being very diluted in each of the two phases, one can assume that they are diluted ideal solutions. In this case, the equilibrium relationship follows Henry's law

$$p = H' x_e \text{ or } P = H' x_e$$

where p is the partial pressure of the compound in the gaseous phase (atmosphere), y is the molar fraction of the pollutant in the gaseous phase (mole/mole), H' is Henry's constant (atmosphere/molar fraction), P is the total pressure (atmosphere), and x_e is the molar fraction at equilibrium in the liquid phase (mole/mole). For a given solute, Henry's constant depends on the temperature and also, to a lesser degree, on the composition of the aqueous phase of scrubbing (Chapter 4).

The value of H' of several odorous compounds is given in Table 4.1 for water as a solvent. These values can be approached from the solubility in water and the partial pressure of pure liquid at the specified temperature. When treating compounds that ionize in water, it is often practical to calculate a modified Henry's constant that takes the dissociation percentages of the compound to be absorbed into account. Thus, one defines $H'_f = fH'$, f being the rate of the nondissociated compound (cf. Section 10.4.2).

10.4.2 Dissociation of Odorous Products in Water

In an aqueous medium, many odorous compounds dissociate giving totally soluble ionic species, essentially ammoniac and amines,

$$R\ Nh_3^+ + H_2O \overset{K_1}{\rightleftarrows} R\ HN_2 + H_3O^+$$

and hydrogen sulfide and mercaptans

$$H_2S + H_2O \overset{K_1}{\rightleftarrows} HS^- + H_3O^+$$

$$HS^- + H_2O \overset{K_2}{\rightleftarrows} S^{2-} + H_3O^+$$

These dissociation reactions are chemical equilibria characterized by their dissociation constants K_1 or K_1 and K_2.

In the case where there is only one dissociation constant as for amines, one gets

$$K_1 = \frac{[R\ NH_2]\ [H_3O^+]}{[R\ NH_3^+]}$$

$R\ NH_2$ fraction: $\dfrac{[R\ NH_2]}{\text{total amine concentration}} = \dfrac{1}{1 + K_1/[H_3O^+]}$

In the case where there are two dissociation constants as for H_2S, one gets

$$K_1 = \frac{[HS^-]\ [H_3O^+]}{[H_2S]} \quad K_2 = \frac{[S^{2-}]\ [H_3O^+]}{[HS^-]}$$

H_2S fraction: $\dfrac{H_2S}{[H_2S] + [HS^-] + [S^{2-}]} = \dfrac{1}{1 + K_1/[H_3O^+] + K_1\ K_2/[H_3O^+]^2}$

HS^- fraction: $\dfrac{HS^-}{[H_2S] + [HS^-] + [S^{2-}]} = \dfrac{K_1/[H_3O^+]}{1 + K_1/[H_3O^+] + K_1\ K_2/[H_3O^+]^2}$

S^{2-} fraction: $\dfrac{S^{2-}}{[H_2S] + [HS^-] + [S^{2-}]} = \dfrac{K_1\ K_2/[H_3O^+]^2}{1 + K_1/[H_3O^+] + K_1\ K_2/[H_3O^+]^2}$

Dissociation constant values are very well known for the first acidity. For the second acidity, the values found in the literature are different according to the authors. For some odorous products, Table 4.2 indicates the values of dissociation constants ($pK_A = -\log K_A$).

10.4.3 Oxidant Action

The addition of an oxidant into a scrubbing aqueous solution (chlorine, hydrogen peroxide, ozone), or the oxidation of the scrubbing solution itself, has a double objective:

1. To improve the pollutant removal by the acceleration of the mass transfer (E value) and/or by increasing the transfer driving force ($C^* - C$) by the destruction of the absorbed odorous compound and therefore the decrease of the value of C. This latter point is particularly important to perform the removal of hardly soluble products or hardly dissociated products at the operating pH, such as mercaptans and disulfides with high yields.
2. To oxidize the absorbed product in order to continuously regenerate the scrubbing solution.

The control of oxidant use requires the knowledge of their action and their by-products as well as the knowledge of the kinetics and the reaction rates. These data make it possible to avoid the formation of new odorous or toxic compounds and to refine the scrubbing reactor design by better evaluating the value of the accelerating factor E.

Only a few studies were reported that deal with the oxidation of odorous compounds with the purpose of improving oxidizing scrubber design. However, a good deal of information can be found in works about the use of chlorine, ozone, and hydrogen peroxide.

10.4.3.1 Chlorine and Its Derivatives

Chlorine can be used either in its gaseous form or as sodium hypochlorite NaClO. Both uses result in the formation of hypochlorous acid HClO:

$$Cl_2 + 2H_2O \overset{K}{\rightleftarrows} HClO + Cl^- + H_3O^+ \quad (K = 1.5\text{--}4.0 \times 10^{-4} \text{ according to } t°)$$

The hypochlorous acid is dissociated as (Doré, 1989)

$$HClO + H_2O \overset{K}{\rightleftarrows} ClO^- + H_3O^+ \quad (K = 1.6\text{--}3.2 \times 10^{-8} \text{ according to } t°)$$

The form under which chlorine (Cl_2, HClO, and ClO^-) will be used depends upon pH and temperature (Figure 10.8). The examination of the redox potential of these species leads to the diagram shown in Figure 10.9.

10.4.3.1.1 Action on Nitrogenous Odorous Compounds

The action of hypochlorous acid on amoniac has been known for a very long time and has been the subject of many studies (Morris,1967; Struppler and

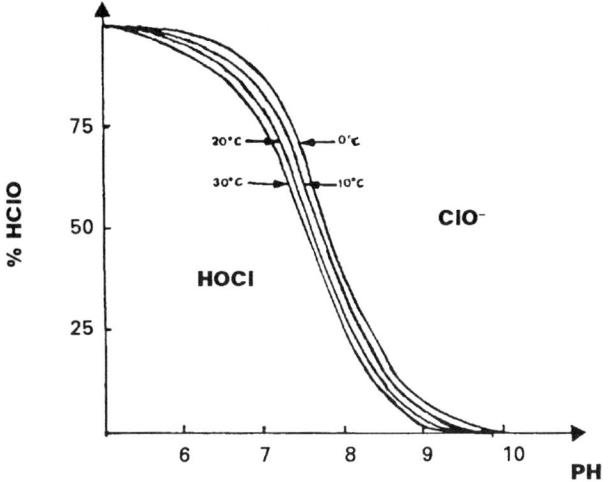

Figure 10.8. Occurrence of HOCl and ClO$^-$ in water as a function of pH and temperature (after Martin, 1979).

Husson,1974; Saunier and Sellier,1979). The complete set of these reactions is shown in Figure 10.10 and deserves the following comments:

1. The final products of the reaction that are the most abundant are NO$_3$, and N$_2$.
2. The kinetic constant k_1 is very large ($k_1 = 5.1 \times 10^6$ M^{-1} s^{-1} at 25 °C) and $k_2 = 3.4 \times 10^2$ M^{-1} s^{-1}. The disappearance of monochloramine NH$_2$Cl and dichloramine NHCl$_2$ occurs within minutes whereas the ap-

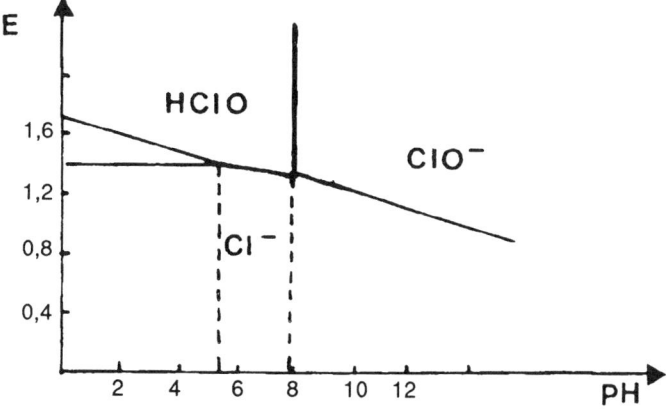

Figure 10.9. Species occurring in a 1 g atom aqueous solution as a function of pH (after Martin, 1979).

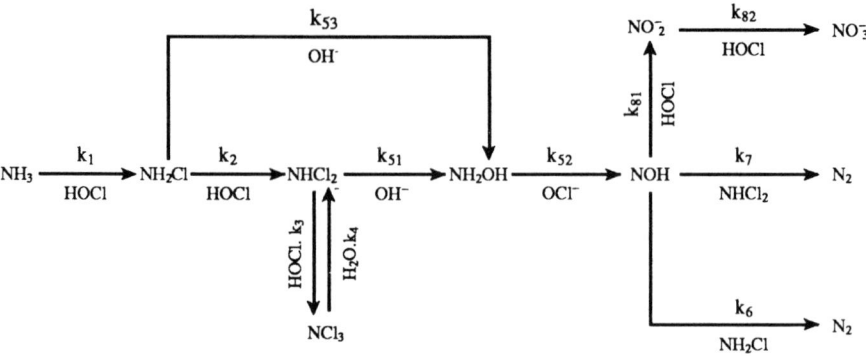

Figure 10.10. NH_3 chlorination reactions diagram.

pearance of the final species (NO_3 and N_2) occurs within hours or even days. The reaction rate depends upon initial conditions, pH, and temperature.

With amines, the overall reaction is similar but stops at the dichloramine step.

$$R\,NH_2 \xrightarrow[\text{HOCl}]{} R\,NHCl \xrightarrow[\text{HOCl}]{} R\,NCl_2$$

For methylamine, Poncin et al. (1984) determined the reaction rates:

$r_1 = 7.33 \times 10^7\ [CH_3NH_2]\ [HOCl]$ mol s^{-1} at pH $= 5.35$ and 5.75

$r_1 = 2.88 \times 10^2\ [CH_3NHCl]\ [HOCl]$ mol s^{-1} at $5.35 < pH < 8.7$

HOCl being the active chlorine form.

10.4.3.1.2 Action on Sulfurous Odorous Compounds

The reactions quoted in the literature transform H_2S into colloidal sulfur or into sulfate:

$$HS^- + ClO^- \longrightarrow S{\downarrow} + OH^- + Cl^-$$
$$S^{2-} + 4ClO^- \longrightarrow SO_4^{2-} + 4Cl^-$$

Waltrip (1985) indicates that the first reaction occurs below pH $= 10$ whereas Cadena and Peter (1988) suggest an upper limit at pH $= 7.5$. A recent work by Bonnin (1990) suggests that the two reactions are superimposed and that the second reaction is favored by a pH increase.

In order to reach the sulfate step, the theoretical chlorine consumption amounts to four moles of Cl_2 per each of mole H_2S. Practically, according to industrial continuous pilot experimentation, with incoming H_2S concentrations ranging from 5 to 10 mg/m^3 and for a controlled pH around 9, removal yields higher than 99.8% and quasitotal transformation into sulfate are reached. The chlorine consumption amounts to 5.2 moles per mole of destroyed H$_2$S and the soda consumption amounts to about 5 moles per mole (Bonnin, 1990).

The action of chlorine on methyl mercaptan is more complex. The first oxidation product is dimethyldisulfide (CH$_3$-S-S-CH$_3$), which is itself very odorous, and the final oxidation step seems to be methylsulfonic acid CH$_3$SO$_3$H. In order to remove this kind of sulfurous compound, the use of excess oxidant is necessary. The removal yields go from 10–15% to 85–95% through the use of excess chlorine at pH > 10. Chlorine and soda consumptions are high (8.5 and 7.6 moles, respectively, per mole of removed CH$_3$SH). Oxidation reactions are much too slow, when they occur, to be useful in the process of removing odorous compounds such as aldehydes, acids, or hydrocarbons.

Treatment of odors by scrubbing with chlorinated water is used industrially to treat gaseous effluents resulting from various manufacturing operations: the pulp and paper industry (Schwab and Kaschke, 1975), meat quartering industries (Schaue et al., 1985), and sewage treatment (Bonnin, 1990 and Chapter 14 of this book, for example).

Most of the processes studied associate an acid pH scrubbing tower with an alkaline pH scrubbing tower to remove nitrogenous and sulfurized compounds. For Abe and co-workers (1979, 1982, 1983), for example, a 1.80 m high packed tower operating with a gas superficial velocity of 1.7 m s^{-1} and a NaClO concentration of 1.5 × 10^{-3} mole L^{-1} removes NH$_3$ (inlet: 2.68 ppm; outlet: 0.001 ppm) and (CH$_3$)$_3$ N (inlet: 0.37 ppm; outlet: 0.000, 5 ppm) at pH 7. A second tower operated at pH 11 removes H$_2$S and CH$_3$SH (yield>96%).

10.4.3.2 Ozone

This very powerful oxidizing agent is widely used in potable water treatment. It is produced in situ and mixed with an oxygen containing gas (air or pure oxygen). It is necessary to transfer this agent from the gaseous phase to the liquid phase when using it in deodorization by scrubbing and oxidation. Two technologies are then feasible:

1. Its introduction using a hydro-injector before the scrubbing operation
2. Its use in a second step where the absorbed pollutant is oxidized, thereby regenerating the scrubbing solution.

The latter operation is carried out using a bubble tower. Pilot plant experiments showed that the first technique leads to a very bad use of the oxidant as it is stripped by the air flux to be treated (Laplanche et al., 1987).

Ozone is also used to directly oxidize odorous compounds in the gaseous phase (First et al., 1974; Ponomarev et al., 1979). Generally, ozone reacts with odorous compounds either directly, by

1. dipolar 1,3 cyclo-addition on double bonds (styrene, allyl- and crotyl-mercaptans)
2. electrophilic substitution on an electron donor reactive site like nitrogen, sulfur, or carbon atoms activated by a donor group (amines, mercaptans, benzene, substituted phenol and cresol)

or by indirect action due to its decomposition into $OH°$, which is a very reactive specie with respect to any organic compound. The production of these hydroxyl radicals is enhanced by higher pH, is autoaccelerated, and may depend on the presence of such inhibitory species as carbonate, bicarbonate, or phosphate (Doré, 1989; Yurteri and Gurol, 1988; Sens et al., 1990).

10.4.3.2.1 Effect on Odorous Nitrogen Derivatives

If the pH is lower than 10, ozone does not react with either NH_3 or NH_4^+. On the other hand, Singer and Zilly (1975) showed that ozone transforms NH_3 into NO_3^- in basic conditions. (See Table 10.4.)

For amines and the different nitrogenous compounds, the first step is an electrophilic attack of the oxidizing agent onto the the nitrogen atom. In aqueous conditions, the products can be classified into four categories:

1. The products of the direct oxidation of the nitrogen atom and the formation of derivatives such as hydroxylamine, oxime, and aminoxide
2. The oxidation products of the α carbon with respect to the nitrogen atom and the formation of products such as amide or formylamine
3. The C-N bond destruction products. The nitrogenous part is transformed into NH_4^+ or NO_3^- and the carbonaceous part gives aldehydes and acids
4. Condensation products involving the original compound and already oxidized derivatives.

The reactions are of the first order with respect to amine and ozone, and the rate constants were measured by Hoigne and Bader (1979). Even though the free amine values are quite high (10^5–10^8 M^{-1} s^{-1}), the actual rate constants depend on the ionization rate and, hence, on the pH value (Figure 10.11). Pilot scale studies showed that it is necessary to find a compromise between the scrubbing efficiency and the ozone-amine reaction rate (6 < pH < 7). The ozone consumption ranges from 1 to 1.5 M per mole of oxidized amine (Laplanche et al., 1988).

10.4.3.2.2. Effect on Sulfurized Odorous Compounds

H_2S is oxidized by ozone into mainly sulfate in aqueous conditions. The oxidation mechanism should proceed through the different oxidation states

Table 10.4. Reactivity and Rate Constants of Some Odorous Compounds with Respect to Ozone. Direct Attack (kO_3); Indirect Attack (kOH)

Products or Family of Products	kO_3 $M^{-1}s^{-1}$ Dissociation Products	kO_3 $M^{-1}s^{-1}$ Nondissociated Products	kOH $M^{-1}s^{-1}$	References
Alcanes		$10^{-3}-1$	10^6-10^9	a
Dibromochloropropane	—		1.47×10^8	b
Chloroform	—	0.1	8.5×10^6	c
Carbon tetrachloride	—	0.005	—	d
Olefins		$1-10^5$	10^8-10^{11}	a
Linoleic acid		10^6	10^9	e
Styrene	—	3.10^5		d
1-1 dichloro ethylene	—	1.1×10^2	4×10^9	e
Trichloro ethylene	—	17	10^8	d
Tetrachloro ethylene	—	0.1		c
Aromatic compounds	—	$1-10^3$	10^8-10^{10}	a
Benzene	—	2	5×10^9	d
Chlorobenzene	—	0.75	$1.5-2.10^9$	f
Naphtalene	—	3×10^3	—	d
Nitrobenzene	—	9×10^{-2}	3×10^9	e
Phenolic compounds	$10^{-5}-10^9$	10^3-10^6	10^9-10^{10}	a
Phenol	1.4×10^9	1.3×10^3	5×10^9	f
4.chlorophenol	0.2×10^9	0.6×10^3		d
4.nitrophenol	1.6×10^7	50		d
Pentachlorophenol	—	3×10^5		d
Resorcinol	3×10^5	3×10^5		d
Aldehydes		$1-10$	10^9	a
Acetaldehyde	—	1.5	—	d
Benzaldehyde	—	2.5	—	d
Ketones	—	10^2-1	10^9-10^{10}	a
Acetone	—	0.03	—	
Alcohols		10^2-1	10^9-10^{10}	a
Methanol	—	2.4×10^{-2}	10^9	e

(Continued)

Table 10.4. *(Continued)*

Products or Family of Products	$kO_3\ M^{-1}\ s^{-1}$ Dissociation Products	$kO_3\ M^{-1}\ s^{-1}$ Nondissociated Products	$kOH\ M^{-1}\ s^{-1}$	References
Ethanol	—	0.37	—	d
Cyclopentanol	—	2	—	d
Glucose	—	0.5	2×10^9	e
Carboxylic acids	—	10^{-5}–10	10^7–10^9	a
Formic acid	100	5	—	d
Acetic acid	3×10^{-5}	3×10^{-5}	2×10^8	d
Succinic acid	3×10^{-2}	—	—	d
Glycocollic acid	1.9	0.17	—	d
N-containing organics	0–1	10^2–10^7	10^9–10^{10}	a
NH_4^+	0	1.4×10^5	—	d
Methylamine	—	1.9×10^7	—	d
Dimethylamine	0.13	4.1×10^6	10^{10}	d
Trimethylamine	0.01	3	—	d
Pyridine	0.01	6.4×10^3	—	d
Alanine	3×10^{-3}	9×10^7	9×10^9	e
S-containing organics	—	10–10^5	10^8–10^{10}	a
Ethylmercaptan	—	2×10^5	—	d
Dipropylsulfide	—	2×10^5	—	d

[a]Barker and Jones (1988).
[b]Glaze and Kang (1989).
[c]Francis (1987).
[d]Hoigne et al. (1983).
[e]Sonntag (1989).
[f]Anbar and Neta (1967).

of sulfur as sulfites and thiosulfates occur at trace levels (Le Sauze et al.,1989). As far as mercaptans and sulfides other than H_2S are concerned, the oxidation mechanisms can be written as (Bailey, 1982)

$$RHS \rightarrow R\!-\!S\!-\!S\!-\!R \rightarrow R\!-\!\overset{\displaystyle O}{\underset{}{\overset{\|}{S}}}\!-\!S\!-\!R \rightarrow R\!-\!\overset{\displaystyle O}{\underset{\underset{\displaystyle O}{\|}}{\overset{\|}{S}}}\!-\!S\!-\!R \rightarrow R\!-\!\overset{\displaystyle O}{\underset{\underset{\displaystyle O}{\|}}{\overset{\|}{S}}}\!-\!OH$$

Mercaptan   Disulfide   disulfide oxide disulfide dioxide sulfonic acid

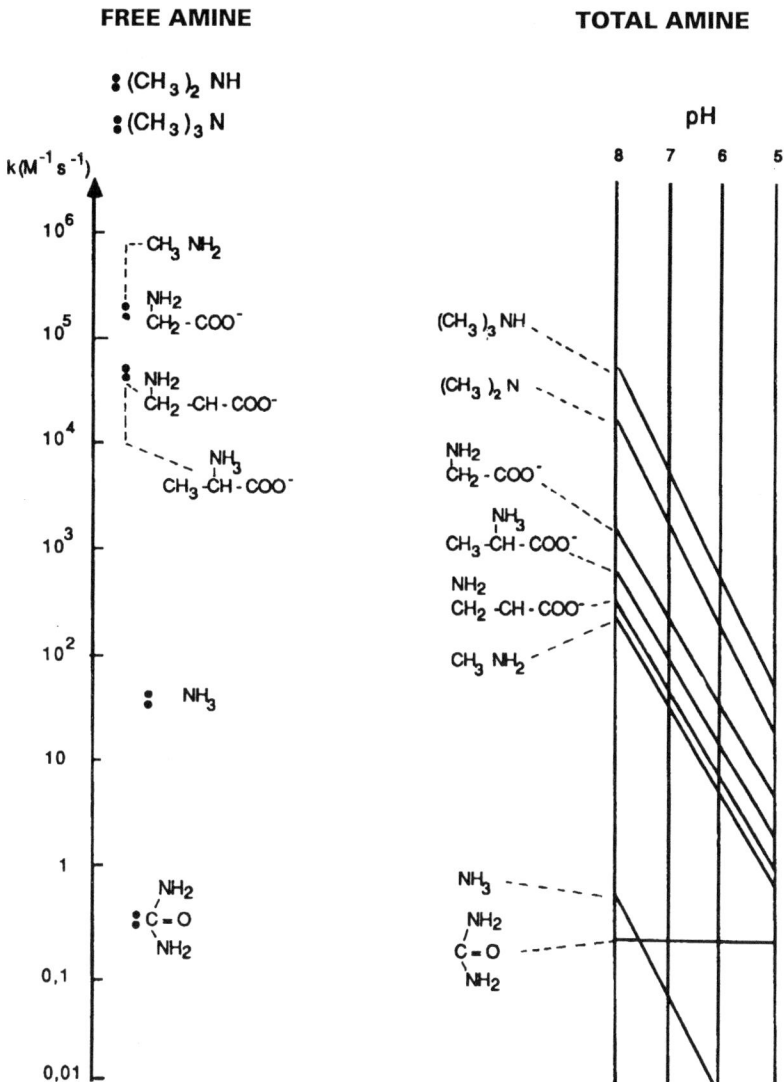

Figure 10.11. Rate constants for amine-ozone reactions after Hoigne and Bader, *Ozone Sci. Eng.* **1**(1),73 (1979).

Hoigne et al. (1985) measured the H_2S oxidation rate. They found that the rate constant amounts to $3 \times 10^4 M^{-1} s^{-1}$ at a pH below 2, then increases by a factor of 10 for each pH unit. Therefore, it depends upon dissociation. For HS^-, it amounts to $3 \times 10^9 M^{-1} s^{-1}$. This very high value explains that the apparent O order kinetics was observed during pilot scale experiments (Le Sauze et al.,1989). Thus, the limiting factor is the supplied ozone and not the ozone-H_2S reaction kinetics.

The consideration of these pilot results suggest that a reasonable operating pH range from 9 to 11 should be applied for H_2S. As far as mercaptans are concerned (CH_3SH), the liquid-gas physical equilibrium is limiting. Therefore, the pH should be as high as possible (equilibrium shift toward ionized species) and the parameters that can modify Henry's constant should be taken into account (solvent and salt effects).

Ozone consumption ranges from 0.8 to 3 M per mole of oxidized H_2S (Figure 10.12). This is due to radical mechanisms and chain reactions (oxidant autodestruction and free radical inhibitors). Ozone can oxidize numerous organic compounds as well. Reaction rates are high with ethylenic hydrocarbons and with some aromatic compounds. They are low with alcohols, aldehydes, ketones, acids, and saturated hydrocarbons. On the other hand, $OH°$ radicals have very high oxidation rates regardless of the nature of the reactant. Some data are shown in Table 10.5.

An example of the use of ozone in deodorization of exhaust gases from wastewater treatment plants is given in Chapter 14.

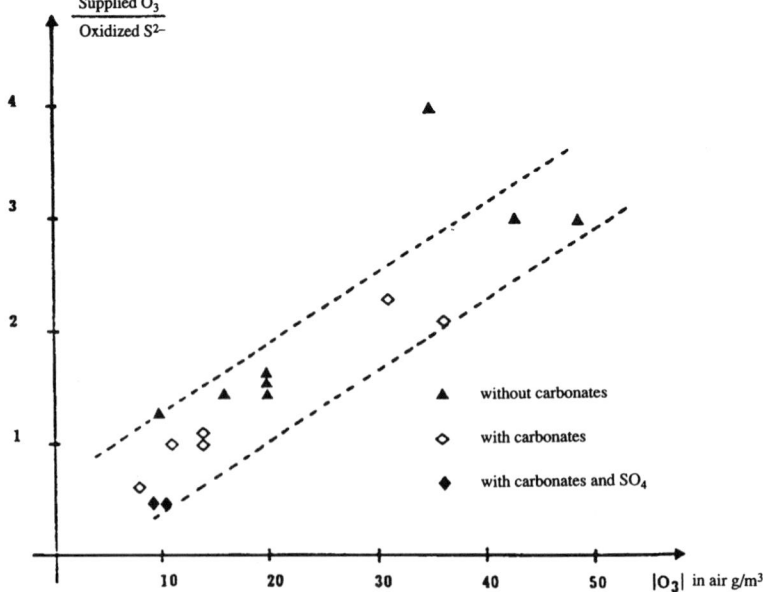

Figure 10.12.   Supplied ozone to oxidized sulfide ratio.

Table 10.5. Comparison of the Effect of Four Oxidizing Agents on Some Typical Odorous Products

	Chlorine			Ozone		
Compound	Operating Conditions	Main By-Products	Reaction Rate Constant	Operating Conditions	Main By-Products	Reaction Rate Constant
NH_3	pH ≈ 5.5	NH_2Cl $NH\,Cl_2$ N_2 NO_3^-	5.1×10^6 M^{-1} s^{-1}	pH > 10	NO_3^-	20 M^{-1} s^{-2}
CH_3NH_2		CH_3NHCl $CH_3N\,Cl_2$	7.3×10^7 M^{-1} s^{-1}	pH = 7	HCHO HCOOH	0.5×10^2 M^{-1} s^{-1}
$(CH_3)_3$ N				pH = 7	$(CH3)_3$ N $-$ O	0.8×10^4 M^{-1} s^{-1}
H_2S	pH < 7.5 pH < 7.5	S⇓, S$_2$O$_3^{2-}$ SO$_4^{2-}$ S⇓, S$_2$O$_3^{2-}$ SO$_4^{2-}$	3.2×10^6 M^{-1} s^{-1}	pH = 2 pH = 10	SO_4^{2-}	3×10^4 M^{-1} s^{-1} 3×10^9 M^{-1} s^{-1}
CH_3SH	pH > 10	$CH_3-S-S-CH_3$ CH$_3$ SO$_3$ H	2.1×10^6 M^{-1} s^{-1}	pH > 9	$CH_3-S-S-CH_3$ CH$_3$ SO$_3$ H	2×10^5 M^{-1} s^{-1}

	Hydrogen Peroxide			KMnO$_4$		
Compound	Operating Conditions	Main By-Products	Reaction Rate Constant	Operating Conditions	Main By-Products	Reaction Rate Constant
NH_3		No reaction			No reaction	
$(CH_3)_3$ N		$(CH_3)_3$ N $-$ O				
H_2S	pH < 6 pH > 7	S⇓, S$_2$O$_3^{2-}$ SO$_4^{2-}$ S⇓, S$_2$O$_3^{2-}$ SO$_4^{2-}$	1.15 M^{-1} s^{-1}	pH < 7.5 pH < 7.5	S⇓, S$_2$O$_3^{2-}$ SO$_4^{2-}$ S⇓, S$_2$O$_3^{2-}$ SO$_4^{2-}$	
CH_3SH	20° 60°	$CH_3-S-S-CH_3$ CH $-$ SO $-$ H	Slow Rapid			

10.4.3.3 Hydrogen Peroxide

Hydrogen peroxide is a powerful oxidizing compound capable of reacting on some odorous products. Its redox potential at pH 0 amounts to 1.77 V. Easy to implement, it is used either in association with a scrubbing operation or directly (e.g., in the case of sewage collectors).

In aqueous conditions, hydogen peroxide does not react with ammoniac. It can react with amines according to the following schematics (Schirmann and Delavarenne, 1979):

$$R'C = N - OH$$
$$\text{Oxime}$$

$$\nearrow$$

$$RNH_2$$

$$\searrow$$

$$R - NHOH \longrightarrow R - NO \longrightarrow R - NO_2$$

Hydroxylamine   nitroso-derivative   nitro-derivative

or

$$R_3N \longrightarrow R_3 N - O$$
$$\text{N-oxide}$$

It is its effect on sulfurous compounds that drew the attention. H_2S oxidation proceeds through several intermediate sulfur oxidation states and the following equations can be written (Le Goallec et al.,1990b):

$$HS^- + H_2O_2 + H^+ \longrightarrow S \Downarrow + 2 H_2O$$
$$2HS^- + 4 H_2O_2 \longrightarrow S_2O_3^{2-} + 5 H_2O$$
$$S_2O_3^{2-} + 2 H_2O_2 \longrightarrow 2 SO_3^{2-} + H_2O + 2H^+$$
$$SO_3^{2-} + H_2O_2 \longrightarrow SO_4^{2-} + H_2O$$

The apparent H_2S disappearance kinetic rate is of the first order with respect to sulfides and hydrogen peroxide and the constant rate is about $1.15 \text{ M}^{-1} \text{s}^{-1}$.

The effect of H_2O_2 was especially pointed out in the purification of steam loaded with H_2S and in air purification. Numerous patents have been filed for pieces of equipment designed to use this product (O'Neill and Kibbel, 1976; Chelu, 1984; Woertz, 1979; Mehl,1985).

Hydrogen peroxide is also used to fight odors that develop in wastewater networks. It oxidizes both organic compounds and sulfur derivatives. Tests carried out on the La Baule (France) network showed overall consumptions of 12 M H_2O_2 per mole of oxidized H_2S for an injection ahead of a pipe portion (Le Goallec et al., 1990a). If injected ahead of the pipe exit, the consumption drops to 6–7 M per mole (Le Goallec et al., 1991). These results must be compared with the theoretical consumption (4 M/M), and to laboratory scale experiments showing an instantaneous consumption of 2 M/M. This is due to the formation of less oxidized intermediate species.

Finally, it should be noted that hydrogen peroxide is often used with metallic catalysts in order to generate very reactive OH^0 radicals.

10.4.3.4 Other Oxidants and Conclusions

Other oxidants can be used to improve odorous compound scrubbing: potassium permaganate, peroxide compounds, etc. However, the coupling of an oxidant to a scrubbing operation should be beneficial to the overall process; for example, it should do the following:

- Improve the purification yield (this is that much more true if the odorous compound - oxidant reaction rate constant is high)
- Reduce the size of the equipment
- Improve the quality of the aqueous effluent (this depends on the nature of the by-products; it should be noted that hydrogen peroxide and ozone do not add any specific pollution)

Table 10.5 shows the effects of chlorine, ozone, hydrogen peroxide, and potassium permanganate on several major odorous compounds. It should also help in the solution of the preceding questions.

10.5 Examples of Use and Implementation

The treatment of odors by physical transfer or simultaneous physical transfer and chemical reactions (acid-basic scrubbing or oxidizing scrubbing) requires the pollutant to be transferred from the gaseous to the liquid phase. Therefore, it must be somewhat soluble in the liquid phase.

Three cases should be consdered:

1. Water soluble pollutants
2. Pollutants hardly soluble in water, but reacting with a water soluble compound
3. Practically water insoluble compounds but soluble into other liquids

10.4.1 Water Soluble Pollutants

In this category, the most frequently encountered pollutants are:

1. Odorous nitrogenous pollutants: ammoniac, amines
2. Some organic pollutants such as acids or alcohols

The absorption step could be carried out using water, but the high volatility of these compounds necessitates the use of large amounts of water. In this case, a simultaneous absorption and acid-alkaline reaction should be used when possible, for the treatment of nitrogenous compounds or acids, for example.

Examples of acidic treatments. Nitrogenous compounds are treated by acid scrubbing. The frequent use of sulfuric acid leads to the formation of sulfates usable for agricultural enrichment. The reaction being almost instantaneous, transit rates through the tower can be very high (7,000–10,000 kg m^{-2} h^{-1} for the gas).

The liquid flowrate depends on the type of packing used as well as the height of the transfer units which varies from 0.2 to 0.6 m, also according to the type of packing. The efficiency depends on the height of packing. It may be very high and generally exceeds 99%.

Examples of alkaline treatment. Organic acids, often very odorous, can be removed as sodium salts through a soda treatment. The reaction kinetics are fast and gas mass flowrates from 6,000 to 8,000 kg m^{-2} h^{-1} can be used. The operation efficiency can be high (99% and higher).

10.5.2 Hardly Soluble Reactive Pollutants

This is the case of sulfurous compounds. The process involves

1. An alkaline reagent (soda) for hydrogen sulfide
2. An oxidizing alkaline reagent for mercaptans

The reaction kinetics can be slow and gas mass flowrates from 2,000 to 5,000 kg m^{-2} h^{-1} should be used. The transfer unit heights are more important and vary from 0.6 to 1.5 m according to the type of packing material.

High yields require significant packing heights. Yields ranging from 80 to 99% are obtained on industrial units.

10.5.3 Water Insoluble Pollutants

This category includes most of the hydrocarbons. When mass transfer technology is competitive, solvent scrubbing is used. The solvent must have

1. A high solvent potential with respect to the pollutant
2. A low vapor pressure at the operating temperature so as not to be stripped by the gaseous flow

The determination of the pressure drop cannot be approached with the preceding data given for the air-water system.

Preliminary pilot tests are always necessary. The pollutants dissolved into the solvent can subsequently be separated from it by distillation if this technology is economically feasible. In some cases, the pollutant-solvent mixture can be recycled. Tar vapor scrubbing by oils makes this recycling possible. In some other cases, the mixture can be reused as fuel.

10.5.4 Some Examples of Treatment Combined with Scrubbing

The use of scrubbers in a series is the current practice. Most often, three cases are encountered:

1. The treatment of a pollutant mixture requiring different treatments
2. The desired recovery of a volatile product
3. The treatment of a dusty gas susceptible to plug up the scrubber, thereby requiring a preliminary treatment

10.5.4.1 Treatment of a Pollutant Mixture

The most frequent case concerns gaseous exhaust from the fermentation or putrefaction of sulfur or nitrogen containing organic waste. The treatment proceeds through a series of 2 or 3 steps: an acid treatment step for the removal of nitrogenous compounds followed by an oxidizing alkaline treatment step, or an oxidation treatment followed by an alkaline treatment.

The design of each tower remains identical as well as the reagent consumption, but pressure drops, hence energy consumptions add up to reach values from 200 to 300 mm water height. These combined treatments concern

1. Meat quartering and tallow processing
2. Urban wastewater treatment plants
3. Organic fertilizer manufacturing
4. Etc.

The pollutant concentrations range from 1 to 10 mg/m³ for nitrogenous compounds and from 10 to 30 mg/m³ for sulfurous compounds in the case of wastewater treatment plants; they are 500 mg/m³ for nitrogenous compounds and 500 mg/m³ for sulfurous compounds in the case of noncondensable products from meat quartering workshop gases.

10.5.4.2 Volatile Compound Recovery

The recovery of a compound from the point of view of reusing purposes requires that the product should be in a solution that is as concentrated as possible (the maximum concentration corresponds to the liquid concentration in equilibrium with the inlet gas when the liquid is recycled). The vapor pressure in contact with the concentrated solution does not allow for a single step high removal.

A multistage operation (two, three, or more stages) makes it possible to reach high gaseous pollutant removal while yielding concentrated liquid solutions. This type of arrangement is found in the treatment of relatively concentrated gases issued from the reactors of the chemical industry. Ammoniac or acetic acid vapors are treated in this way.

10.5.4.3 Dusty Gas Treatment

Packed towers may be plugged up by dusty gases and preliminary dust removal may be essential. If the gas is not too humid and if the particles are not too plugging, sleeve filters can conveniently be used upstream from the tower, as this type of filter is very efficient for particle removal.

In the opposite case, a preliminary treatment using wet dust removers (venturi for instance) is absolutely necessary.

10.5.5 Scrubber Operation, Trouble Shooting, Maintenance, and Fine Tuning

Besides dust plugging, several problems can disturb proper packed tower operation. Plugging may come from crystallization due to a too high salt concentration in the liquid. An uncontrolled liquid flowrate increase may flood the column and prevent the gas circulation. Regular monitoring and maintenance are therefore highly recommended.

Continuous monitoring should be applied to

1. Pressure drop over the packing and the entrainment eliminator (preferably separately so as to better locate the trouble source)
2. Pressure and flowrate on the recycle pump
3. Reagent concentration through pH monitoring in the case of acid-alkaline treatment or Rh in the case of an oxidizing treatment or possibly the liquid density
4. Liquid level by a probe (variable in the case of an water unsaturated air treatment leading to evaporation).

These measurements can be easily automated, leading to a more accurate operation of the equipment.

The identified defects should be rapidly corrected as follows:

- Cleaning in the case of an increase in the pressure drop after the checking of the liquid flowrate
- Water makeup to reach nominal level
- Reagent addition
- Exhausted product removal
- Etc.

10.6 Conclusion

Odor treatment by scrubbing, when feasible, brings an efficient means to air quality maintenance. This process should be considered any time there is an odor problem. It should be compared to the other processes described in this book, on the basis of efficiency and cost.

The main advantage of scrubbing resides in its ability to reach very high yields for selected problems. Its major drawback comes from the creation of an associated liquid pollution that is not always easily eliminated.

Liquid recycling should always be envisaged.

These combined advantages and drawbacks can bring an economically and technically valuable answer by providing scrubbing as a powerful tool for the resolution of specific odor problems.

10.7 Notations

A	theoretical exchange surface area
A	column cross section (m^2)
a	actual exchange surface area m^2 m^3
C	concentration in the liquid (M/m^3)
C^*	concentration in the liquid at the interface (M/m^3)
C_E	concentration in the liquid in equilibrium with the gaseous phase
C_T	total concentration (M/m^3)
D, D_G, D_L	diffusivity in the gaseous phase, in the liquid phase (m^2/s)
D_C	column diameter (m)
d	packing particle diameter (m)
E	transfer acceleration coefficient
F	packing factor
G, L	massic flowrates (kg/s)
G', L'	massic flowrates per unit surface area ($kg/m^2 s$)
G_m, L_m	molar flowrates (M/s)
G'_m, L'_m	molar flowrates per unit surface area ($M/m^2 s$)
g	gravitational acceleration
H	Henry's constant ($atm/m^3/M$)
H'	Henry's constant (atm/M fraction)
H''	Henry's constant (dimensionless)
H_G, H_L	height of a film transfer unit
H_{OG}, H_{CL}	height of an overall transfer unit
Ha	Hatta number
J	molar flux with respect to a plane where no mass transfer occurs ($M/s\ m^2$)
K	liquid/vapor equilibrium constant

K_1, K_2	equilibrium constants
k_G, k_L	film transfer coefficients
K_G, K_L	global transfer coefficients
k	kinetic constant of a chemical reaction
m	slope of the equilibrium curve
N	transferred amount per unit transfer surface area (M/m^2 s)
N'	transferred amount (M/s)
N_G, N_L	number of transfer units (film)
N_{OG}, N_{OL}	number of transfer units (overall)
P_e, P_s	inlet, outlet total pressure
p_e, p_s	inlet, outlet partial pressure
p^*	interfacial partial pressure
p^E	equilibrium partial pressure
R	ideal gas constant
r	reaction rate
T	temperature °C or K
t	time (s)
u	velocity (m/s)
U_{SL}, U_{SG}	specific velocities (m/s)
U_E, U_T	flooding and operational velocities (m/s)
V	reactor volumn (m^3)
x_e, x_s, x	liquid phase molar fractions
x_e^*, x_s^*, x^*	interfacial liquid phase molar fractions
x_{eE}, x_{sE}, x_E	equilibrium liquid phase molar fractions
y_e, y_s, y	gaseous phase molar fractions
y_e^*, y_s^*, y^*	interfacial gaseous phase molar fractions
y_{eE}, y_{sE}, y_E	equilibrium gaseous phase molar fractions
Z	column height
z	transfer direction
δ_G, δ_L	film thickness, on the gas side and the liquid side (m)
β	liquid volumic fraction = $(1 - \varepsilon)$
ε	void fraction
ρ_L, ρ_G	volumetric masses kg/m^3)
μ_L, μ_G	viscosities (Pl)

Bibliography

Abe, K., Hag Wara, J., Machida, W. (1979). Conditions for oxidation treatment of ammonia and trimethylamine by sodium hypochlorite: Taili Oscn Gakkaishi 14:11–12, 474–478. After CA 93 (26) 244496 c.

Abe, K., Machida, W. (1982). An effective process for hypochlorite treatment of malodor in waste gas. *Bull. Chem. Soc. Jpn.* **1**(11), 3677–88.

Abe, K., Uehara, Y., Yamada, H., Kimora, M. (1983). Treatment of offensive odor substances by hypochlorous acid and hypochlorite; *Kogai to taisaku* 19, (4), 337–340. After CA 99(10) 76024 a.

Anbar, M., Neta, P. (1967). A compilation of specific biomolecular rate constants for the reactions of hydrates electrons hydrogen atoms and hydroxyl radicals with inorganic and organic compounds in aqueous solution. *Int. J. Appl. Radiat. Isotopes* **18**, 493.

Bailey, P. S. (1982). *Ozonation in Organic Chemistry*, Vol. 11, Chap. VII-4. Academic Press, New York.

Barker, R., Jones, A. R. (1988). Treatment of malodorants in air by UV/Ozone technique. *Ozone Sci Eng.* **10**(4), 404.

Bonnin, C. (1990). Centre de recherche de la compagnie générale des eaux (personal communication).

Cadena, F., Peters, R. W. (1988). Evaluation of chemical oxidizers for hydrogen sulfide control. *J.W.P.C.F.* **60**, 7, 1259–1263.

Chelu, G. (1984). Procédé de désodorisation d'air pollué. European Patent 0057624 of 01/08/1984.

Colburn, A. P. (1939). The simplified calculation of diffusional processes. General consideration of two film resistance. *Trans. Am. Inst. Chem. Eng.* **35**, 211.

Copigneaux, P., 1986. Distillation, absorption, les colonnes garnies. *Les Techniques de l'Ingénieur J.* 2626.

Coulson, J. M., Richardson, J. F. (1980). *Chemical Engineering*, Vol. 2. Pergamon Press, New York.

Danckwerts, P. W. (1951). Significance of liquid film coefficients in gas absorption. *Ind. Eng. Chem.* **43**, 1460.

Doré, M. (1989). *Chimie des oxydants et traitement des eaux*, Lavoisier, Technique et Documentation, Paris.

Eckert, J. S. (1961). Design techniques for designing packed towers. *Chem. Eng. Prog.* **57** (9), 54.

Elmghari Tabib, M., Laplanche, A., Venien, F., Martin, G. (1982). Ozonation of amines in water. *Water Res.* **16**, 223.

First, M. W., Schilline, W., Govan, J. H., Quinby, A. H. (1974). Control of odors and aerosols from spent grain dryers. *Proc. Spec. Conf. Control. Technol. Agric. Air Pollut.* 133–146, Pittsburgh.

Francis, P. D. (1987). Oxidation by UV and ozone of organic contaminents dissolved in deionized and raw mains water. *Ozone Sci. Eng.* **9**(4), 369–390.

Glaze, W. H., Kang, J. W. (1989). Advanced oxidation processes. Test of a kinetic model for the oxidation of organic compounds with ozone and hydrogen peroxide in a semi batch reactor. *Ind. Eng. Chem. Res.* **28**(11), 1580–1587.

Higbie, R. (1935). The rate of absorption of a pure gas into a still liquid during short period of exposure. *Trans. A.I.Ch.E.* **35**, 365.

Hikita, H., Asai, S. (1964). Gas absorption with *m, n*th order irreversible chemical reaction. *Intern. Chem. Eng.* **4**, 332.

Hoigne, J., Bader, H. (1979). Ozonation of water, selectivity and rate of oxidation of solutes. *Ozone Sci. Eng.* **1**(1), 73.

Hoigne, J., Bader, H., Haag, W. P., Staehelin, J. (1983). Rate constant of reactions of ozone with organic and inorganic compounds in water, part II: Dissociating organic compounds. *Wat. Res.* **17**, 185–194.

Hoigne, J., Bader, H., Haag, W. P., Staehelin, J. (1985). Rate constant of reactions of ozone with organic and inorganic compounds in water part III. *Wat. Res.* **19**, 8, 993–1004.

Laplanche, A., Wei, Y., Martin, G., Langlais, B. (1987). A process of washing and ozonation to deodorize an atmosphere contaminated by amines. *Proceedings of 8th Ozone World Congress,* Zurich, F11-F25.

Laplanche, A., Le Sauze, N., Wei, Y., Martin, G., Langlais, B., Paillard, H. (1988). 8èmes Journées information Eaux, Poitiers 28-30, September, 1988.

Le Goallec, O., Traineau, N., Martin, G., LaPlanche, A. (1990a). Elimination des sulfures par l'eau oxygénée dans le réseau d'assainissement de La Baule; *T.S.M.* (in press), 1990a.

Le Goallec, O., Laplanche, A., Martin, G. (1990b). Action du peroxyde d'hydrogène sur l'hydrogène sulfuré: Application aux réseaux d'assainissement. *Rev. Sci. l'Eau* (in press), 1990b.

Le Sauze, N., Laplanche, A., Martin, G., Paillard, H. (1989). A process of washing and ozonation to deodorize an atmosphere contaminated by sulfides. *Proceedings, Wasser Berlin,* P(4). 1, 12, April 1989.

Martin G. (1979). *Le problème de l'azote dans les eaux;* Lavoisier, Techniques et Documentation, Paris.

Mehl, E. (1985). Ver fahren zur Reimigung von Gasen aux Pyrolyseanlagen von Abfallstoffon; Deutsch Patertat DE 34 12581 A1 of 24/10/1985.

Morris, J. C. (1967). Kinetics of reaction between aqueous chlorine and nitrogen compounds. In *Principles and Applications of Water Chemistry,* edited by S.P. Faust and J.V. Hunter. John Wiley, New York.

Morris, G. A., Jackson, J. (1953). *Absorption Towers.* Butterworths, London.

O'Neill, E. T., Kibbel, W. (1976). Animal waste odor treatment. U.S. Patent 3966450 of 29 June 1976.

Perry, R. H., Green, D. W. (1984). *Perry's Chemical Engineers' Handbook,* 6th ed. McGraw Hill Book Company, New York.

Poncin, J., Le Cloirec, P., Martin, G. (1984). Étude cinétique de la chloration de la méthylamine par l'hypochlorite de sodium en milieu dilué. *Environ. Technol. Let.* **5**, 263–274.

Ponomarev, V. A., Yacobi, V. A., Esip, V. P., Sukhanov, S. V., Sharagoni, N. M. (1979). After CA 90 (16) 126643 P.

Rachamandran, P. A., Sharma, M. V. (1980). Simultaneous absorption of two gases. *Trans. Inst. Chem. Engrs.* **58**, 242.

Saunier, B. M., Sellier, R. E. (1979). The kinetics of breakpoint chlorination in continuous flow system. *J.A.W.W.A.* 164–172.

Schaue, A., Hockun, F., Thiele, W., Goette, H. (1985). The absorption of odors from the waste gases from a carcass processing plant. *Luft - Kaeltetech.* **21**(2), 94–97.

Schirmann, J. P., Delavarenne, S. Y. (1979). *Hydrogen Peroxide in Organic Chemistry,* Chap. 7. Edition et Documentation Industrielle, Paris, pp. 147–154.

Sherwood, T. K., Shipley, G. H., Holloway, F. A. L. (1938). Flooding velocities in packed columns. *Ind. Eng. Chem.* **30**, 765.

Schwab, H., Kaschke, V. Y. (1975). Slime and odor elimination in process water of the paper industry. *Papier* **29**, 10A, 43–51.

Sens, M. L., Le Sauze, N., Laplanche, A., Langlais, B. (1990). Effets des anions sur la décomposition de l'ozone dans l'eaux. *Rev. Sci. Eau* **3** (3).

Singer, P. C., Zilly, W. B. (1975). Ozonation of ammonia in wastewater. *Wat. Res.* **9**, 127.

Sonntag, C. V. (1989). The chemistry behind the upgrading of water with UV light. *Proceedings Wasser* Berlin, Vol. 1.1

Struppler, N., Husson, G. P. (1974). Étude sur la chloration des eaux chlore résiduel et formation de chloramines. *J. Fr. Hydro* **15**, 31–46.

Toor, H. L., Marchello, J. M. (1958). Film-penetration model for mass and heat transfer. *A.I.Ch.E.J.* **4**, 97.

Trambouze, P., Van Landeghem, H., Wauquier, J. P. (1984). *Les Réacteurs Chimiques.* Technip., Paris.

Van Krevelen, D. W., Hoftijzer, P. J. (1948). Kinetics of gas liquid reactions. Part I: General theory. *Recueils Trav. Chim. Pays Bas.* **67**, 563.

Waltrip, F. (1985). Elimination of odor at six major waste treatment plant. *J.W.P.C.F.* **57** (10), 1027–1032.

Whitman, W. G. (1923). A preliminary confirmation of the two film theory of gas absorption. *Chem. Met. Eng.* **26**, 149.

Woertz, B. B. (1979). Hydrogen sulfide abatement in geothermal steam, US Patent 4 151 260 of 24/04/1979; US Patent 4 163 044 of 31/07/1979.

Yurteri, C., Gurol, A. D. (1988). Ozone consumption in natural waters: effects of background organic matter, pH and carbonate species. *Ozone Sci. Eng.* **10**, 277.

11

Treatments with Gas-Solid Transfer-Adsorption

P. Le Cloirec, G. Dagois, and G. Martin

11.1 Introduction

Adsorption of molecules on solid surfaces is a very common phenomenon. The implementation of this phenomenon gave rise to a number of industrial processes for the transformation or the treatment of fluids including heterogeneous catalysis and purification of liquids or gases. The latter application is developed in this chapter, specifically in the practical case of the removal of odorous substances by adsorption.

After having given the general principles of adsorption, the materials used in the purification of gases by gas to solid transfer will be described. The use and the characteristics of activated carbon will be especially developed. Additionally, other adsorbants presently under development (zeolites, resins) will be presented as well. The performances and the modeling of reactors will be presented prior to examples of industrial implementation.

11.2 The Basic Concepts of Adsorption

Adsorption can be simply considered as an increase in the concentration of a substance at the fluid-solid interface. This transfer as well as the stabilization of the molecule are carried out through different steps that are described in this paragraph.

11.2.1 General Consideration About Transfer and Adsorption in a Porous Material

The overall adsorption concept of a compound into a porous material can be simulated through the diagram presented in Figure 11.1. It includes the steps (Kast and Otten, 1987) described in Fig. 11.1. Steps 1 and 2 of Fig. 11.1 correspond to the odorous air transport through the granular bed. The boundary layer corresponds to a region where the adsorbable molecule concentration is high. The thickness of this layer depends on the turbulence of the gaseous stream entering the reactor. This concept will be detailed in the section dealing with the modeling of the overall process (section 11.5). The transfer in the material (step 3) corresponds to a diffusion in the porous volume as well as a diffusion on the internal surface of the material. This relatively new concept has been developed in numerous studies, especially those concerning liquids (Weber and Liu, 1980, Fettig and Sontheimer, 1987). Adsorption in its strictest sense will be developed in section 11.2.2. The exothermal character of adsorption is often neglected. However, in the case of very concentrated gases (blowdowns, storage vents, . . .), this heat generation can be high locally; the temperature can reach values high enough to ignite flammable adsorbing materials (Ladousse, 1989).

11.2.2 The Molecule-Adsorbing Solid Interactions

This adsorption phenomenon corresponds to step 4 of the model previously described. It is generally accompanied by an orientation of the molecules at the interface and, therefore, by a decrease in the molecular disorder

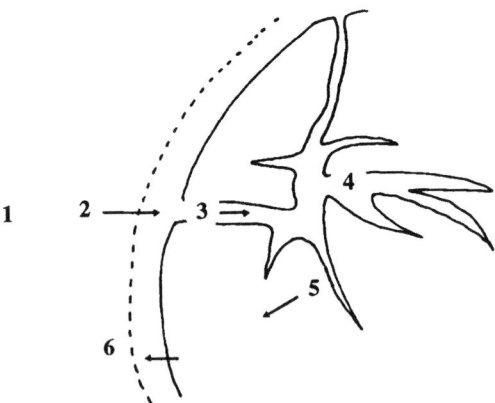

Figure 11.1. The different steps of the adsorption of a molecule into a porous material. **1.** Odorous molecule in the gaseous phase. **2.** Mass transfer through the material boundary layer. **3.** Molecule transfer in the porous material. **4.** Surface adsorption with possible heat generation (exothermal phenomenon). **5.** Thermal conduction through the porous material. **6.** Thermal conduction through the boundary layer.

(Gradsztajn et al.,1970,1971; Le Cloirec et al.,1988). A system is thereby obtained with a lower free energy and, hence, a higher stability. The interactions between the solid and the molecules can be either chemical or physical.

11.2.2.1 Physisorption

The interaction phenomena between the solid surface and the adsorbable molecules are based on a set of attractive–repulsive electrostatic forces (through free charges). They can be due to ionic forces, dipole-dipole interactions, dispersion forces, or London–van der Waals' forces (London,1930) and to hydrogen bonds (Stumm and Morgan,1981). Generally, these adsorbate-surface bonds are weakly exothermal and reversible because of the low interaction energy value. 5–40 kJ mol^{-1} are assigned to van der Waals bonds, whereas dipole-dipole interactions have an energy lower than 5 kJ mol^{-1} (Montgomery, 1985).

11.2.2.2 Chemisorption

In this case, the adsorbate-solid bonds should be considered to be close to covalent bonds. They can be site specific or functional surface group specific (Donnet,1970); these bonds produce an important amount of heat and they are irreversible because of the high energy of the bond. For a covalent bond, the energy is higher than 40 kJ mol^{-1}. It is assumed that a monolayer is formed at the surface of the solid (Montgomery, 1985). Of course, this type of interaction is a function of the chemical nature of the adsorbant surface. As developed further, the treatment or the impregnation of the solid can enhance this type of chemical interaction.

11.2.3 Adsorption Isotherms

A relatively simple method to reveal the adsorptive potential of a material with respect to a pollutant consists of establishing the adsorption isotherm. This is the amount of adsorbed material as a function of the pollutant concentration (or the partial pressure) in the gas to be treated. An example of this type of isotherm is presented by Figure 11.2. The adsorption capacity is not only a function of the concentration but also of the nature of the compound to be adsorbed. Generally, it is assumed that a compound is more adsorbable if its molecular weight is higher (Cheremisinoff and Ellerbush,1978; Legros,1980).

Adsorption isotherm modeling applies to diluted concentrations, hence it is especially suitable for the study of the adsorption of odorous molecules occurring at very low concentrations (cf. Chapter 9). A simple approach

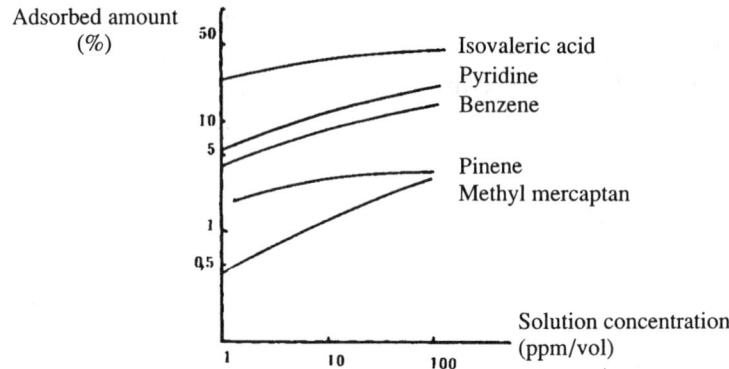

Adsorbed amount (%)

Isovaleric acid
Pyridine
Benzene

Pinene
Methyl mercaptan

Solution concentration (ppm/vol)

Figure 11.2. Adsorption isotherm for different odorous molecules (Waller, 1974).

consists of assuming that the adsorption kinetic rate is a function of the adsorbing surface. The written expression of this assumption is

$$A + \sigma \rightleftarrows A\sigma$$

where A represents the adsorbing molecule and σ the porous solid surface. The equation representing the adsorption process can then be written as

$$\frac{dq}{dt} = K_1 C (q_m - q) - K_2 q \tag{1}$$

where C is the concentration (or pressure) of the solute, q is the amount of adsorbed compound, q_m is the maximum amount of adsorbed compound (saturation), K_1 and K_2 are the forward and backward first-order kinetic constants.

At equilibrium, the Langmuir equation is then obtained:

$$q_e = a\, q_m\, C_e / (1 + a\, C_E) \tag{2}$$

where $a = K_1 / K_2$. This equation is not applicable on a wide range of concentrations. Redlich and Petterson (1959) suggested an empirical modification of Eq. (2):

$$q_e = a\, q_m\, C_e / (1 + a\, C_e^{\gamma}), \text{ with } \gamma < 1 \tag{3}$$

More recently, Baudu et al.(1990) suggested an equation allowing the description of the adsorption isotherm over a wide range of equilibrium concentration ($1 < C_e$ in ppm $< 1{,}000$):

$$q_e = a\, q_m\, C_e^{\alpha} / (1 + a\, C_e^{\beta})$$

with α and $\beta < 1$.

Table 11.1. Influence of the Humidity on the Adsorption Capacity of an Activated Carbon with Respect to Odorous Molecules

Material Capacity (%)

Pollutant	Concentration mg/m³	Treated Air Humidity (20%)		
		5 g/m³ (25%)	11 g/m³ (60%)	17 g/m³ (90%)
Acetaldehyde	80	0.23	0.17	0.13
	1050	2.10	1.80	1.50
Trimethylamine	100	10	4	1.5
	1000	16	9	7
Methyl mercaptan	100	19	9	3.5
Hexane	200	10	9	5
	1700	16	13	8

The Freundlich equation gives a very good description of the isotherms:

$$q_e = K C_e^{1/n} \tag{4}$$

The adsorbed amount (q) is a function of the equilibrium concentration and of different parameters as well. The isotherm equation is relative to an adsorbant. Competition may occur, as shown by the Table 11.1, which gives the influence of the humidity on the adsorbtion of odorous molecules. High values of humidity tend to decrease the adsorption capacity. These few figures show the importance of operating with a gas as dry as possible in order to use the adsorbant at its maximum adsorbing capacity (Perret, 1982).

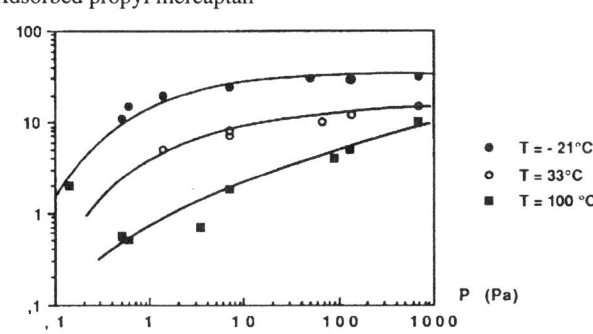

Figure 11.3. Influence of temperature on the adsorption capacity.

The size of the particles is also an important factor in adsorbtion, since the solid-gas interface is essential. A reduction in the size of the adsorbant particles favors the adsorption. Boki and Tanada (1977; 1983) showed that the porosity and the pore diameter make it possible a selection of the adsorbed molecules. H_2S, for instance, is trapped in the smallest micropores whereas methyl mercaptan is adsorbed in larger micropores.

The adsorbtion measurements must be performed at constant temperature. As described by Waller (1974), the adsorption capacity varies as an inverse function of temperature (Figure 11.3).

11.3 The Main Adsorbing Materials

11.3.1 Activated Carbon

11.3.1.1 Characteristics

Activated carbon is a product manufactured from mineral or organic hydrocarbons. It is characterized by a very high specific surface area as well as a high porosity. The pores have diameters ranging from a few angstroms to a few tens of thousands angstroms. Additionally, activated carbon presents an opened or closed superstructure. For an opened superstructure, the channels are approximately a few microns in diameter.

Two main processes are used in the manufacturing of activated carbon: the physical process and the chemical process.

11.3.1.1.1 "Physical" Process

In this process, the raw material is charred prior to its activation. It can be described as follows:

	Amounts
Raw material carbonization $T = 500$–600 °C $t = 5$–6 h	1000
Charred material	250
Activation Steam CO_2 $T = 800$–1000 °C $t = 24$–72 h Activated carbon	100

To these different processes, granulating, sieving, and conditioning are added. These operations remove undesirable elements (ashes, dust), thereby decreasing the final yield to 5.1%.

11.3.1.1.2 "Chemical" Process

In this process, carbonization and activation are simultaneous. It can be described as follows:

	Amount
Raw material charring/activation	1,000
$ZnCl_2$ or P_2O_5	
$T = 400–600 \degree C$	
$t = 5–24$ h	
Activated carbon	400

Sieving and conditioning should be then performed, decreasing the final yield to approximately 25%.

The main difference between the two processes relies on the value of the final yield:

- 5% in the case of the physical process
- 25% in the case of the chemical process.

However, other differences exist in the structures, which will be discussed later in this chapter.

Appended treatments must be added to the two main processes (physical or chemical).

Agglomeration. Activated carbon can be naturally granulated. This is the case when the raw material is resistant enough and directly yields granulated products whose mechanical properties are satisfactory. In the opposite case, granulation must be implemented. This process consists of the addition of a binder to the raw material prior to its carbonization. This is the case of peat and mineral coal.

Impregnation. This process is implemented after the carbonization step. It consists of the exposure of the carbonized material to organic or mineral products. This makes it possible to combine physisorption and chemisorption, hence making the activated carbon more selective or more efficient in the adsorption of a target pollutant or family of pollutants. Impregnation with an organic compound has been developed (Le Cloirec and Martin, 1983) in order to improve the adsorption of primary amines and ammoniac as well as H_2S.

11.3.1.2 Main Characteristic Parameters

In order to characterize an activated carbon, the following parameters must be known:

1. Its specific surface area
2. Its pore volume
3. Its physical characteristics
4. Its surface chemical complexes

11.3.1.2.1 Specific Surface Area

The specific surface area is experimentally determined by applying the Brunauer-Emmet-Teller model (or BET method):

$$\frac{P}{V_a (P_0 - P)} = \frac{1}{V_m C} + \frac{(C-1)}{V_m C} \frac{P}{P_0} \tag{5}$$

where P is the equilibrium pressure, P_0 is the saturation pressure, V_a is the adsorbed volume at equilibrium pressure, V_m is the adsorbed volume necessary to reach a monolayer in the porosity, and C is the constant related to the energy difference between liquefaction and adsorption.

The most commonly used operating gas is nitrogen, which covers 4.39 m^2/mL of adsorbed gas. If P/V_a (P_0-P) is plotted as a function of P/P_0, a straight line is obtained for $0.05 < P/P_0 < 0.35$. This gives access to the value of V_m and, hence, to the specific area.

11.3.1.2.2 Pore Volume

The pore volume is determined by the penetration of mercury at a known pressure. The pressure and the pore radii are related through the following equation:

$$R = 2\gamma \cos \alpha/P$$

where R is the pore diameter (in angstroms).

For mercury, the surface tension g amounts to 480 dynes/cm and the contact angle is close to 140° for solids. This equation makes it possible to follow not only the total pore volume, but also the pore diameter in which the mercury penetrates. Figure 11.4 exemplifies this principle for two activated carbons:

1. Picatif: coconut based, "physically" processed
2. Picaflo : wood based, "chemically" processed

The considerable difference in the pore distribution function should be noted, the physical processing hardly yielding any macropores (pore radii $> 10^4$Å) and the chemical processing not giving any mesopores (pore radii ranging from 6×10^2 to 8×10^3 Å). In both cases, the pore diameter partition function is flat in the corresponding diameter ranges.

11.3.1.2.3 Simple Indexes Related to the Applications by Activated Carbon Manufacturers

These two parameters are not sufficient to know the fixation capacity of the studied material with respect to the pollutants. This is the reason for the

Pore volume in %

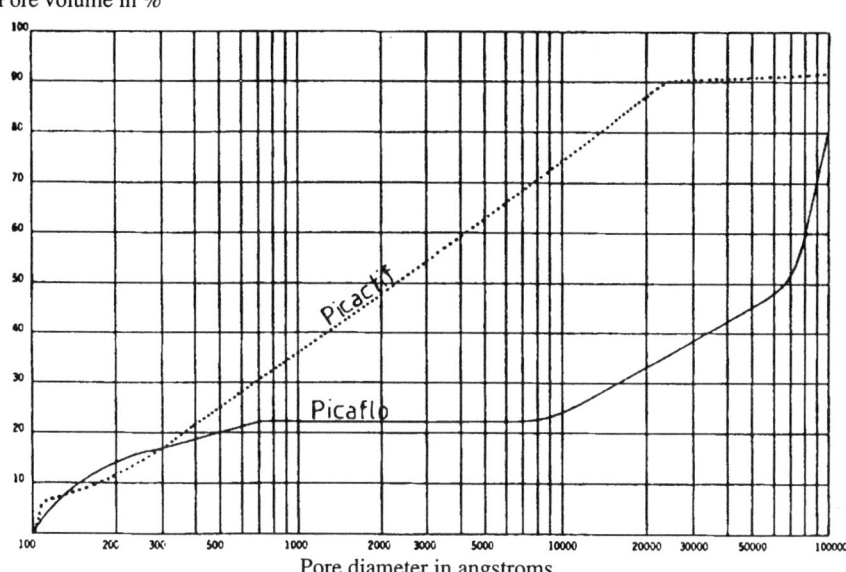

Pore diameter in angstroms

Figure 11.4.   Porosity distribution.

development of simple indexes related to the applications by activated carbon manufacturers, in accordance with end users, such as

1. Carbon tetrachloride adsorption/retention capacity
2. Benzene adsorption capacity
3. Butane adsorption/desorption capacity

All these tests are performed in gaseous phase, by circulating an air stream containing the studied pollutant through a temperature controlled granulated activated carbon bed. The adsorption capacity is expressed as mass of product per mass of adsorbant.

11.3.1.2.4 Physical Characteristics

Besides the adsorption characteristics, other parameters are determinant in activated carbon implementation.

- *Packed bed density:* This characteristic determines the mass of adsorbant per unit volume of adsorber.
- *Granulometry or mean particle diameter:* this is determinant in the adsorption kinetics, as well as in the adsorber headloss.
- *Hardness:* Activated carbon is possibly subject to several constraints such as handling, transport, or lifting. Hardness is certainly important in the possibility for the product to generate fine particles.

11.3.1.2.5 Surface Functional Groups

In addition to their structural reactivity, activated carbons are characterized by their surface reactivity. This is due to the occurrence of either acidic or alkaline oxygenated functional groups located at the edge of the condensated polyaromatic rings.

These functional groups can be characterized by acid/basic titration. Four acidic groups are distinguished:

- *1st group:* Carboxylic acids, measured by $NaHCO_3$ titration
- *2nd group:* lactones, measured by the difference between $NaHCO_3$ titration and Na_2CO_3 titration
- *3rd group:* phenols, measured by the difference between Na_2CO_3 titration and NaOH titration
- *4th group:* carbonyls, measured by the difference between NaOH and C_2H_5ONa titration

as well as an "alkaline" group: the pyrones.

11.3.1.3 Implementation

11.3.1.3.1 Acivated Carbon Presentation

The material presentation depends upon its manufacturing process and its final use. It is presently possible to find the following on the market:

1. Powders with a mean particle diameter ranging from 8 to 80 μm
2. Naturally granulated products, with particle diameters ranging from 200 μm to 6 mm
3. Extruded materials and rods, with diameters ranging from 0.8 to 5 mm and lengths ranging from 5 to 20 mm
4. Pellets with diameters approximating a few tens of a mm
5. Fibrous or woven material constituted of either 100% activated carbon, or powder or pellets deposited on a textile support.

The efficiency of an operation depends on the activity of the material, as well as its presentation. For instance, a granulated material with a 1 mm mean particle diameter is more efficient than a granulated material with a 4 mm mean particle diameter, but the headloss of the former material is higher than the latter material for a similar bed depth. This might be prehibitory in certain applications. This headloss characteristic may be limiting as it is in the case of gas masks, since the inhaling power of an individual is limited.

11.3.1.3.2 Conditioning

Activated carbon implementation depends on the packing, which is itself a function of the application. In the case of full scale industrial filtration and for flowrates of several thousands of m³/h, a fixed deep bed implementation is most often preferred. It consists of providing a few tenths of a second contact time between the pollutant and the activated carbon, which amounts to a few tenths of a m/s superficial velocity, in beds a few tenths of a meter deep. In all cases, the filtration is performed from top to bottom. A bottom-up filtration mode would expand the material and the adsorption efficiency would then decrease considerably. A turbulent tangential filtration mode is currently being studied and its performances are promising. In this type of implementation, the carbon layer thickness is very low (a few mm). The regeneration problems still remain to be solved.

11.3.1.4 Performances

Performances of an activated carbon are expressed as the mass of pollutant fixed per mass of adsorbing material. Two cases should be distinguished:

1. The odorous compound is recovered;
2. The odorous compound is not recovered, but only trapped by the adsorbant.

11.3.1.4.1 Odorous Product Removal and Recovery

This case is most frequently encountered in the industries where solvents are used. Then, odorous compound concentrations are very high, up to several grams of solvent per m³ of air. Solvent fixation capacities per kg of adsorbant are very high as well. This is exemplified by Table 11.2 where saturation adsorption capacities of five different activated carbons are given for five usual solvents. These values were obtained for a fluid concentration $C_0 = 30$ mg/m³. This table clearly shows that activated carbons may have very different adsorptive capacities (all these carbons have a similar mean particle diameter). Since this operation leads to the recovery of the adsorbate (solvents), it is important to know if desorption can be obtained within economically acceptable conditions (steam at 110 °C). Table 11.3 gives desorbable quantities for the same adsorbants and adsorbates such as those in Table 11.2. From these two tables (11.2 and 11.3), Table 11.4 can be calculated as the percentage of desorbability. All of these results show that it is impossible to predict performance unless dynamic tests are previously carried out.

In the case of solvent recovery, it is also important to know the steam consumption necessary to complete desorption. In Table 11.5, steam consumptions are given for the same set of solvents and activated carbons as in Tables 11.2, 11.3 and 11.4. These consumptions are given in kg of steam per kg of recovered solvent.

Table 11.2. Examples of Maximum Adsorption Capacities for Several Solvents.

		Saturation Capacity in kg Solvent Per Ton of Activated Carbon			
Carbon	Carbon Tetrachloride	Perchlorethylene	Trichlorethylene	Toluene	Monochlorobenzene
Coconut 1	725	715	720	470	620
Coconut 2	810	880	785	500	640
Coconut 3	755	830	730	440	515
Extruded peat	690	730	670	400	500
Extruded wood	1000	1150	990	600	760

Table 11.3. Examples of Desorption Capacities for Several Solvents.

		Desorption Capacity in kg Solvent Per Ton of Activated Carbon			
Carbon	Carbon Tetrachloride	Perchlorethylene	Trichlorethylene	Toluene	Monochlorobenzene
Coconut 1	560	440	620	320	430
Coconut 2	680	590	720	360	460
Coconut 3	600	510	620	340	340
Extruded peat	510	300	490	230	260
Extruded wood	1000	1150	990	600	760

Table 11.4. Regeneration Percentage for Several Activated Carbons.

	Desorption Capacity in kg Solvent Per Ton of Activated Carbon				
Carbon	Carbon Tetrachloride	Perchlorethylene	Trichlorethylene	Toluene	Monochlorobenzene
Coconut 1	77	61	86	68	69
Coconut 2	84	67	92	72	72
Coconut 3	79	61	85	77	66
Extruded peat	74	41	73	57	52
Extruded wood	100	100	100	100	100

Table 11.5. Steam Regeneration: Examples of Steam Consumption.

	Steam Consumption in kg/kg of Desorbed Solvent				
Carbon	Carbon Tetrachloride	Perchlorethylene	Trichlorethylene	Toluene	Monochlorobenzene
Coconut 1	3.5	7.3	4.8	9.8	7.6
Coconut 2	4.7	6.0	4.5	9.0	7.3
Coconut 3	7.4	8.6	7.0	12.6	15.0
Extruded peat	6.3	7.0	8.7	12.8	13.9
Extruded wood	3.4	3.1	4.1	6.0	5.5

Table 11.6. Examples of Diverse Organic Compounds Removal by Activated Carbons (Impregnated or Not).

Gases and Vapors	Activated Carbon	Activated Impregnated Carbon
Acetaldehyde	+	
Amyl acetate	+ + +	
Butyl acetate	+ + +	
Ethyl acetate	+ +	
Isopropyl acetate	+ +	
Methyl acetate	+ +	
Acetone	+ +	
Acetylene	−	
Acetic acid	+ + +	
Butyric acid	+ + +	
Caprylic acid	+ + +	
Hydrogen cyanide	−	±
Hydrogen bromide	+	±
Hydrogen chloride	+	±
Hydrogen fluoride	−	±
Hydrogen iodide	+	±
Formic acid	+	
Nitric acid	+	±
Palmitic acid	+ + +	
Propionic acid	+ + +	
Sulfuric acid	+ +	
Acroleine	+ +	

Gases and Vapors	Activated Impregnated Carbon	Activated Carbon
Crotonic aldehide	+ + +	
Ammoniac	−	±
Hydrogen sulfide	+	±
Sulfuric anhydride	+	±
Benzene	+ + +	
Bromine	+ + +	
Butane	+	
Butyl (chloride)	+ + +	
Butyl (ether)	+ +	
Butylene	+	
Butyne	+	
Camphor	+ +	
Carbon disulfide	+ +	
Carbon tetrachloride	+ + +	
Chlorine	+ +	
Chloroform	+ + +	±
Cyanogen chloride	−	
Cresol	+ + +	
Decane	+ +	
Diethylketone	+ + +	
Ethyl chloride	+ +	
Ethylene	−	

Compound	Rating		Compound	Rating
Amyl alcohol	+ + +		Ethyl (ether)	+
Butyl alcohol	+ + +		Eucalyptol	+ +
Ethyl alcohol	+ + +		Carbonic acid	−
Isopropyl alcohol	+ + +		Heptane	+ +
Methyl alcohol	+		Hexane	+ +
Buthyl aldehide	+ +		Hydrogen sulfide	− ±
Formic aldehide	−		Hydrogen arsenide	− ±
Indole	+ +		Ozone	− ±
Iodine	+ +		Kitchen smell	+ +
Iodoform	+ +		Misc. smells	+ +
Isopropyl (chloride)	+ +		Pentane	+
Isopropyl (ether)	+ +		Pentene	−
Menthol	+ +		Perchloethylene	+ + +
Methyl (chloride)	+ +		Phenol	+ + +
Methylene (chloride)	+ + +		Phosgene	+ + ±
Methyl ethyl ketone	+ +		Propane	−
Methyl (ether)	+		Propylene	−
Methyl isobutyl ketone	+ + +		Propyl mercaptan	+ +
Methyl mercaptan	+ +		Propyne	−
Naphtalene	+ + +		Pyridine	+ +
Nitrobenzene	+ +		Pyridine (derivatives)	+ + +
Nonane or octane	+ +		Chlorin'd and benzenic solvents	+ + +
Carbon monoxide	−		Tetrahydrofuran	+ +
Toluene	+ + +		Trichloroethylene	+ + +
Xylene	+ + +			

11.3.1.4.2 Odorous Compound Removal Without Recovery

When the odorous pollutant occurs either at trace concentrations (a few mg/m³ of polluted air) or at high concentrations, recovery has no economical interest since it hardly raises solvable technological problems. In all cases, the problem is to fix a maximum amount of pollutant in a minimum volume of adsorbant.

According to the nature of the pollutant, adsorption can be implemented on a nonimpregnated activated carbon. However, in some cases the material is impregnated so as to combine adsorption and chemisorption in order to increase the yield of the operation. Table 11.6 summarizes activated carbon (impregnated or not) maximum adsorption capacity for a certain number of organic compounds frequently occuring in industry.

At this stage, it is important to specify a fundamental issue related to activated carbon adsorption capacity. The use of this compound is related to the volume of adsorbant in a column. This volume must be as small as possible to solve the problem. Therefore, it is essential to reach a maximum efficiency for a given volume. The increase in adsorption capacity (mass per mass) of a given carbon is obtained by an increase of the activation time, the effect being an increase in the porosity. However, this is accompanied by a decrease in the specific density of the material. Figure 11.5 shows the CCl₄ adsorptive capacity of a coconut based carbon as a function of its density. This leads to a volumic adsorptive capacity (Figure 11.6), which exhibits an optimum value. This value is a characteristic of the material, which depends upon the manufacturing process. It is therefore different from one

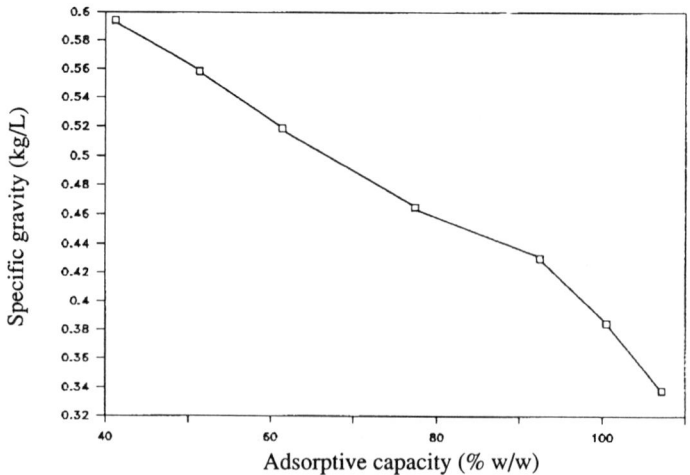

Figure 11.5. Relationship between an activated carbon specific gravity (kg/L) and its adsorptive capacity (% w/w).

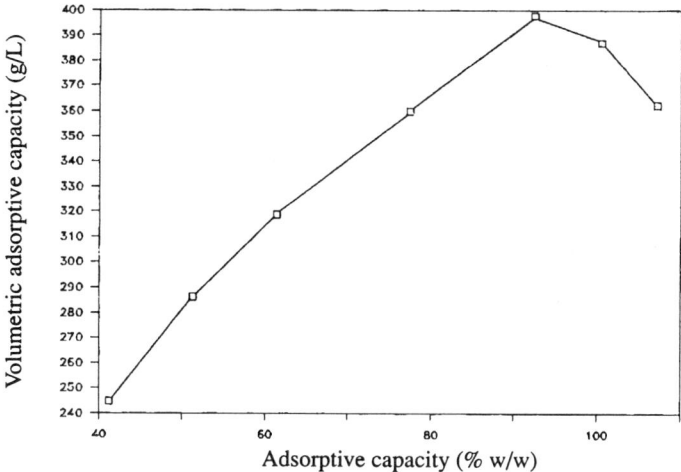

Figure 11.6. Relationship between an activated carbon volumic adsorptive capacity (g/L) and its adsorptive capacity (% w/w).

activated carbon to the other. In the case of impregnated carbons, the adsorptive capacity becomes a function of the impregnating compound. In Figure 11.7, the adsorptive capacity for a cyanide is shown as a function of the impregnating compound carbon content. Beyond a certain content, the activated carbon is saturated, that is to say, the impregnating compound occurs as multilayers, the lower layer no longer being efficient.

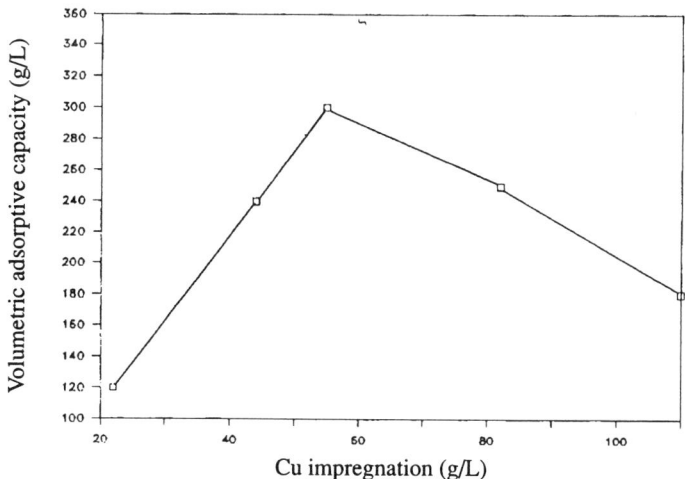

Figure 11.7. Relationship between an activated carbon HCN adsorptive capacity (g/L) and its impregnation (g/L).

11.3.1.5 New Activated Carbon Forms

New activated carbon presentations have recently appeared on the market. They are fibers, foams, sieves, or membranes. Their characteristics are quite interesting as well as their high specific surface area (1,500 m^2/g). The adsorption rate can be 30–50 as fast as for the granulated form. These materials have not been developed on a large scale as yet. However, experimentation is underway (Le Cloirec, 1990), and there is the possibility of a promising future for these "new" activated carbons.

11.3.2 Natural Adsorbants

Activated carbon is the most frequently used material for the adsorption of odorous compounds. However, a large number of other materials are used as well. The purpose of this section is to propose a nonexhaustive list of natural materials and to give their characteristics and performances in deodorization treatment (Table 11.7).

11.3.2.1 Soils

Earth can be used as packing material in adsorbing columns for odorous molecules (Bohn,1975). Several studies have been carried out on this type of material. Dodorik and Noguchi (1984) obtained 94% reduction of the odorous intensity in the treatment of fat processing gases. 78–99% efficiency was reached for molecules such as H_2S, and 84% efficiency for NH_3. These experiments were performed with a 70 cm deep dry soil packing and with a 28 m/s velocity (Sato and Nakao,1986; Sueda,1986).

The adsorptive capacity determination showed that soils have a relatively low fixation capacity. Nishida and Inone (1986) found dynamic retention capacities of $2–8 \times 10^{-3}$ mg/g of dry soil for dimethyl sulfide and 20–30 mg/g of dry soil for NH_3. The adsorption capacity seems to increase with increasing molecular weight and number of oxygenated or sulfurous groups of the molecule (Bohn, 1975). Modeling is possible through the use of the Freundlich's equation or the mass transfer laws (see section 11.5.1).

Table 11.7. Characteristics of Some Natural Adsorbants (after Harper and Purnell, 1990).

Adsorbants	Specific Surface Area (m^2/g)	Pore Type ([1])
Activated carbon	800–1500	Micropores
Silica gel	300–800	Micro & mesopores
Zeolites		Micro/mesopores
Activated alumina	100–300	Mesopores

[1]Micropores < 2 μ; mesopores (2–50 μm).

The odorous molecules removal by soil is attributed to physical and chemical interactions between these molecules and the soil itself. A large proportion of the molecules are oxidized upon acidification, the soil playing the role of a catalyst. This mechanism is akin to the oxidation of H_2S into $S°$ by activated carbon.

11.3.2.2 Peat

Even though this material is commonly used in the biological treatment of odorous gases (see Chapter 12), peat has hardly been studied as an adsorbant. However, it was pointed out that biological removal of odorous compounds using peat as support material is due to two phenomena: physical-chemical transfers (adsorption-absorption) followed by biodegradation. Since peat has an acidic pH, compounds such as aminated molecules and ammoniac are easily fixed. In the case of adsorption, the most likely mechanism is peat surfacic function neutralization (Martin et al., 1989). These functions are strong acidic sites, which were titrated by the method of Bœhm and co-workers (1965, 1966). An example of peat pH evolution after ammoniac adsorption is given in Figure 11.8.

Martin et al. (1989) gave some values of peat adsorptive capacity:

- 20–75 mg NH_3/g dry peat for equilibrium concentrations (C_e) from 5 to 200 g NH_3/m^3 air
- 0.6 mg NH_3/g dry peat for equilibrium concentrations (C_e) of 10 mg NH_3/m^3 air

These results are in accordance with those of Togashi et al. (1986) who reported 10 mg NH_3/g dry peat in a dynamic system with highly concentrated gaseous effluents.

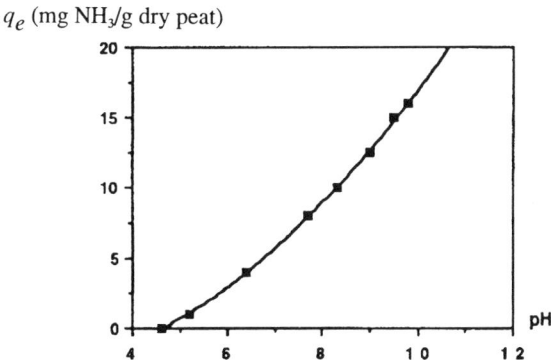

Figure 11.8. Trapped NH_3 as a function of peat pH.

11.3.2.3 Compost

This material is of the same category as soil and peat and can be used for odor removal as well. Odor intensity reduction from 4,000 to 52 was found in the case of fermentation gas treatment (Nogushi et al.,1984). The following order of adsorption was found in the case of specific molecules (Terashima,1984):

$$NH_3 >> MeSH > H_2S >> Me_2S$$

The adsorptive capacities found were similar to those found for soils. In the case of dry composts, Freundlich's equation is usable for the modeling of batch process. In the case of wet compost, a mass transfer term must be added, corresponding to an absorption phenomenon (Martin et al., 1989).

11.3.2.4 Sawdust

As another natural material, sawdust has been tested as an adsorbing material. Its adsorptive capacity was found to be very low (Ose et al.,1977). Additionally, NH_3 and H_2S removal by wet sawdust (90% and 97%, respectively) were not attributed to solid-molecule, but to absorption into the intersticial liquid.

11.3.2.5 Zeolites

Like activated carbons, zeolites are materials under extensive investigation, even though their use as odorous compound adsorbants is the most recent. The implemented zeolites are generally clinoptilolite, mordenite, and montmorillonite. For these materials, pore diameters range from 3 to 7.5 nm. Of course, this fact limits the use of zeolites to the trapping of small odorous molecules. The adsorption mechanisms seem to indicate that micropores are involved. The Dubinin-Artakhov equations are usable for the determination of the microporous volume occupied by odorous compounds (Tanada et al.,1977; Tanada and Boki, 1980). In the case of styrene, the adsorption rate was found to be a function of the pore diameter (Miyoshi et al., 1978).

Besides trapping by adsorption, ion exchange phenomena can occur in zeolites. Kasaoka et al. (1981) showed that dimethylsulfide removal may occur on type Y zeolites by an ion exchange phenomenon. In this case, the material was previously loaded with metallic ions (Ag^+, Cu^{2+}, . . .).

The following order of adsorption was found during the determination of adsorptive capacities (Nogushi et al., 1984):

$$NH_3 > H_2S > CH_3SH$$

For NH_3, the adsorptive capacity was determined to be around 6.3 mg N/g clinoptilolite (Kœlliker et al., 1980). A correlation was found between the adsorbtive capacity and the surface area of the adsorbate such as H_2S, amines or organo-sulfurous compounds (Tanada and Boki, 1980). Additionally, molecules such as SO_2 can be trapped by zeolites as well (Tsuchiya, 1980).

Zeolite implementation is carried out in columns. Humidity was found to be detrimental to adsorption. However, it is possible to operate on a wide range of temperature: 40–200 °C (Kasaoka et al., 1981). Kasaoka et al. uses zeolites in a 20 cm deep bed. Under such conditions, 50% removal is reached for 3–4 mg $N-NH_3/m^3$ gas influent concentration, and a 600–1000 m/h operating velocity. Frazier (1986) implemented a 1.2 cm deep bed to treat a diffuse odorous atmosphere containing formaldehyde, ammoniac, and sulfurous hydrogen. Removal yields reached 97% after 1 h of operation.

These materials can be used in mixtures containing activated carbon, silica gels, or clays as well.

11.3.2.6 Clays, Silica Gel

Clays are used as support material for activated carbon (Tanaka et al., 1976). A batch test for the removal of NH_3 by a 1–9 mixture of activated carbon and clay, respectively, gave a capacity of 32%. In the case of gas mixtures containing amines, mercaptans, H_2S, and ammoniac, columns were packed with activated carbon and acid washed clays. A good efficiency was found with a 30 cm bed depth and a 1,000 m/h operating velocity.

For silica gel, only laboratory experiments were performed. For a biogas containing H_2S, CO_2, and methane, capacities ranging from 0.45 to 0.55 mg H_2S/g silica gel were obtained in continuous tests (Tse Chuan Chou et al., 1986).

11.3.3 Synthetic Adsorbants

Synthetic adsorbants are used two different ways: either as packing in gas chromatography columns (see Chapter 8) or as adsorbants for odor removal. Table 11.8 shows some of these materials along with their characteristics. In this paragraph, only resins used as packing in industrial columns implemented for deodorization will be concerned. These adsorbants can be categorized into two classes: macroreticulated adsorbing resins and ion exchange resins.

11.3.3.1 Adsorbing Resins

Different types of resins are used for odorous molecule removal. In addition to the macroreticulated conventional XAD type Amberlite resins (Table 11.8), iron sulfate and aluminum-sodium sulfate impregnated polyethylene resins are manufactured (Shigematsu and Suzuki,1987). Urea based or formaldehyde monomer resins are also found (Verbestel,1980). However, the most common structures for these resins are based on styrene-divinylbenzene polymers.

Table 11.8. Characteristics of Some Synthetic Adsorbants (after Harper and Purnell, 1990).

Adsorbant	Specif. Surf Area (m²/g)	Maximum Temperature (°C)	Pore Type[1]	Polymer Type[2]
Carbonax	14	500	Microporous	
Tenax	19	375	Macroporous	DPPPO
Chromosorb				
101	<50	275	Macroporous	STY-DVB
102	300–400	250	Mesoporous	STY-DVB
103	15–25	275	Macroporous	STY
104	100–200	250	Macroporous	Acrylate
105	600–700	250	Mesoporous	Aromatic
106	700–800	225	Mesoporous	STY
107	400–500	225	Mesoporous	Acrylate
108	100–200	225	Macroporous	Acrylate
Porapack				
P	50–100	250	Macroporous	STY-EVB
Q	500–600	250	Mesoporous	EVB-DVB
R	550–750	250	Mesoporous	NVP
T	250–350	190	Mesoporous	EGDMA
N	225–350	190	Meso-macroporous	CVP
Amberlite				
XAD1	700–400	250	Mesoporous	STY-DVB
XAD4	498		Micro-mesoporous	STY-DVB
XAD7	326		Micro-mesoporous	Acrylate

[1]Microporous, <2 nm; mesoporous, 2–50 nm; macroporous, >50 nm.
[2]DPPPO: diphenyl p phenylene oxide; CVP: C-vinylgnolidene; STY: styrene; DVB: divinylbenzene; EVB: ethylvinylbenzene; EGDMA: ethylene glycodimethacrylate.

These materials have been implemented especially in the case of pollution by ammoniac, hydrogen sulfide, or sulfur dioxide. Shigematsu and Suzuki (1987) showed that these compounds were adsorbed from 88 to 95% in batch tests. The pollutant concentrations were at a few mg/m³ levels. In the case of a gas containing 1.8 mg/L of phenol, XAD resins were capable of totally removing this compound. In this case, regeneration is performed with methanol.

11.3.3.2 Ion Exchange Resins

Odorous pollution removal is performed by the use of strong cationic resins. Thus, Osterburg et al. (1987a and 1987b) used cation exchange resins loaded with silver or palladium to remove isopropanol from a gaseous effluent. In the case of ammoniac or amine containing effluents, these same resins can be used successfully.

In the case of mixtures containing H_2S, CH_3SH, or NH_3, Akita and Hamagushi (1981) implemented ion exchange resins. The regeneration process includes a washing step by a diluted solution of sodium hypochlorite.

11.4 Regeneration

Because of the cost and the adsorptive capacity of these materials, regeneration is often necessary. This regeneration is generally applied to activated carbon. Three types of processes can be implemented: the thermal (and most widely used) process, the chemical process, and the biological process.

11.4.1 Thermal Regeneration

11.4.1.1 Generalities

Activated carbon having adsorbed different compounds has its porosity clogged by these compounds, thereby considerably hindering its adsorptive efficiency. In order to recover the adsorptive capacity, it is necessary to proceed to a "scrubbing" of the porosity: the regeneration process. The thermal methods consist of an increase of the temperature so that adsorbed compounds can desorb. Two cases should then be distinguished:

1. The adsorbed compounds are volatile, hence desorbing at low temperature.
2. The adsorbed compounds are barely volatile. It is necessary to heat the carbon to a high temperature. It is then important to know the thermal stability of the adsorbate if its recovery is envisaged.

Information is obtainable on these two cases through thermogravimetric analysis: before performing a regeneration, it is important to analyze the adsorbed compound's desorption from the activated carbon. This process consists of submitting a sample of saturated carbon to a nitrogen flow whose temperature is increased. The sample mass variation is recorded as a function of temperature. Figure 11.9 shows an example of a thermogram. A first peak appears at 100 °C on the derivative curve, corresponding to the adsorbed water. The second peak occurs at 200 °C. In this case, regeneration should be performed at about 300 °C. Under these conditions, the adsorbed compounds can be desorbed without being carbonized. In the second thermogram, presented in Figure 11.10, two types of compounds are adsorbed on the activated carbon surface besides water. The regeneration should then be performed at a high temperature, over 700 °C. Actually, the temperature must exceed 400 °C to desorb fixed compounds. In such a case, the organic compounds are no longer stable and they are decomposed by carbonization at the surface of the activated carbon. The adsorbing surface is then free of adsorbate, but an amorphous carbon layer is formed at the surface of the activated carbon. The regeneration process should allow for the activation of this new carbon layer, hence the chosen regeneration temperature value at 700 °C in the presence of steam.

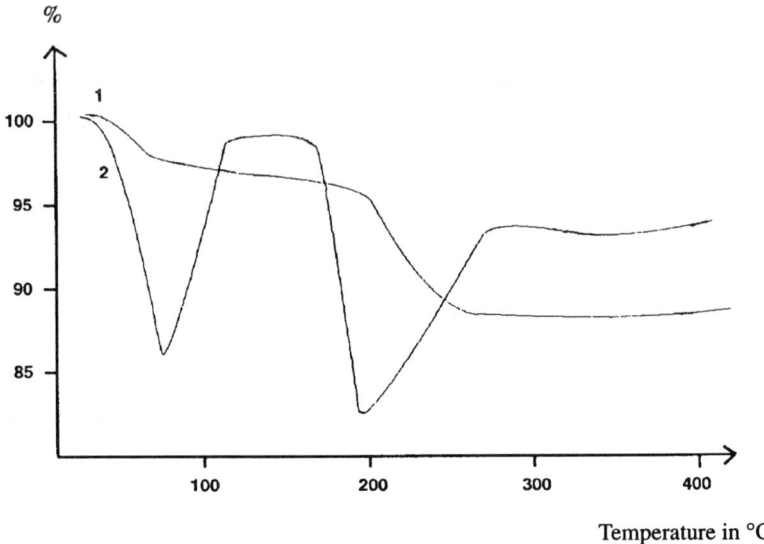

Figure 11.9. Example of thermogram. Activated carbon 1.

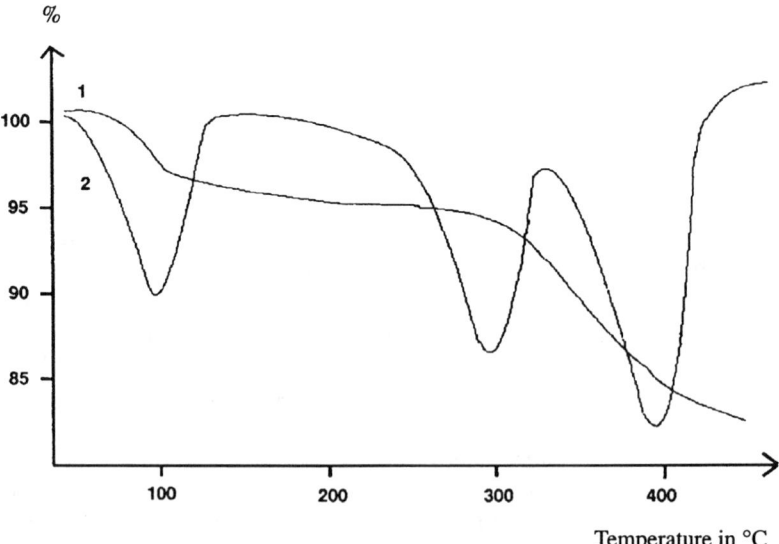

Figure 11.10. Example of thermogram. Activated carbon 2.

11.4.1.2 Different Thermal Regeneration Processes

11.4.1.2.1 Steam Regeneration

This is the most commonly used process in the case of solvents. It makes it possible to regenerate activated carbons saturated with volatile compounds. Its disadvantage is the formation of hydrochloric acid resulting from the decomposition of chlorinated solvents. This acid is corrosive for the tank shell.

11.4.1.2.2 Hot Air Regeneration

This process allows the use of higher temperatures than the former process. It avoids problems related to decomposition due to the presence of water but since some adsorbants are sensitive to oxidation, its use is limited.

11.4.1.2.3 Nitrogen Regeneration

This process allows the recovery of fragile adsorbates sensitive to oxidation. Its temperature application domain is limited to 300 °C, thereby limiting its application to the regeneration of adsorbants loaded with volatile adsorbates.

11.4.1.2.4 CO_2-Steam Regeneration or "Thermal" Regeneration

This is the most commonly used process. It allows the use of a whole range of temperatures. However, adsorbate recovery is not economically feasible.

11.4.1.2.5 Joule Effect Regeneration

Most recently, a new regeneration process has been used. It consists of submitting the carbon to the passage of an electric current. The carbon can be either granular (Ritoul, 1983) or fibrous (Le Cloirec et al., 1990a, 1990b, 1990c). The activated carbon structure and the material resistance tests show that this structure is akin to a semiconductor. Laboratory scale experiments on solvent desorption are very promising. The advantage of such a process is that it can be implemented in situ.

11.4.2 Chemical Regeneration

In order to avoid manipulation of the adsorbant prior to their thermal regeneration, studies were carried out on the in situ chemical desorption.

11.4.2.1 Acid-Basic Washing

Mineral acids are used conventionally for the desorption of adsorbed compounds. Stecker (1962) used hydrochloric acid for the regeneration of activated carbon saturated with aminated alcohols (ethanol amine, di- and tri-ethanolamine). The acid concentration ranges from 5 to 35% w/w. The process can be performed at temperatures between 70 and 130 °C. Under these conditions, 95–100% yields are reached. The adsorbant is not rinsed, the residual acid forming a salt with aminated alcohols during the following adsorption cycle. In the case of an activated carbon saturated with alkaline compounds, regeneration with sulfuric acid can be envisaged as well. However, it is important to choose the dilution so as to avoid a surface oxidation which would strongly alter the porous structure (Lemarchand, 1981).

Basic washing allows the desorption of acidic adsorbates. The sodium hydroxide concentration varies from 0.5 to 25% w/w, the contact time ranging from 30 min. to 2 h for velocities of about 1 m/h. Water rinsing is necessary after treatment. The optimum temperature range is from 25 to 90 °C. This type of regeneration gives good quality activated carbon (Benzaria and Zundal, 1973).

11.4.2.2 Solvent Extraction

Sulfurous compound adsorption on activated carbon forms elemental sulfur at the adsorbing surface (Legros, 1980). Regeneration of such a saturated material is possible through the use of CS_2. Wendlandt and Henrich (1969) developed such a process. The following regeneration process is implemented on 20% w/w saturated activated carbon by elemental sulfur:

1. Nitrogen adsorber fill up to replace the air
2. CS_2 fill up of the adsorber, and contact occurs in 1 h
3. Solvent blowdown followed by material scrubbing by steam

Other solvents were used as well, such as ethyl acetate (Himmelstein, 1976) or formaldehyde (Popper et al., 1978). These solvents desorb acetic acid and carbonaceous solvents, respectively. After the solvent application, water rinsing is carried out to eliminate all traces of solvents.

11.4.2.3 Oxidative Regeneration

Different oxidation processes can be used on adsorbed molecules. Pure oxygen or air performs this oxidation between 100 and 300 °C and under 5 bars, as shown by pilot scale experiments (Cheremisinoff and Ellerbusch, 1978).

Some more powerful oxidizing agents were tried in laboratoratory scale experiments (ozone, hydrogen peroxide, persulfate, etc, . . .). Two main problems hindered industrial implementation of these processes. The occurence of a strong redox system (oxidizing agent: activated carbon) makes the implementation hazardous. Additionally, the oxidizing agent tends to dramatically change the adsorbant pore structure, thereby changing the adsorption capacity and the adsorption kinetic rate.

11.4.3 Biological Regeneration

An alternative to thermal or chemical regeneration is to scrub the porosity by the use of bacteria. Of course, this type of regeneration process is possible only if the adsorbates are biodegradable. This possibility must be compared with gas biological scrubbing, as described in Chapter 12.

11.4.3.1 Principle

The saturated material is contacted with a bacterial suspension. The bacteria are then fixed onto the material. The resulting biomass utilizes the adsorbates as substrate, thereby cleaning the adsorbant porosity. In order to reach optimum conditions, it is often necessary to circulate a nutritive solution so as to balance bacterial nutrition. A basic schematic of this type of regeneration is presented by Figure 11.11 (Gaïd, 1980; Le Cloirec et al., 1986).

11.4.3.2 Implementation

This type of operation is generally carried out in submerged filters. This system can only be operated with two filters working alternatively in the adsorption mode and the regeneration mode. The regenerated filter is isolated, fed with nutrients, and possibly seeded with either specific strands or with sludges from a wastewater treatment plant. This seeding increases the start-up rate of the regeneration. The nutrient solution is fed through a closed loop. The content of this solution is a function of the adsorbate nature. Thus, in the case of a carbonaceous adsorbed substrate, the necessary nutrients will be ammoniacal nitrogen and a phosphorus source (potassium phosphate). Oligo-elements are brought about by the dilution water (Wallis and Bolton, 1982; Kim et al., 1986). The system can be operated either in the aerobic or anaerobic mode. Recently, a dentrification process was implemented in the case of sulfurous adsorbed compounds. This process uses

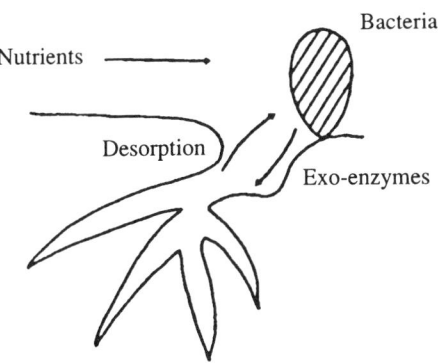

Figure 11.11. Schematics of biological regeneration.

Thiobacillus dentrificans, oxidizing sulfur, and sulfurous compounds into sulfate (Le Cloirec, 1990). The regenerated material is of good quality. The regeneration yield, that is, the adsorptive capacity of the regenerated material as compared to the original adsorptive capacity, ranges from 70–95% (Wallis and Bolton, 1982). Regeneration time is generally between 20 and 50 days. The knowledge of this time makes it possible to design the adsorber so that the breakthrough time is slightly longer than the regeneration time. A 1.2 to 1.5 ratio between breakthrough time and regeneration time is commonly used. This includes the regeneration, washing, and drying times.

11.4.3.3 Process Limits

Presently, two phenomena limit the use of biological regeneration for adsorbers and, more specifically, for activated carbon. After regeneration, it is necessary to eliminate the biomass present at the material surface. Water is used for this purpose. However, it is difficult to completely remove the biomass, which tends to plug the material porosity with time. A research on acidic washing and enzymatic washing is necessary to develop a more thorough washing sequence. Connesson (1988) successfully used cellulase to break exopolysaccharides that form bridges between the bacteria and the support material, thereby eliminating the whole biomass. The second limit of this process is the plugging of the material porosity by the bacterial metabolism by-products. This may explain that regeneration is not complete, and that the material ages rapidly. Furthermore, a mineralization may occur at the material surface, with the formation of calcium, iron, and copper carbonate, which plug the material porosity. This reduces the access by the adsorbing compounds to the inner porosity. An acidic washing, or the use of a descaling agent, insures the complete scrubbing of the porosity.

11.5 Modeling and Design

After a first approach to choose the adsorbing material for the removal of odorous molecules, it is necessary to put it into action. This implementation requires the ability to design the plant and to predict its behavior, in other words to size, model, and to simulate the adsorber operation.

11.5.1 Adsorber Modeling Concepts

The modeling of adsorption of a molecule onto a porous material implemented in the form of a filtration bed relies on several concepts that can be defined as follows (Kast and Otten, 1987):

1. The gas flow pattern in the bed
2. The molecule transfer into the porous particle

3. The adsorption itself, that is, the interaction between the adsorbate and the adsorbant surface
4. Heat transfer from the particle then from the bed. This phenomenon is often negligible enough to be taken into account.

This set of concepts on adsorption has the purpose of simulating and predicting the adsorber's breakthrough curve. Several types of operation must be considered. The most common are the contact filtration, the fixed bed reactor, and the fluidized bed reactor, which will not be developed here.

11.5.2 Contact Filtration

This system consists of putting the gas into contact with the adsorbants in a batch reactor. When equilibrium is achieved, the solid phase is separated from the gas. The modeling is very simple as it relies on a mass balance:

$$M (q_e - q_0) = G (y_e - y_0)$$

where M = adsorbant mass, q_0 = initial adsorbant loading (w/w), q_e = equilibrium adsorbant loading (w/w), G = solution mass, y_0 = initial adsorbate concentration in the solution (w/w), and y_e = equilibrium adsorbate concentration in the solution (w/w). At equilibrium, the following relationship can be written (as in the case of Freundlich's equation):

$$q_e = k (y_e)^{1/n}$$

This modeling applies to closed system deodorization.

11.5.3 Fixed Bed Continuous Reactor

11.5.3.1 Batch Reactor Transfer Models

Before investigating the analysis of a continuous reactor, it is necessary to model the laws of solute transfer. Only isotherm models will be described. However, several cases should be distinguished according to the origins of the transfer rate limitations:

1. External surface area:

$$\frac{dq}{dt} = k_1 C (q_m - q) - k_2 q$$

   and with Langmuir's equation at equilibrium:

$$q_e = a q_m C_e/(1 + a C_e)$$

2. External diffusion:

$$\frac{dq}{dt} = \frac{k_f a_v}{m} (C - C_i)$$

where a_v is the specific surface area—for a spherical particle, $a = 6/d_p$, C_i is the interface concentration, k_f is the film transfer coefficient (extra-particle transfer), m is the adsorbent mass. The film transfer coefficient (k_f) is a function of Sherwood's number:

$$Sh = k_f \, d_p / D_L$$

This dimensionless number is directly related to Schmidt's and Reynolds' numbers (Sc and Re, respectively); for instance,

$$Sh = 2.0 + 0.6 Sc^{1/3} Re^{1/2}$$

where $Sc = \mu/D_L$ and $Re = d_p \, \rho \, U/\mu$. Different correlations were suggested between these dimensionless numbers by other authors (Trambouze et al., 1984; Perry and Green, 1984).

3. Internal (intraparticle) diffusion:

$$dC_s/dt = k_p \, a \, (C_{si} - C_s)$$

where C_s = surfacic "concentration."

11.5.3.2 Continuous Flow Modeling

In order to simplify the modeling of a continuous adsorber, some assumptions are necessary:

1. The adsorber operates in isothermal conditions and the pollutant concentrations are low (diluted system). Therefore, exothermic adsorption energies are negligible.
2. Packing particles are assumed to be spherical.
3. Adsorber headloss is low.
4. The maximum adsorption capacity is independent of the depth (z) in the column.
5. The gas velocity is constant.

The modeling relies on three equations: the headloss, the mass balance, and the heat balance. In the case of the adsorption of odorous molecules contained in air, isothermal conditions can be assumed, as stated earlier. Hence, the following equations are obtained:

11.5.3.2.1 Headloss per Unit of Column Length

Several equations are suitable as, for example, Leva's equation:

$$\frac{\Delta P}{H} = 2 f \frac{U_0^2 \, (1 - \varepsilon)^{3-n}}{d_p \quad \varphi^{3-n} \, \varepsilon^3}$$

n and f are functions of the Reynolds' number $Re = d_p U_0 P/\mu > 10$ for turbulent and transitory regimes and $n = 1$ and $f = 100/Re$ for laminar regime

($Re < 10$). The shape factor can be approximated through the following relationship:

$$\varphi = 4.87 V^{2/3}/a$$

where V = particle volume and a = particle surface area. The following values are given for comparison purposes: $\varphi = 1$ for spherical particles and $\varphi = 0.74$ for activated carbon particles with an effective diameter ranging from 1 to 1.5 mm.

It is also possible to use the following equation:

$$\frac{\Delta P}{H} = 180 \frac{\mu \ U_0(1 - \varepsilon)^2}{d_p^3 \ \varepsilon^3} + 1.75 \frac{U_0^2 \rho}{d_p} \frac{1 - \varepsilon}{\varepsilon^3}$$

This relationship is applicable whatever the value of Reynolds' number. However, U_0 is prevalent in a laminar regime whereas, in a turbulent regime, U_0^2 becomes important.

11.5.3.2.1 Mass Balance

The general relationship is

$$-D_L \frac{\partial^2 C}{\partial z^2} + u \frac{\partial C}{\partial z} + \frac{\partial C}{\partial t} \frac{1 - \varepsilon}{\varepsilon} \frac{\partial q}{\partial t} = 0$$

with

$$q = \frac{3}{R^3} \int_0^R qr^2 dr$$

The resolution of this equation assumes the knowledge of the function $q(C,t)$. Several approximations can be made and several processing methods carried out. Assuming that the reactor works as a plug flow reactor, the following relationship can be derived:

$$u \frac{\partial C}{\partial z} + \frac{\partial C}{\partial t} \frac{1 - \varepsilon}{\varepsilon} \frac{\partial q}{\partial t} = 0$$

If equilibrium is assumed along the adsorption front, the equation yields its propagation velocity along the adsorption bed:

$$W = \left(\frac{\partial z}{\partial t}\right)_e = \frac{U}{1 + \dfrac{1 - \varepsilon}{\varepsilon} \dfrac{\partial qe}{\partial C}}$$

If the transfer obeys a kinetic rate law of the kind $q = f(C,t)$, this law is used to "accelerate" W.

11.5.3.3 Transfer Units

Assuming that the external film transfer is limiting, the following equation is obtained:

$$- U_0 \frac{\partial C}{\partial z} = Ka\,(C - Ce)$$

K is the overall mass transfer coefficient. C_e is related to q_e through a relationship of the Freundlich's kind. The solution to this equation relies on the knowledge of the equilibrium relationship, as well as the operating line (Figure 11.12). It makes it possible to calculate the number of transfer units (NTU) and the height of the transfer unit (HTU). Thus the total height of the adsorber becomes

$$Z = \text{NTU} \times \text{HTU}$$

The integration of the transfer equation and the Freundlich's equation then yields the outlet concentration (C).

The mathematical manipulation of these equations leads to the following relationship (Clark, 1987):

$$C = \frac{[C_0^{n-1}]^{1/n-1}}{1 + A_e^{-rt}}$$

A and r being the model parameters. This equation allows the calculation of the outlet concentration C at any time t as a function of the inlet concentration C_0 to, therefore, draw breakthrough curves. This kind of equation has been successfully applied to the case of solvent removal by activated carbon beds (Le Cloirec, 1990; Baudu, 1990). An example is presented in Figure 11.13, for the removal of dichloromethane by granular or woven fibrous activated carbon. Additionally, it is interesting to note that the values of the A and r parameters are correlated to the gas contact time in the reactor.

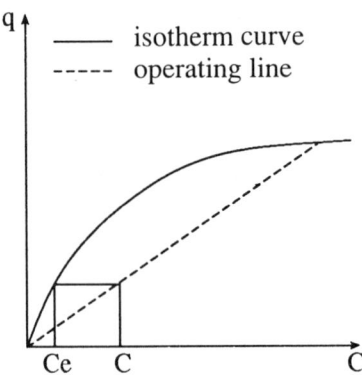

Figure 11.12. Use of the adsorption isotherm and the operating line for the determination of NTU and HTU.

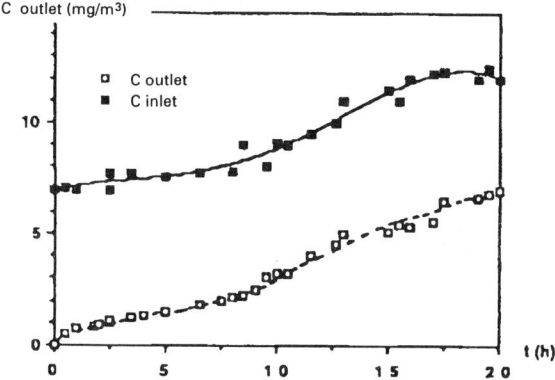

Figure 11.13. Use of a model in the case of the removal of CH_2Cl_2 by activated carbon.

11.5.3.4 Filtration Time

The mass of adsorbed molecules on the adsorbant is

$$(H - \delta) \, S \, q_0 + S \, q_0 \, f \, \delta$$

where f is the used fraction of the adsorption front δ. The filtration time is therefore

$$t = \frac{H - \delta \, (1 - f)}{U_0 \, C_0}$$

which is simply expressed as

$$t = a \, H - b$$

Experimentation is used to determine the values of a and b in practical conditions.

11.5.3.5 Modeling Limitations

The solution of these equations is quite complicated and requires the use of sophisticated computing equipment. Furthermore, they can be applied only to systems containing only one pollutant or a global pollution measurement.

11.5.4 Design Elements

The filtration equipment to be implemented is a function of three independant parameters:

1. The flowrate of polluted air
2. The nature of the pollutant(s) to be removed
3. The pollutant's concentration

In this section, design elements are specified, such as the filtration surface area, the adsorbant bed depth, and the power of the compressor or the fan to be installed.

11.5.4.1 Surface Area

The relationship between the air flowrate Q and the adsorber's surface area S is $Q = SU_0$, where U_0 is the air velocity (filtration rate). The choice of the value of the filtration rate U_0 is critical since it depends upon the pollutants concentrations and adsorbabilities. Generally, the filtration rate is set to 600 m/h (0.166 m/s) but in the case of very adsorbable pollutants, this rate can reach 1,000–2,000 m/h (0.278–0.55 m/s). Figure 11.14 represents the relationships to be used for the determination of the filtering surface area.

11.5.4.2 Bed Depth

The filtration rate and the pollutant concentrations determine the required contact time (Θ) that finally gives the adsorber bed depth. This is due to the relationship

$$\Theta = V/Q = H/U_0$$

Figure 11.15 represents this relationship. Generally, the contact time is set between a few tenths of a second and a few seconds (0.1–5 s).

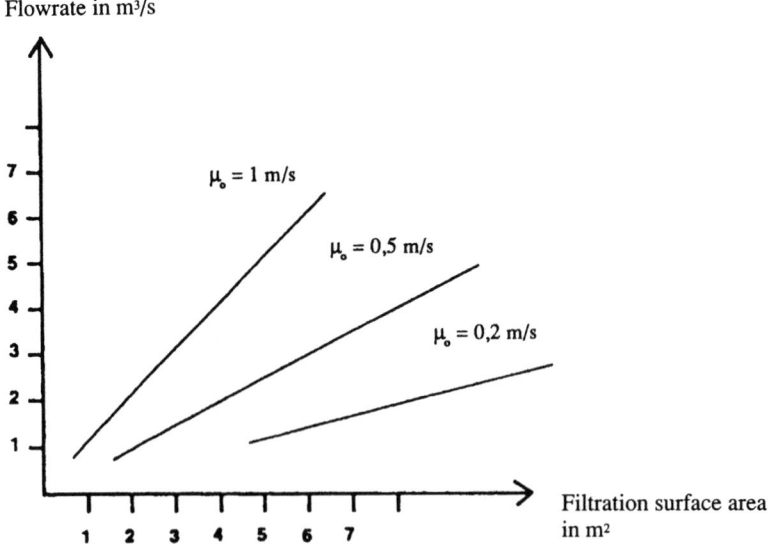

Figure 11.14. Relationships between the filtration surface area, the air flowrate, and the filtration rate.

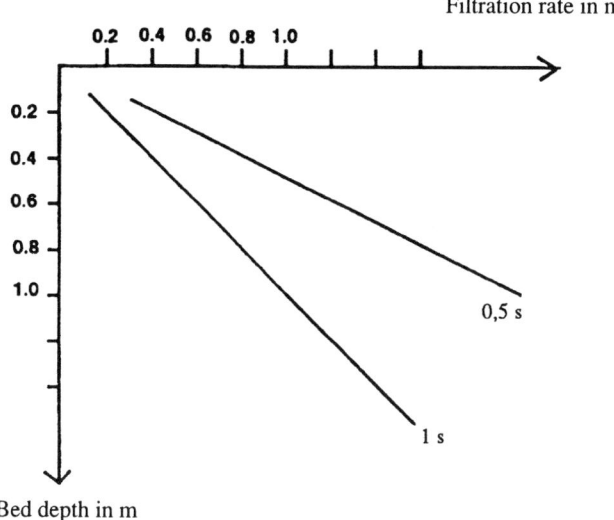

Figure 11.15. Relationship between the filtration rate, the bed depth, and the contact time.

11.5.4.3 Headloss

One of the main operating parameters of an adsorber is the headloss. This can be calculated using the relationships presented in Section 11.5.3.2. Figure 11.16 represents an example of the relationship between the headloss and the filtration rate for different activated carbons commonly used in deodorization by adsorption. Thus, for activated carbon C and for a filtration rate of 0.1 m/s and a bed depth of 1.2 m, the headloss amounts to 24 cm water head.

11.6 Industrial Examples

Two examples are given in this paragraph. They concern a low and high pollutant concentration, respectively.

11.6.1 Perchlorethylene - IPA at Low Concentration in the Air

The air to be treated has the following characteristics:

Flowrate	30,000 m³/h
Perchlorethylene	65 mg/h (2.17 g/m³)
IPA	3.5 kg/h (0.12 g/m³)
Temperature	20 °C

NC	14 x 30	= A	:	NC	6 x 12	= D
NC	12 x 20	= B	:	NC	4 x 8	= E
NC	8 x 16	= C				

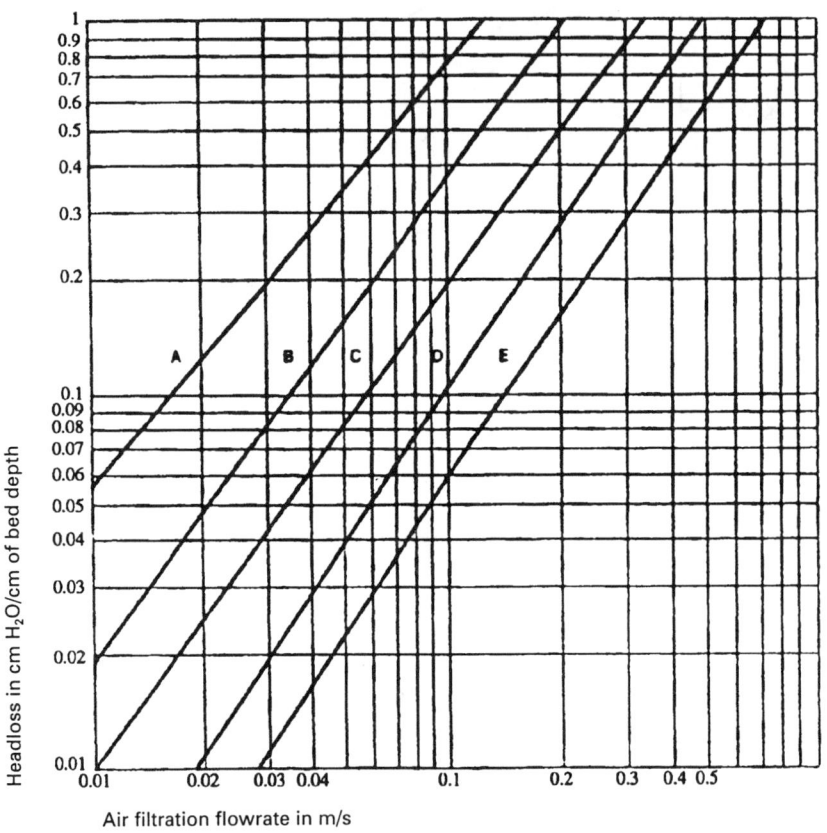

Figure 11.16. Adsorber headloss as a function of the activated carbon and the filtration rate.

The specifications for the treated gas are

Perchlorethylene	0.2 g/m³
Perchlorethylene + IPA	0.3 g/m³

The solution to this problem gave the following treatment plan:

Number of adsorbers	6
Adsorbers in operation	5

Air flowrate per adsorber	6,000 m²/h
Adsorber diameter	2,500 mm
Filtration rate	0.339 m/s
Activated carbon bed depth	1,110 mm (1,010 + 100 mm)
Air-activated carbon contact time	3.3 s
Activated carbon mass per adsorber	2,070 kg
Activated carbon saturation rate	10%
Trapped solvent mass per adsorber	207 kg
Adsorber operating time 15 h	
Headloss	560 mm water head
Required steam per regeneration	620 kg
Number of daily regenerations in continuous operation	8
Drying fan flowrate	6,000 m³h
headloss	560 mm water head
Condensation water content	20 g solvent/litre

The adsorber flowsheet is given by Figure 11.17. The overall plant flowsheet is represented by Figure 11.18.

11.6.2 Freon and Chlorothine at High Concentration in the Air

The air to be purified has the following characteristics:

Air flowrate	7,500 m³/h
Freon	90 kg/h (12 g/m³)
Chlorotine	7.5 kg/h (1 g/m³)
Temperature	20 °C

The air must not contain more than 0.3 g/m³ of either freon or chlorotine.

The implemented purification plant is based on an activated carbon adsorber that is steam regenerated. Figures 11.19 and 11.20 represent the flowsheets of the adsorber and the plant, respectively. The plant has the following design and operational characteristics:

Number of adsorbers	3
Adsorbers in operation	2
Air flowrate per adsorber	3,750 m²/h
Adsorber diameter	2,200 mm
Filtration rate	0.274 m/s

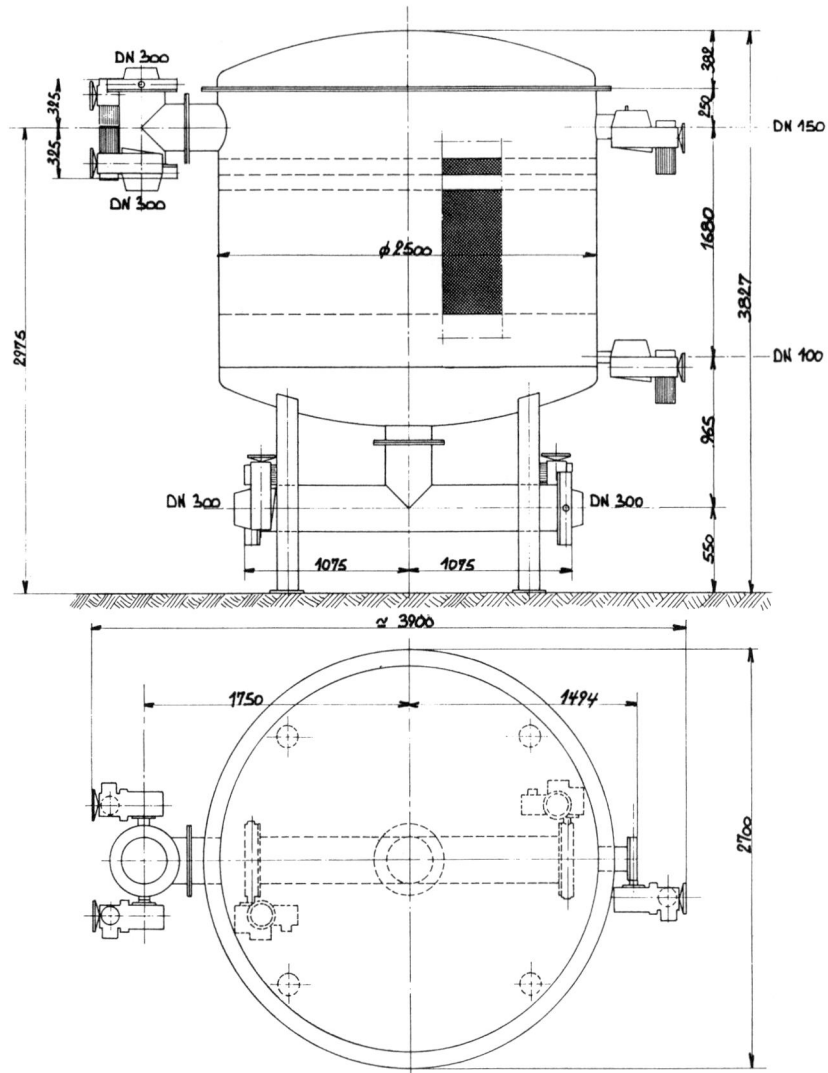

Figure 11.17. *Example 1*: Schematics of an adsorber.

Activated carbon bed depth	1,350 mm (1,250 + 100 mm)
Air-activated carbon contact time	4.9 s
Activated carbon mass per adsorber	1,950 kg
Activated carbon saturation rate	10%
Trapped solvent mass per adsorber	195 kg
Adsorber operating time	4 h

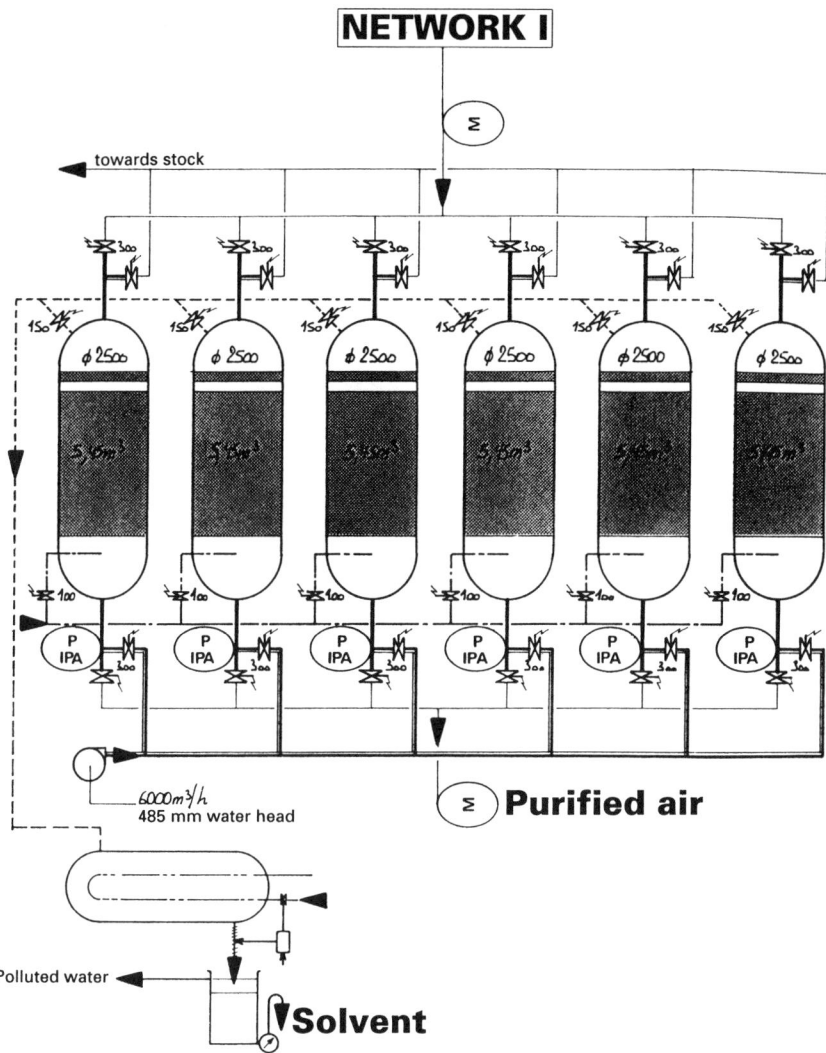

Figure 11.18. *Example 1*: Treatment plant schematics.

Headloss	520 mm water head
Required steam per regeneration	975 kg
Number of daily regenerations in continuous operation	12
Drying fan flowrate	3,750 m³/h
headloss	520 mm water head
motor power	11 kW
Condensation water content	1 g solvent/L

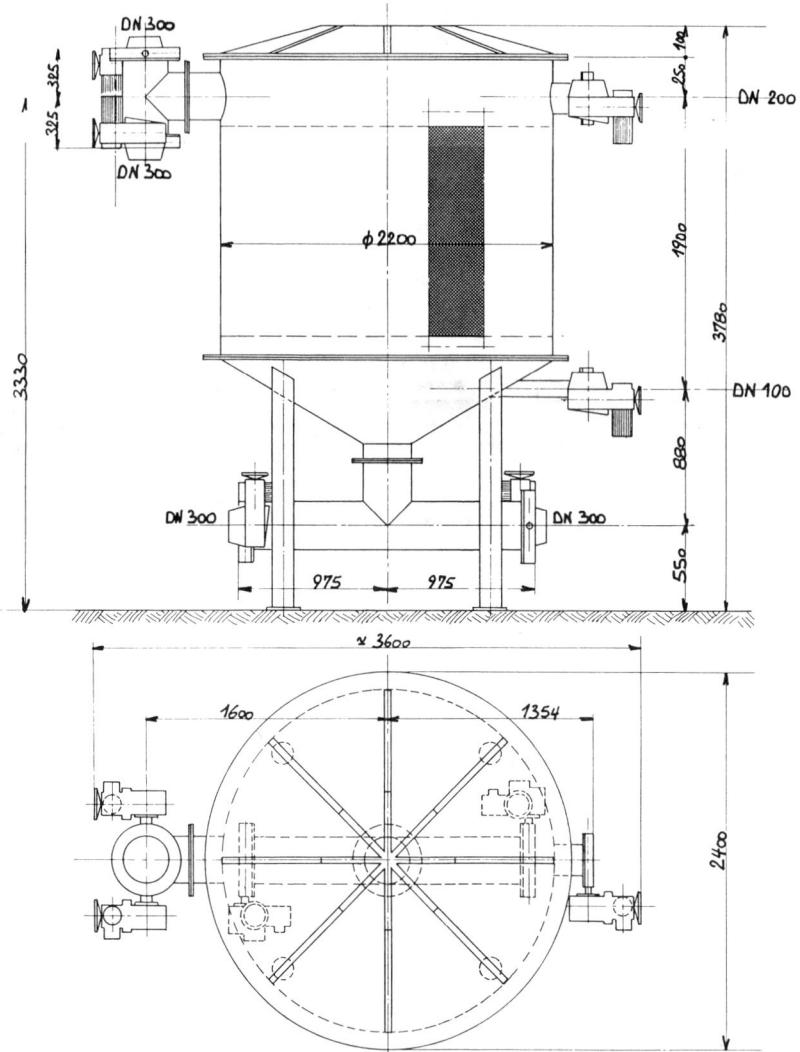

Figure 11.19. *Example 2*: Schematics of an adsorber.

11.7 Conclusion

Odorous molecule adsorption is a very interesting technique. It allows a great flexibility of use and a proper design for a variety of cases. Granulated activated carbon is used more and more frequently for this application. However, new forms of activated carbon (fibers, woven and nonwoven material, foams, . . .) as well as materials such as zeolites and resins are begining to be used with promising results.

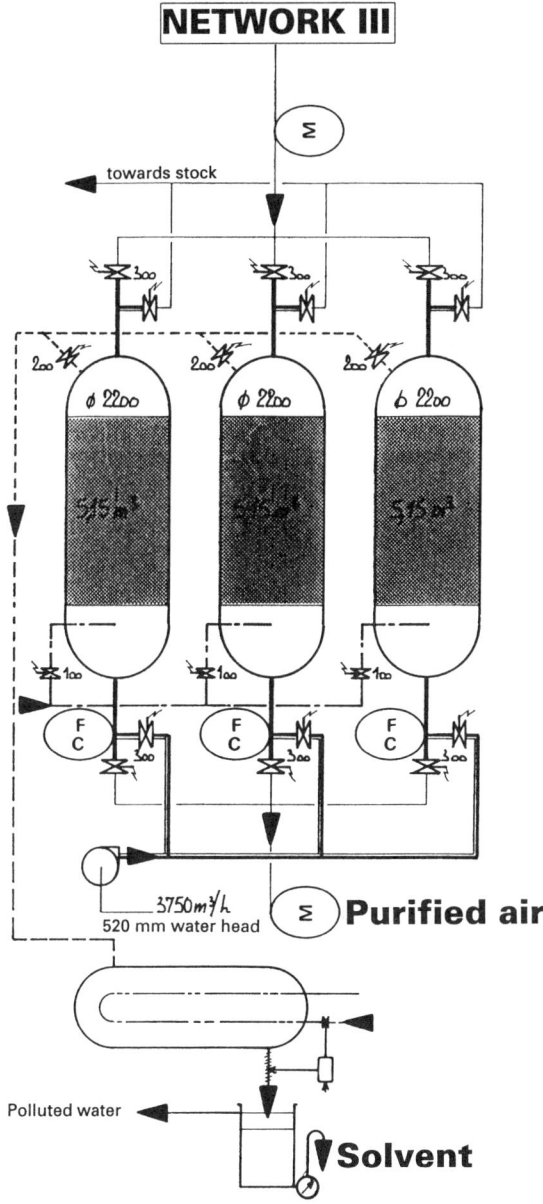

Figure 11.20. *Example 2*: Treatment plant schematics.

The adsorption mechanisms knowledge as well as modeling and simulation are well advanced in the case of gas-solid reactors. These data and knowledge allows to suggest designs for the removal of odorous molecules from air.

11.8 Notations

a	Specific surface area
b	Constant
C	Concentration
C_e	Equilibrium concentration
C_i	Interface concentration
d	Diameter
d_p	Particle diameter
G	Solution mass
H	Bed depth
k	Constant
k_f	Mass transfer coefficient (film)
k_1	Adsorption constant
k_2	Desorption constant
M	Adsorbate mass
m	Adsorbant mass
n	Constant
P	Pressure
P_0	Saturation pressure
q	Adsorbed amount (loading)
q_e	Equilibrium capacity
q_m	Maximum capacity
Q	Flowrate
Re	Reynolds' number
S	Cross section surface area
Sc	Schmidt's number
Sh	Sherwood's number
t	Time
U	Velocity (filtration rate)
U_0	Empty bed velocity
V	Volume
V_a	Adsorbed volume
V_m	Required volume for the formation of a monolayer
z	Bed depth
α	Constant
β	Constant

γ	Constant	
ε	Void fraction	
Θ	Contact time	
μ	Viscosity	
ρ	Volumetric mass	
φ	Shape factor	

Bibliography

Akita, M., Hamagushi, M. (1981). A deodorization system employing ion exchange resin for an adsorption method. *Kagaku to Kogyo* **55**, 80–84.

Bansal, R. C., Donnet, J. B., Stoeckli, F. (1988). *Active Carbon*. Marcel Dekker Inc., New York.

Baudu, M. (1990). Interactions de molécules organiques avec un charbon actif, Thèse de l'Université de Rennes 1.

Benzarl, J. R., Zundal, C. (1973). Brevet USA no. 8, 3720626.

Boehm, H. P., Diehl, E., Heck, N. (1965). *Proceedings of the Conference on Ind. Carbon Graphiste,* London, pp. 369–379.

Boehm, H. P. (1966). *Advances in Catalysis,* Vol. 16, edited by D. D. Eley, H. Pines, and P. B. Weisz. Academic, New York, pp. 179–274.

Bohn, H. L. (1972). Soil absorption of air pollutants. *J. Environ. Quality* **1**(4), 372–377.

Bohn, H. L. (1975). Soil and compost filters of malodorant gases. *J. Air Pollut. Control Ass.* **25**(9), 953–955.

Boki, K. (1977). *Nippon Kagaku Kuishi* **32**, 482–493.

Boki, K., Tanada, S. (1983). Comparative adsorption of hydrogen sulfide and methyl sulfide inside micropores of activated carbon. *Environ. Technol. Lett.* **4**, 411–416.

Brunauer, S., Emmet, P. H., Teller, E. (1938). *J. Am. Chem. Soc.* **60**, 309–319.

Cheremisinoff, P. N., Ellerbusch, F. (1978). *Carbon Adsorption Handbook*. Ann Arbor Science, Ann Arbor, MI.

Chou, T. C., Lin, T. Y., Awang, B. J., Wong, C. C. (1986). Selective removal of H_2S from biologics by a packed silica adsorber tower. *Biotechnol. Progress* **2**, 4, 203–209.

Clark, R. M. (1987). Evaluating the cost and performance of Full Scale granular activated carbon systems. *Environ. Sci. Technol.* **21**, 6, 573–580.

Connesson, C. (1988). Étude des polysaccharides bactériens: analyse et formation. Thèse de l'Université de Rennes 1, no. d'ordre 296.

Dodorik, T., Noguchi, Y. (1984). Deodorization of waste gases from cost steel plants by soils. *Chuko to Tunko,* 385, 9–13.

Donnet, J. B. (1970). *Bull. Soc. Chem.* 3353–3366.

Donnet, J. B., Hueber, F. (1962). *Bull. Soc. Chim.* **8–9**, 1727.

Fettig, J., Sontheimer, H. (1987). *Kinetics of Adsorption on Activated Carbon,* parts I, II, III, 113, 4, 764–810.

Frazier, S. E. (1986). Filter element and method for removing odors from indoor air. US patent no. 4 604 110.

Gaïd, K. (1980). Mode d'élimination de polluants sur filtre, Thèse Docteur ès Sciences, Université de Rennes 1, Série B, no. d'ordre 344, no. de série 200.

Gradsztajn, S., Conard, J., Benoît, H. (1970). *J. Phys. Chem Solids* **31**, 1121–1135.

Gradsztajn, S., Vivien, D., Conard, J. (1971). *J. Phys. Chem. Solids* **32**, 1507–1520.

Harper, M., Purnell, C. J. (1990). Alkylamonium montmorillorite as adsorbents for organic vapors from air. *Environ. Sci. Technol.* **24**, 1, 55–62.

Himmelstein, K. J. (1976). US patent no. 3965036.

Kasaoka, S., Susaoka, E., Funahara, M., Assano, K. (1981). Adsorption of dimethyl sulfide with various ion exchange Y type zeolite. *Nippon Kagaku Kuishi* **12**, 1945–1950.

Kast, W., Otten, W. (1987). The breakthrough in fixed bed adsorbers: methods of calculation and the effects of process parameters. *Intern. Chem. Eng.* **29**, 2, 197–211.

Kim, B. R., Chian, E. S. K., Cross, W. H., Cheng, S. S. (1986). Adsorption, desorption and bioregeneration in an anaerobic granular activated carbon reactor for the removal of phenol. *J. Wat. Pollut. Control. Fed.* **58**, 1, 35–70.

Koelliker, J. K., Miner, J. R., Hellickson, M. L., Nakaue, H. S. (1980). A zeolite packed scrubbers to improve poultry home environment. *Trans. ASAE* 157–168.

Ladousse, A. (1989). Service environnement, Elf Aquitaine Lacq (personal communication).

Le Cloirec, P. (1985). Étude des interactions soluté charbon actif, modélisation de réactions biotiques et abiotiques, Thèse Doctorat ès Sciences, Université de Rennes 1, Série B, no. d'ordre 423, no. série 235.

Le Cloirec, P. (1990). Dégradation biologique de produits soufrés. Internal Report, Groupement de recherche de Lacq, Elf Aquitaine.

Le Cloirec, P., Baudu, M., Martin, G. (1990a). Dispositif de traitement de fluides, patent.

Le Cloirec, P., Baudu, M., Martin, G. (1990b). Élimination de solvants sur filtre du charbon actif (to be published).

Le Cloirec, P., Baudu, M., Martin, G. (1990c). Les nouveaux charbons actifs (to be published).

Le Cloirec, P., Martin, G. (1983). Procédé de désodorisation d'effluents gazeux sur charbon actif, French patent no. 8310934.

Le Cloirec, P., Martin, G., Bernard, T. (1986). Bioregeneration of granular activated carbon: an investigation by radiochemical compounds and microbreakdown. *Water S.A.* **12**, 4, 169–172.

Le Cloirec, P., Martin, G., Gallier, J. (1988). 1H NMR investigation on saturated and unsaturated activated carbon, quantification and dynamics of protons. *Carbon* **26**(3), 275–282.

Legros, E. F. (1980). Élimination des odeurs par adsorption sur charbon actif. *Trib. Cebedeau* **435**(33), 81–90.

Lemarchand, D. (1981). Contribution à l'étude des possibilités de rétention de matières organiques en solution dans l'eau potable sur charbon actif. Thèse de doctorat d'ingéneiur, Université de Rennes, Série B, no. ordre 133, no. de série 96.

London, F. (1930). Properties and applications of molecular forces, 2. *Phys. Chem.* **B11**, 222–251.

Martin, G., Le Cloirec, P., Lemasle, M., Cabon, J. (1989). Rétention de produits odorants sur tourbes. *Proceedings of the 8th World Clean Air Congress,* The Hague, The Netherlands, Vol. 4, pp. 373–378.

Miyoshi, J., Tanada, S., Boki, K. (1978). Studies on adsorptive removal of styrene gas. *Sangyo Igaka* **20**, 6, 374–375.

Montgomery, J. M. (1985). *Water Treatment Principles and Design.* Wiley Interscience, New York, pp. 174–197.

Nishida, K., Inone, H. (1986). Removal of odor components by soil. *PPM* **17**(9), 37–47.

Nishins, H., Aibe, T., Ogino, F. (1981). Deodorization method, US Patent no. 4 256 728.

Noguchi, F., Nakamura, R., Veda, Y., Otor, K. (1984). Adsorption removal of malodorous gases by zeolotite tuffs. *Nippon Kogyo Karischi* **100**, 1162, 1150–1156.

Ose, Y., Nagase, H., Sato, T. (1977). Deodorization with sawdust. *PPM* **8**(5), 14–20.

Osterburg, G., Gluzek, K. H., Neier, W. (1987a). Process for deodorizing isopropyl alcohol. German patent, DE 360 82 02.

Osterburg, G., Gluzek, K. H., Reith, W., Neier, W. (1987b). Deodorizing isopropyl alcohol. German patent, DE 36 082 10.

Perret, R. (1982). Étude de l'amélioration des performances des charbons actifs pour traiter des effluents gazeux malodorants. Commission of the European Communities, Technical coal research.

Perry, R. H., Green, D. W. (1984). *Perry's Chemical Engineer's Handbook.* McGraw Hill, New York.

Popper, K., Camirand, W. M., William, G. S., Metchi, E. P. (1978). US patent no. 4 073 747.

Radeke, K. H., Woff, J. H. (1989). Separation of organic components from gases. German patent DD 23 4797.

Redlich, O., Petersen, B. L. (1959). *J. Chem. Phys.* **63**, 1024–1025.

Ritoul, J. C. (1983). A heating apparatus for heating solid, particulate material. European patent no. 0 104 749.

Sato, E., Nakao, H. (1986). Odor central at a sewage treatment plant. *Kogai to Taisaku* **22**, 10, 27–34.

Shigematsu, A., Suzuki, K. (1987). Thermoplastic resin composition. European patent no. 2 203 928.

Stecker, H. A. (1962). US patent no. 3 024 203.

Stumm, W., Morgan, J. J. (1981). *Aquatic Chemistry,* 2nd ed. John Wiley and Sons, New York, pp. 599–684.

Sueda, H. (1986). Efficiency of soil deodorization method. *Suido Korn* **22**, 8, 33–35.

Tanada, S., Boki, K. (1979). Adsorption of various kinds of offensive odor substances on activated carbon and zeolite. *Bull. Environ. Contam. Toxicol.* **23**, 4–5, 524–530.

Tanada, S., Boki, K. (1980). Adsorption and removal of industrial offensive malodotous substances. *Kushima Bunri Daigaku Kivo* **20**, 9–18.

Tanada, S., Boki, K., Omori, Y., Imaki, M. (1977). Adsorption of methylsulfide on porous adsorbants (activated carbon, Zeolites and silicates). *Tokushima bunri Daigaku Kiyo* **17**, 67–74.

Tanaka, H., Tanaka, T., Tanaka, T. (1976). Composite adsorbent. US Patent no. 3 960 771.

Terashima, Y., Urabe, S., Ito, H. (1984). The utilisation of composed sewerage sludge and odor treatment. *Asian Environ.* **6**(2), 18–26.

Togashi, J., Suzuki, M., Hirai, M., Skodu, M., Kuboku, H. (1986). Removal of ammonia by a peat biofilter with and without nitrifier. *J. Ferment. Technol.* **67**(5), 425–432.

Trambouze, P., Van Landeghem, H., Wanquier, J. P. (1984). *Les réacteurs chimiques*. Technip, Paris.

Tse Chuan Chou, Theng Yueh Liu, Bong Joe Huang, Chung Cheng, Wang (1986). Selective removal of H_2S from biogas by a packer silicon adsorber tower. *Biotech. Progress* **2**(43), 203–209.

Tsuchiya, K. (1980). Development of Zeolite adsorption paper and its effect. *Kagaku Gijut-sushi Mol.* **18**(2), 89–92.

Verbestel, J. B. (1980). Apparations and method for treatment of lignocellulosic materials basic on formaldehyde resins. Belgian patent no. 879322.

Waller, G. (1974). Adsorption. *Chem. 2 Industry* **1**, 853–856.

Wallis, D. A., Bolton, E. E. (1982). Biological Regeneration of activated carbon. *Alche J.* **219**(78), 64–70.

Weber, W. J., Liu, K. T. (1980). Determination of mass transport parameters for fixed bed adsorbers. *Chem. Eng. Commun.* **6**, 49–60.

Wendlandt, H. G., Henrich, L. (1969). US patent no. 2 961 413.

Yoshino, T. (1979). Ion exchange and odor Removal. *Akushu No Kenkyer* **8**, 37, 44.

12

Biodeodorization

G. Martin and G. Besson

12.1 Introduction

In previous chapters, the mechanisms of olfaction and odor quantification have been presented as well as physical-chemical gas purification processes. It is important to emphasize that most gas nuisances are the result of

1. Biomass decomposition: sulfur, nitrogen, and carbon containing compounds
2. Industrial anthropogenic compound production (this has to do with V.O.C. such as hydrocarbons, chlorine derivatives, sulfur and nitrogen oxides, . . .)

Biopurification can certainly play a role in removing these compounds. It may be that certain substances are degradation resistant and require the use of special techniques, as will be seen.

As our knowledge stands at the present time, biodeodorization is mainly concerned with organic aliphatic compounds and nonorganic or aliphatic sulfur and nitrogen containing compounds. For example, new processes make it possible to treat cyclic aromatic compounds as well as chlorinated solvents, which, like aromatic hydrocarbons, have an especially resistant nature. In analogy with water treatment, the interest of combining oxidation (for example, ozone) and biological processes can also be mentioned.

Most units work by continuous reactor at steady state; nevertheless, operation shut downs are frequent. It is thus important to specify functioning in both situations.

12.2 Gas Purification Principles

12.2.1 Processes

Gas purification implies contact between the biomass and substratum. The compounds to be degraded are generally in low concentrations. These undesirable compounds are more or less hydrosoluble, that is to say, more or less absorbable. The product-biomass contact can be accomplished in many ways:

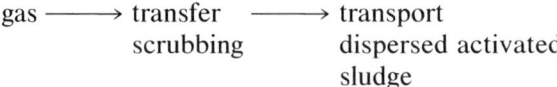

In the case of compounds with low hydrosolubility, transfer and purification are done in a water-solvent emulsion, using the solvent for extraction and transfer.

Scrubbing followed by biopurification leads to a process known as *bioscrubbing* (N.B.: scrubbing can be done with settled water)

gas ⟶ transfer and ⟶ regeneration
      biopurification   or periodic
      by fixed   scrubbing of
      biomass   biofilter

This methodology of bringing microorganisms into contact with noxious compounds is described as *biofiltration:*

gas ⟶ transfer to ⟶ periodic
      adsorbing   regeneration
      support   of support
      material   material

At the present time, this technique has not been used; it is being considered for certain activities, but will not be discussed within the framework of this chapter.

A majority of cases are treated by bioscrubbing or biofiltration. In bioscrubbing, the compounds are generally extracted with water, which transports them to the purification microorganisms. The microorganisms may be present in the water used for scrubbing. Thus, purification may be done in a separate reactor.

In a recent variation of bioscrubbing (French patent Besson/Lebault, September 1989, licensed to Société MURGUE-SEIGLE), the liquid used for scrubbing is an emulsion of water and a solvent with low vapor partial pressure and low biogradability. This allows the treatment of pollutants with very low solubility and a very low toxicity threshold, since the solvent phase slowly diffuses the pollutant into the aqueous phase.

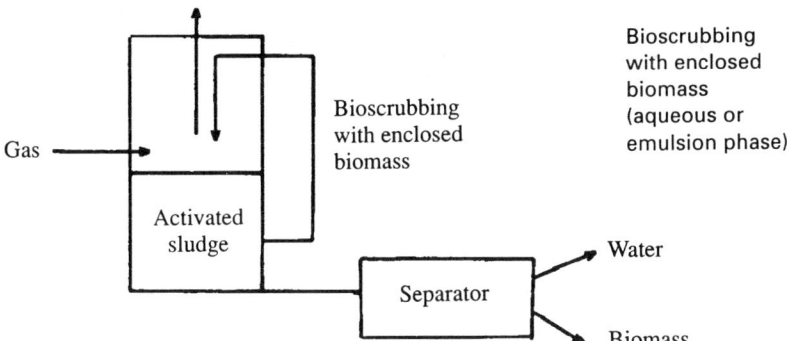

Figure 12.1. Basic flowchart of a bioscrubber.

In biofiltration, the gas to be purified contacts material on which micro-organisms are present. A fixed or movable bed can be used, but the material is often fixed.

The main difference between these two technologies essentially concerns the physical-chemical properties of the compounds to be removed. If the compounds have low hydrosolubility (they are then often adsorbable), one would be well advised to use biofilters or an emulsion bioscrubber. In the latter case, the structure of the supporting and transfer material should be specified. For instance, solvents such as xylenes and gases such as H_2S in acidic conditions are purifiable using emulsion bioscrubbers.

In the case of rather hydrosoluble substances, bioscrubbing will be interesting. As an example, one can think about volatile acids, ketones, and ammoniac (in acidic to neutral conditions).

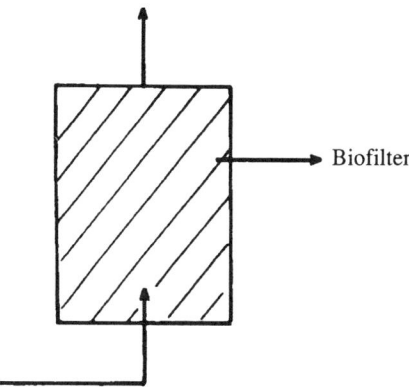

Figure 12.2. Basic flowchart of a biofilter.

In all processes, purification leads to metabolic by-products or even to biomass production. Then the role of the water–biomass separator is crucial and there is a risk of fouling in biofiltration.

12.2.2 Biomass and Nutritional Balance

In biologically transforming various substrates, it is advisable to specify the biomass to be favored and its requirements for each situation. In all cases, catabolism (transformation of a compound producing by-products and energy) and anabolism leading to the production of biomass will be specified.

More or less biodegradable organic compounds such as aldehydes, acids, ketones, arenes, and amines are absorbed in the water and adsorbed on the supporting material and contacted with the biomass. These substrates are often transported with air and aerobic heterotrophic purification can be implemented. Microorganisms such as *pseudomonas* (e.g., Klaus and Kutzner, 1980), *corynebacterium* (Ohta and Sato, 1985), *antherobacter, acinetobacter, Rhodotorula,* and *aspergillus* can develop.

The symbolic balance

$$OM + NH_4^+ + PO_4H_2^- + OE + O_2 \rightarrow C_5H_7NO_2 + CO_2$$

shows the requirements in N and P to fulfill biopurification. The biomass growth leads, for instance, in bioscrubbing to the necessity of implementing a downstream separator.

Some compounds, such as aldehydes and acids, are hydrosoluble and bioscrubbing is feasible. As far as low hydrosolubility solvents are concerned, emulsion bioscrubbing is recommended.

Klaus and Kutzner (1980) give examples of bacteria susceptible to degrade odorous compounds:

Butyraldehyde	P Putida Bu2
Diethylamine	pseudomonas spec. D1
Ethylisocyanate($CH_3CH_2N = C = S$)	Bacillus spec. is
Mercaptoethanol	P. aeruginosa Me
Thiophenol	P. fluorescens TP
Indole	alcaligenes Spec. In 2
Scatole	Nocardia Spec. Sk

These authors specify some values of μ^0 and K_M.

In the case of *nitrogenous compounds,* such as ammoniac, several routes are possible in a biological treatment scheme: Ammoniac is integrated in the cells by biological assimilation with biomass production, for instance, in an aerobic heterotrophic biological process. Then organic matter (for instance, methanol, glucose, etc.), phosphates, and oligo-elements must be provided to fulfill the assimilation. The biomass growth that is produced is significant.

Autotrophic nitrification requiring dissolved CO_2 and CO_3H^- is also feasible. The following mass balance expresses nitrification:

$$NH_4^{+3} + 1.83O_2 + 1.98CO_3H^- \rightarrow 0.021\ C_5H_7NO_2 + CO_2 + 1.04\ H_2O + 0.98\ NO_3^- + 1.88\ H_2CO_3$$

Biomass growth is smaller than in the case of heterotrophic assimilation. Water issued from biotreatment may indeed contain nitrates. In our process [Besson and Martin (French Patent No. 2259-122) (1986)], wet peat is used to fulfill the nitrification by associating an exogenous and/or endogenous heterotrophic denitrification. Nitrogen also undergoes nitrification, denitrification, and the purification of ammoniac-containing gases releases nitrogen into the air.

Thus, nutritional balance depends on the chosen method. In the case of the nitrification denitrification process, nutritional balance requires the addition of CO_2, P, and OE. In the assimilation process, the addition of organic material (OM), phosphates, and oligo-elements (OE) is indispensible.

When dealing with *sulfur compounds*, many solutions can be considered. In the case of H_2S, for example, sulfur is a cellular microelement and can be eliminated by an NH_3 or organic compounds removal process if the contamination level is low. If H_2S is the major or unique pollutant, it must be transformed: H_2S can be oxidized into S, SO_2, and SO_4^{2-} by various microorganisms. The energy can be produced by photochemistry or chemical oxidation. In this way, Depeyre et al. (1987) developed photobiopurification of H_2S containing gases. *Chromatium* bacteria are photolithotrophic.

Another possible method is chemical-autotropic biological oxidation, which requires CO_2, N, P, and OE (Dalouche, 1989). An aerobic biofilter using peat or maërl (patent pending) (Maërl is a natural algal calcium carbonate used for remineralisation. The scientific name of the algae is *Lithotamnium Calcareum*), or a packed bioscrubber with purification at the tank bottom. The microorganisms in use are *thiobacillus*, such as *thiobacillus thioxydans, thiobacillus thioparus*, and *Intermedius*.

The following mass balance has been established by Dalouche (1989):

$$0.018(HS^- + S^{2-}) + 1.5O_2 + 0.078NH_4^+ + 0.078HCO^{3-} + 0.312CO_2 \rightarrow 0.836SO_4^{2-} + 0.078C_5H_7NO_2 + 0.382H^+ + 0.358H_2O$$

For a gas or nonaerated liquid phase, *thiobacillus denitrificans*, for example, can be used with NO_3^- addition (Dalouche, 1989; Lyn, 1986: Cooper et al., 1975).

The significant difference between the biochemical cycles of nitrogen and sulfur should be emphasized: nitrogen may be in a gaseous form (N_2) while sulfur is solid. Under such circumstances, techniques for treating sulphurous products should be oriented toward oxidized soluble species (SO_4^{2-}) biological production, using water for by-product removal from the reactor.

12.2.3 Modeling Elements

Biopurification depends on physical-chemical transfer phenomena and biological reaction. According to implementation (bioscrubbing or biofiltration) and biomass type, the process may be different but at this point, one should be reminded of the general phenomena likely to be involved.

Liquid-gas transfer kinetics (see also Chapter 10). The double boundary layer theory offers an explanation for gas absorption in water. Gas flux rate (polluted air) in the packing material, which is supposed to be invariant with solute removal, is written

$$Q_G dC_g = K_L a(C_e - C_L)A \, dh$$

which formalizes the nonreacting mass transfer occuring in the *dh* height of the packing material (see Section 12.5 for notation).

The C_e concentration follows from Henry's law $[C_e = HC_g]$. Thus

$$U_{g0} \, dC_g = K_L a(C_e - C_L) \, dh$$

Under certain conditions, such as

- H_2S or NH_3 dissolution
- Transfer in presence of chemical or enzymatic catalysts (e.g., Fe^{2+} for H_2S)
- Gas scrubbing with biomass

reaction occurs within the interfacial liquid film and an acceleration coefficient is introduced (E).

The third relationship refers to the gas to liquid mass transfer during a period of time *t* for a liquid volume V_L and whether or not the liquid is stationary:

$$Q_g dt \, dC_g + V_L dC_L = 0$$

Gas - solid transfer kinetics (see also Chapter 11). Solute adsorption onto the solid surface will play an essential role in biofiltration.

In a gas-solid mass transfer operation to an adsorbent with physisorption, gas diffusion, superficial fixation, and porous diffusion can be involved. Gaseous solutes concerned in the present study have a high diffusivity and the limiting phenomenon is probably the superficial fixation. Once fixed on the surface and if the solute is biodegradable, on one hand and, since the biomass only exists within macropores on the other hand, then diffusion into micropores will not occur. (It should be noted that diffusion into micropores could happen in contact adsorption with subsequent bioregeneration.)

Physical laws dealing with mass transfer and adsorption are numerous however, only Langmuir's mechanism will be discussed:

$$+ \frac{dC_g}{dt} + U_0 \frac{dC_g}{dz} + \frac{1}{\varepsilon} \frac{dC_s}{dt} = 0$$

C_g and C_s being related through the mass transfer kinetic law.

In the case of a Langmuir fixation the following relationship formalizes adsorption and desorption:

$$\frac{dC_s}{dt} = K_1 C_g (C_m - C_s) - k_2 C_s$$

The maximum sorption capacity C_m satisfies the following equilibrium relationship:

$$K_1 C_{ge} (C_m - C_{se}) = K_2 C_{se}$$

hence

$$C_{se} = (K_1 C_{ge} C_m)/(K_2 + K_1 C_{ge})$$

The nature of the solid phase and, hence, its superficial structure determines the maximum adsorption capacity. Physisorption involves interactions (dipole, dipole - electron, . . .) determined by the structure of the material. When we chose to work with wet peat, we tried to prove the importance of the chemical functions revealed during wetting [Martin et al., (1989) and Dalouche (1989)]. Figure 12.3 shows the interest in controlling the wetting process.

Microbiological kinetics. A purifying biomass is either fixed onto the solid or dispersed in the water in a flocculated form. The solid can be inert: soil, peat, activated carbon, plastic packing material. In this case the microorganisms are stationary and constitute a biofilm or more or less localized and dispersed aggregates. But it is assumed that the supporting material is involved in the fixation mechanism and not in the biological metabolism.

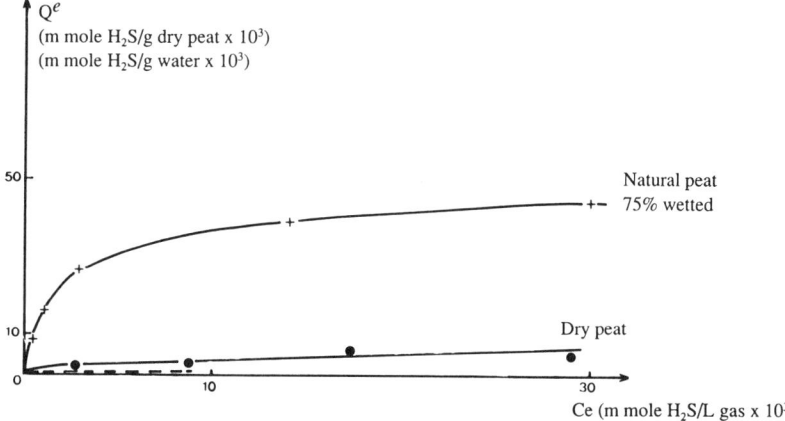

Figure 12.3. Gas adsorption by wet peat.

Biofilters wherein the supporting material is also a substrate are developed as well (Dalouche, 1989). For instance, a bed of *lithotamnium calcareum* (or maërl) allows the fixation and development of *thiobacillus thioxydans* that transform H_2S into SO_4^{2-} and H^+. The carbonates are degraded and serve as substrate in the autotrophic bacteria development. In this case, the packing material is slowly consumed and must be supplemented.

According to the operating mode and the gas concentration to be purified, dispersed biomass or biofilm processes can be implemented. This determines the kinetic rate. The elementary laws of biokinetics explain the correlation between substrate (S) consumption and biomass production:

$$+\frac{dX}{dt} = (\mu - b)\, X$$

$$-\frac{dS}{dt} = (\mu/y)\, X$$

The growth rate coefficient μ depends upon the mechanisms brought into play and the nature of the involved microorganisms.

The mass balance of a biological transformation always results from the catabolism and the anabolism. If several substrates S_i are limiting, μ can be written according to various models among which the Herbert Reynolds and Monod model is discussed:

$$\mu = \mu_0\ \text{II}\ S_i/(K_{Mi} + S_i)$$

Deficiency in one of the substrates inhibits growth. The main substrates are as follows: C, N, and P; Ca, Mg, and K micro-elements; and Mo, Cu, etc. oligo-elements. Biological reactor modeling requires knowledge of the viable biomass partition (e.g., on the supporting material). An easy access to this information is constituted by the measurements of VSS, ATP, TPC, or, better yet, specific microorganism counts. Then substrate transfer is written as corresponding to the local evolution of the biomass.

The maximum growth rate coefficient μ^0 depends on the nature of the microorganisms. For example, autotrophic nitrifying bacteria have a $\mu^0 \approx$ 0.1 day^{-1}. For ordinary heterotrophic microorganisms μ^0 approximates 1 hr^{-1}.

12.2.4 Conclusion

In this rapid presentation of the main phenomena susceptible to being involved in biopurification, the complexity of the process has been demonstrated. It requires both mass transfer and biological assimilation. In the following discussion, precisions will be given on the respective weight of these latter processes as a function of the implemented technology.

12.3 Applied Biopurification

Industrial activities susceptible of being concerned with odorous product emissions are numerous and can be grouped in general categories:

1. *Agricultural industries:* Fermentation; cheese industry; yeast manufacturing; coffee roasting; slaughterhouses; biomass production (maggots, worms, . . .); pectin manufacturing; etc.
2. *Purification and biological refuse treatment:* Sewage treatment plant; sewers; compost; solid waste processing
3. *Heavy and light chemical industries:* Petrochemistry; thiol, amine, and aldehyde industries; fine chemical industry; perfume industry (essential oils)
4. *Iron and steel industries:* Foundries

Situations and levels of compound removal are extremely varied. The following is a discussion of practical application relating to industry, output to be treated, involved compounds, contamination levels, available space, and required technology.

Biological gas purification depends on two main consecutive processes:

1. Solute absorption and/or adsorption in water and a liquid or solid wetted support
2. Biological solution regeneration; generally, an aerobic process

Schematically, if absorption (and adsorption) and biological regeneration are done in the same unit with a solid support, one speaks of biofilters. In the case where transfer to water and regeneration are done in the liquid phase (often in two different units), one speaks of bioscrubbers.

Biofilters or bioscrubbers will be used depending on the situations encountered and the nature of the compounds to be removed. If the pollutant to be removed is a compound with low solubility, the use of a bioscrubber using water as scrubbing liquid should be avoided. If the pollutant is a toxic compound with a low concentration, the use of a bioscrubber using only water as scrubbing liquid and biofilters in which toxins may accumulate should be avoided. If the pollutant is a highly concentrated sulfur compound (200–$2,000$ mg/m^3, for example) a biofilter with a replaceable support is used. These few case examples allow a choice to be made between the different applicable processes described below.

12.3.1 Biofilters

12.3.1.1 Generalities

12.3.1.1.1 Biological Gas Purification

Biological gas purification has been in use for about 30 years. The first patent filed in the United States was Pomeroy's (1957). In Europe, the first biofilter was apparently built in Switzerland in 1964. Since then, research and implementation have multiplied. The "VDI" Commission in the Federal Republic of Germany has established installation recommendations (VDI guideline no. 2377).

In France, research has developed at the request of the Ministry of the Environment and meat packing companies, the signed contract conditions having stimulated interest in biofilter application in slaughterhouses, in particular.

12.3.1.1.2 Malodorous Compounds

When air carrying malodorous compounds goes through wetted material, a transfer of these compounds by absorption and possibly by adsorption occurs. The fixed microorganisms on the material transform the fixed compounds purifying the supporting material.

If the biomass is evenly distributed with a concentration x per unit volume of reactor, an elementary plug flow tank reactor model can be written in the case where the void fraction ε is constant and the packing is not degraded and in the case of a unique process where transfer and consumption compensate each other:

$$\Sigma Q_g dC_g = E\,K_L\,a\,A\,dh(C_s - C_1) = \frac{\mu}{y} \times A\,dh$$

hence $$\eta = \frac{C_{g0} - C_g}{C_{g0}} = \frac{\mu}{Y} \cdot \frac{1}{C_m} = \frac{\mu}{Y} = \frac{x}{\theta}$$

assuming that the growth rate μ is constant (nonlimiting substrate).

As far as biomass production is concerned, assuming that it remains stationary, one writes

$$\frac{dx}{dt} = (\mu - b)\,x$$

The growth rate μ depends on the species concerned. It involves the nutritional equilibrium when one wishes to develop a particular activity. For example, to remove NH_3 by assimilation a solution with assimilable carbon, phosphates, calcium, magnesium, etc., in a ratio close to $DBO_5/N/P \approx 100/5/1$ must be provided to the reactor.

In our process (Besson and Martin, 1986), we attempt to remove NH_3 after biological transformation in N_2 form and hence, with nitrification and denitrification. The use of carbonates in reactors can thus be justified.

In the case of biopurification of gas containing H_2S, we (i.e., G. Martin as well as Société MURGUE-SEIGLE) implement oxidation in SO_4^{2-} and H^+ by autotrophic microorganisms. Carbonates are necessary in this instance as well.

Several remarks should be made at this point:

- The biomass in the biofilter can expand and the yield evolves with the massic load.
- The biomass does not involve a unique species. In certain cases, a specific biomass can be developed, but the cost may be higher than in nonrestricted operations.
- The biomass is not often evenly distributed.
- The growth rate μ varies considerably with the microorganisms involved, particularly heterotrophic bacteria as compared to autotrophic bacteria.
- To optimize the efficiency necessitates an improved growth, that is to say, a modification of the porosity of the supporting material.
- One can attempt to work with a constant biomass by controlling the limiting factors (method recommended by SAPS and ourselves).
- In certain cases, an increase of head loss due to decreased porosity can be observed, which necessitates either additional steps in the process (e.g., scrubbing) or a change in the material itself.

12.3.1.1.3 Packing Material

Packing material is involved in the process in several ways. It must be humid (50–80%), have an affinity for the solute to be removed (NH_3 fixation on wet peat has already been shown), and for the bacterial support. In certain operations, it can be consumed, have a high void fraction ($\varepsilon = 0.5$), and remain as a 1 m thick layer without undergoing packing.

Frequently used materials include peat, bark, compost, limestone, plastic materials, and activated carbon. De Savorin (1986) claims the use of activated carbon as a support under various conditions.

In our operations, having the objective of removing nitrogen containing compounds, we use a peat based mixture as support. Gases entering biofilters have a relatively high temperature, for instance, 15 °C for the air coming from a sludge treatment building or 30–40 °C for sludge cooking condensates. Deodorization operations can be carried out in psychrophilic, mesophilic, or thermophilic conditions. The addition rate of water must be adjusted according to the operating temperature.

The liquid and gas circulations in the packed reactors are very often countercurrent. In the biofilters especially devoted to H_2S removal we can operate in a countercurrent mode, with water recycling. Biodegradation of the support material necessitates periodic washing.

12.3.1.1.4 Examples of Operating Conditions

The figures in Table 12.1 correspond to our experiments and confirm those reported by Brauer (1986). Removal efficiency with respect to treated compounds will be specified in the following examples.

Table 12.1. Examples of Operating Conditions

Material	ε	U_0 (m h^{-1})	θ_s	$\Delta P(CE)$
Peat Compost Bark	0.4–0.6	100–200	15–30	1.5–3
Maërl	0.5	400–500	3–6	10
Plastic	0.9	300	5	0.05

12.3.1.2 Research and Implementation Examples

Research done by Fisher and Bardtke (1984) and Bardtke et al. (1987) demonstrate organic compound purification on compost: xylene, dichlormethane, dimethyl isopropylamine, and butanol. The purification yield is given according to the material as follows:

$$\eta = 1 - e^{-kt}, \; t = \frac{V_{\text{packing}}}{C_{\text{gas}}} \; \varepsilon = \varepsilon \, \theta$$

where θ = empty bed contact time (EBCT) ($\theta = V_{\text{packing}}/Q_{\text{gas}}$), k depends on the material and substrate being treated. It varies between 6×10^{-3} and 10^{-2} S^{-1}, for instance, for xylenes on compost. Figure 12.4 illustrates the suggested implementation. Working speeds described by the authors are rather slow.

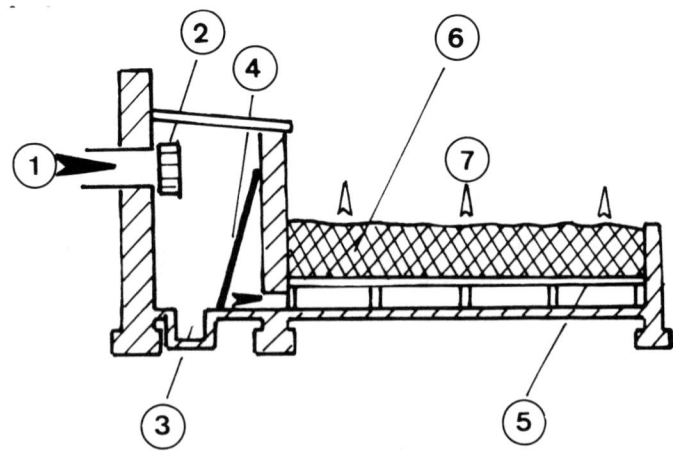

Figure 12.4. Biofilter schematics according to Bardtke et al. (1987). **1**—Gas to be purified. **2**—Ventilator. **3**—Pump. **4**—Dust filter. **5**—Floor. **6**—Filter. **7**—Purified gas.

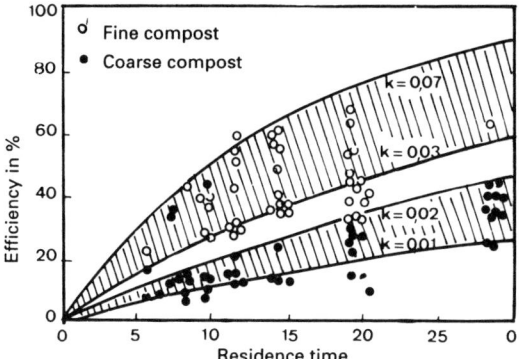

Figure 12.5. Butanol removal efficiency as a function of residence time, according to Bardtke et al. (1987).

The efficiency of the process is equally dependant on the support solvent and the operating load (e.g., in g/m³ j). Butanol is satisfactorily purified (Fig. 12.5); other substrates require a lower loading (Fig. 12.6).

The organic solvent's water solubility and the type of packing are determining factors in purification using fixed bacteria systems. Thus Kirchner et al. (1989) removed 80% of pollutants such as acetone, propionaldehyde, naphtalene, and toluene using single strain cultures using wetted wall (wetted packed) bioreactors with spatial rates approximating 1,000 mh⁻¹ for pollutant concentrations ranging 5–35 ppm.

Tabasaran et al. (1979) deodorized a biological gas containing CH_4, CO_2, N_2, O_2, and H_2 as main constitutuents and H_2S, NH_3, acetone, benzene, and

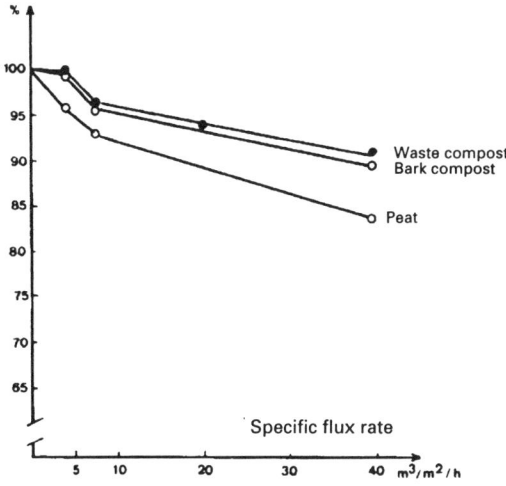

Figure 12.6. Removal rate as a function of loading, according to Bardtke et al. (1987).

mercaptans using a biofilter whose support layer is made of 10–20 cm of sand with an effective diameter ranging 6–8 mm and a wet compost layer of 40–60 cm thickness. In these conditions, CH_4 can be biologically consumed if O_2 is provided in sufficient quantity and if the velocity is low enough.

The results reported by Koch et al. (1982, 1983) on biodeodorization are quite interesting and are summarized in Table 12.2. They were obtained using composts at velocities ranging 130–200 m/h and for contact times ranging 4–8 seconds.

The work currently carried out at Rennes by Martin et al. and the Société MURGUE-SEIGLE are aimed at the improvement of nitrogen loss in the purification of ammoniac containing gases and the implementation of carbonate based reactors (maërl) specifically designed for autotrophic operation (NH_3 and H_2S).

Dalouche et al. (1989) make it possible to exemplify the purification of ammoniac containing gases using peat. A heterogenous biomass is developed, which contains both autotrophic and heterotrophic strains. This biomass is kept constant by preconditioning the peat. Then NH_3 is converted to N_2. (See Fig. 12.7.)

Figure 12.8 shows the profile of H_2S removal along a wetted maërl bed after washing (Dalouche, 1989).

The use of peat is suggested by several authors for gas purification when H_2S is the major pollutant. Furusawa et al. (1984) examined the influence of contact time in the oxidation of H_2S into S and sulfate in a laboratory scale pilot using peat. However, under usual operating conditions, H_2S is transformed into elementary sulfur. We did observe this inconvenience in industrial scale field experiments on H_2S rich gases. It led to the implementation of a pretreatment step (scrubbing or maërl biofilter).

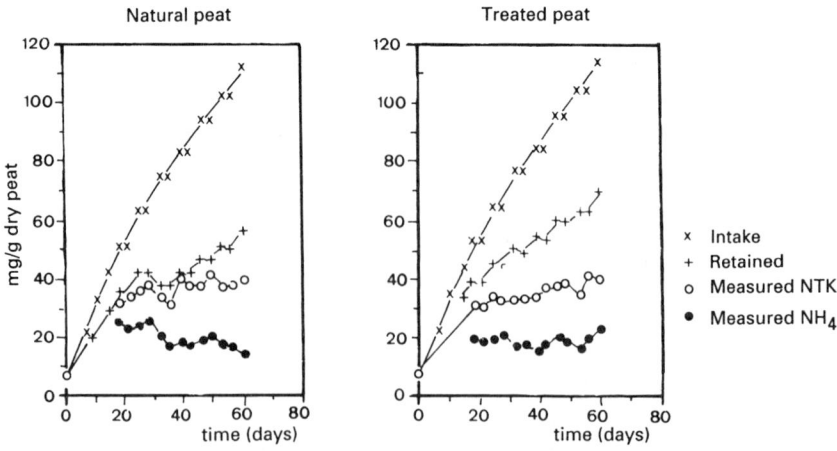

Figure 12.7. Evolution of peat composition for ammoniacal and total nitrogen. Comparison with ammoniac concentrations actually retained on the biofilter.

Table 12.2. Gas Purification Data According to Koch et al. (1983). Parameter nos.: **1:** Ammoniac; **2:** Dimethylformamide; **3:** Hydrazine; **4:** Mercaptan; **5:** Triethylamine; **6:** Formaldehyde.

Vers. No.	Concentration (ppm) (Dräger) Raw air Parameter No.				Purified Air Parameter No.	Olfactometry Raw Air	Olfactometry Purified Air
1							
2							
3	30	60–100	70	30		86	20
4	100–145	400	160	30		84	4
5	18	40	51	25		60	25
6	27	56	200	30		122	24
7	110	250	125	30	Sp.	108	4
8	110–120	250	38	30	Sp.	111	22
9	38	110	65	30		87	8.4
10	60	145		5	2.5 / 1.5	unable to measure (condensation)	

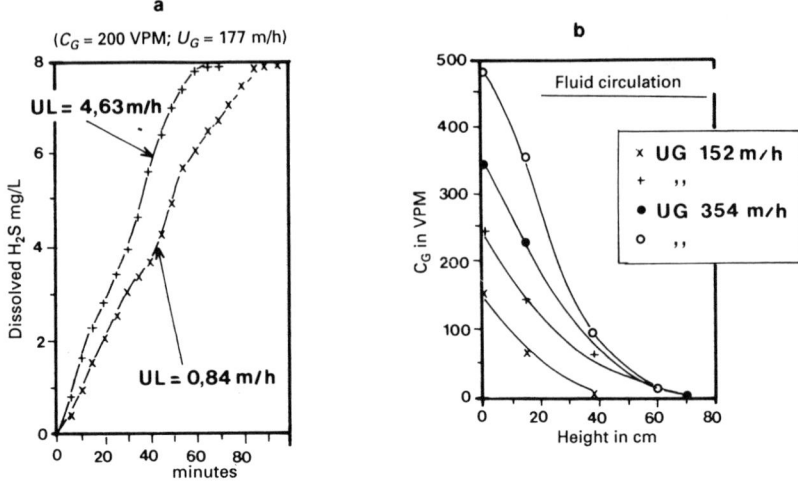

Figure 12.8.   Profile of H_2S removal along a maërl bioreactor.

Wada et al. (1986) showed H_2S degradation into sulfates accompanied by acidification (pH 4.5–2.5) by *thiobacillus intermedius* using peat. These results enable the SAPS to implement several plants:

1. *Peat biofilters* for the purification of gases coming from different industries such as fermentation, meat quartering, tallow melting, organic fertilizer manufacturing, and waste water treatment plants. The most frequent pollutants were

   - Sulfur containing compounds (H_2S and mercaptans) at 10–200 mg m^{-3} loadings.
   - Nitrogen containing compounds (ammoniac and amines) at 10–300 mg m^{-3} loadings.

   With adequate nutrient addition, removal yields of 90–99% are possible with velocities of approximately 200 m/h.

2. *Maërl biofilter or main support degradable by sulfurous gases* originating from organic synthesis, in particular hydrogen sulfide/carbon sulfide mixtures containing approximately

   - 150 mg/m^3 in H_2S (with peaks of 700 mg/m^3)
   - 900 mg/m^3 in CS_2 (with peaks of 4,000 mg/m^3)

   Removal results are

   - 99% on H_2S and
   - 90% on CS_2

   with velocities ranging from 300 to 400 m/h.

Technology implemented for gas biopurification using biofilters is very similar to that used in bacteria filters or beds. With our process (Dalouche et al., 1989), where it is necessary to wash the packing periodically, no doubt we are close to bacteria filter systems used in water purification, which require periodic backwashing. The systems that we are developing by attempting to maintain the biomass are closer to bacteria bed reactors.

Instead of working with a fixed biomass, it is possible to use maërl moving beds or biological disks for biopurification. For example, McKim Matthew (1985) reports a biological contact scrubber using disks.

Biofilter design and management requires considerable control of microbiological processes (metabolism, pH, activity rate, . . .) along with knowledge of materials, kinetics, and pollutant chemistry. The implemented process must be adapted to the pollutants to be treated; thus, it is unreasonable to use the same support and structure to remove NH_3, H_2S, and xylene, for instance. Furthermore, and depending on the process control scheme of the biofilter and/or the microbiological processes brought into play in the case of NH_3 removal, for example, an NO_3^- outflow can be obtained (Jol and Dragt, 1989) or complete purification in the form of N_2 (Dalouche et al., 1989).

12.3.2 Bioscrubbing

In a bioscrubber, the transfer of the solute into the liquid is done in a suitable liquid-gas contacter (spray towers, packed towers, venturi scrubbers, etc.) and the solubilized compounds are degraded in a tank at the bottom of the contacter, which ensures adequate residence time. A separator on the scrubbing liquid line allows the recyling of all or part of the scrubbing liquid.

For a theoretical study on scrubbing, see Chapter 10 and, in the case of H_2S, the work done by Dalouche (1989).

Flowchart

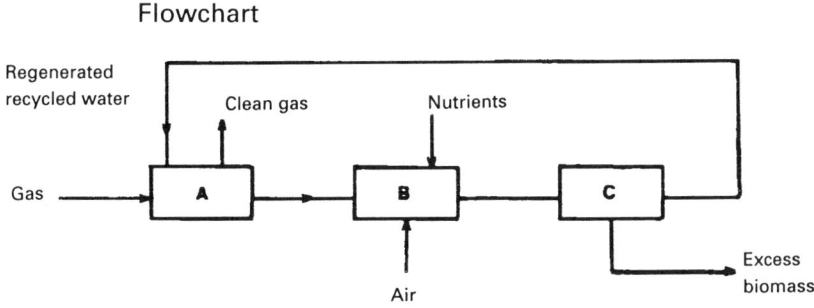

Figure 12.9. Bioscrubber flowchart. Biological emulsion. **A:** Mass transfer unit; **B:** Biological water purifier; **C:** Water-biomass separator.

The chemical metallurgical industry has given this technology a privileged role. Its implementation has been described in Germany in the VDI Guide No. 3478. It has been applied on foundry gases in Daimler-Benz painting workshops in 1982. The main pollutants were formol, ammoniac, phenols, amines, and cracking products. The amount of pollution was estimated at 120–150 mgC m^{-3}. The purification was performed with two bioscrubbers in parallel [see Brauer (1986) and Kölher (1987)]. Glanferd et al.(1979) report implementations in the food industry. The use of scrubbers for the gas purification of a Shell Mold plant was proposed by Kanezaschi and Okada (1980). The gases are contacted in a countercurrent mode with an activated sludge containing powdered activated carbon (0.2–10 g/L). The vertical column is constituted of perforated trays.

Van Geelen and Van der Hoek (1977) examined the design of bioscrubbers and suggest the use of Cloisonyle and Tellerettes as packing material. The treated gases originated from chicken or pig farms. For instance, for a 6,000 m^3/h ventilation rate, the air circulates at 1 m/s through a 1.67 m^2 scrubber. The water circulates at 6 m/h. Since the packing height is 0.5 m, the contact time is 0.5 s.

A laboratory scale experiment using a bioscrubber for the purification of H$_2$S containing gas is reported by Depeyre et al. (1987). The use of photosynthetic bacteria (*Chromatium*) in this process is interesting, but the necessary contact time seems high (10^{-4} mole H$_2$S/h/L). A 1,000 m^3 reactor would be required to treat 10,000 m^3/day of gas at 1–2% H$_2$S, due to the low biomass concentration (10^7–10^8 germs/cm^3) during these experiments.

The use of bioscrubbers operating with water is restricted to the treatment of the following:

1. Soluble, low toxicity pollutants as they can accumulate in the liquid phase, thereby inhibiting biodegradation
2. Low, relatively constant pollutant flux, as the aqueous phase does not provide a sufficient buffer capacity.

In these conditions, a water operated bioscrubber can insure

1. an 80–90% removal yield for pollutants such as butanol, ethylhexanol, glycol at a concentration of 500 mg/m^3 (in these cases, the pollutant is present at constant concentration)
2. a 15–20% removal yield for such pollutants as xylene with an equivalent load (in this case, the pollutant is poorly soluble and toxic at low concentration)

To widen the field of application for bio-scrubbing in case of

• Important charge variations
• Pollutants that are hardly soluble in water
• Toxic pollutants at low concentrations

a process known as "reversed phase" has recently been put into practice using a water–solvent emulsion as a scrubbing liquid in water–solvent proportions varying from 10–50%.

A good solvent is a liquid that has low volatility, is slightly or not biodegradable, insoluble in water, nontoxic, and being a good solvent for the pollutants to be removed. Thus, products such as

- Silicone oils
- Hydrocarbon oils

have been utilized as solvents.

This process has made it possible to degrade pollutants reputed to be hardly biodegradable such as aromatic hydrocarbons (toluene, benzene, xylene, styrene, etc.), chlorinated solvents like hexachlorocyclohexane, or polychlorobiphenyls that have a treated air flux per m^3 of reactor of 30–400 m^3/h depending on the type of material used (stirred tank reactor or venturi).

The removal yield reached with a flux of 400 $m^3/m^3/h$ for air containing 1 g/m^3 of toluene is superior to 92% and reaches 80% for the same quantity of chlorinated solvents with a flux of 30 $m^3/m^3/h$.

12.3.3 Biopurification at the Source

In a certain number of situations, one can attempt to eliminate odors by biologically treating the source of odorous gases, consequently without having to resort to a purification bioreactor. Obviously, one way to fight the problem at the source efficiently is to change the environmental conditions (O_2, redox potential, pH, etc.) in such a way that odor production and/or emission is impossible. Below are several examples of biological treatment.

Ishizaki et al. (1984) report the solid sludge deodorization by actynomecetes; Ohta et al. (1979) discuss the deodorization of bird feces.

Wastes particular to the food industry such as manure, droppings, etc., can be deodorized without NH_3 loss by species of *Cornybacterium* and *Micrococcus*. Fatty acids, amino acids, and H_2S are removed according to methods reported by Ohta et al., 1984 and Ohta and Sato, 1985. In deodorization with aeration scum formation appears to be linked to various microorganisms where *Cornybacterium* is in the majority (Ohta and Toshiyuki, 1987).

Tannery wastes in ponding operations contain, among other soluble compounds, a high proportion of sulfur compounds such as HS^-, $S_2O_3^{2-}$, SO_4^{2-} (Cooper et al., 1975). In ponds a strong H_2S and mercaptans odor develops due to bacteria (e.g., *Thiocapsa*) and algaes (*Chlorella englena, Chlamydomonas, Scenedesmus, diatoma, navicula, . . .*). Red photosynthetic bacteria (e.g., *Thiocapsa*) have been developed by aeration to remove sulfur compounds; NO_3 addition has been attempted as well.

Odor emissions are avoided in purification systems by biological activator additions (proxeronine and proxeroninase) according to Heinicke, 1987.

12.4 Conclusion

Bioscrubbing or biofiltration are economical techniques that are currently being developed. Their maintenance is more or less complex according to the process. Choice of transfer material or fixed bacteria is important and must still be more accurately specified. Reactor surveillance is simple. A chemical analysis technology may often be sufficient for self-surveillance, but optimization requires more thorough methods.

The most noteworthy advantage of biofilters, apart from their running costs, concerns process steadiness, which is advantageous in noncontinuous operation. It should be pointed out that packing material reuse is possible in the case of compost, peat, and bark use.

Bioscrubbers of superior technical quality are, no doubt, less steady but more compact; they concern more hydrosoluble compounds.

12.5 Notations

$\bar{\varepsilon}$	Void fraction of material $\bar{\varepsilon} = \varepsilon/(1 - \varepsilon)$.
a	Transfer volume
A	Right section of a reactor
h	Reactor height
Q_G	Gas flux
U_{g_0}	Gas flux rate in empty tank
K_L	Total gas-liquid transfer coefficient
ΔP	Head loss
H	Henry's constant
k	Kinetic constant
K_{mi}	Monod's constant of substate i
OE	Oligo elements
OM	Organic matter
Q_L	Liquid flux
t	Time
TPC	Total plate counts
V_L	Liquid volume
X	Biomass
μ	Biomass growth rate
μ_0	Maximum growth rate
Y	Cellular yield
S	Substrate

b	Microorganism mortality constant
VSS	Volatile suspended solid
ATP	Adenosine triphosphate
C_m	Massic load
θ	$h/U_0 = V_R/Q_g$ (Average time)

Bibliography

Bardtke, D., Fisher, K., Sabo, F. (1987). Air purification with biofilters; field of application and design criteria. *80th annual meeting Air Pollution, Control association*. New York, 22–26 June, 1987.

Besson G., Lebeault, J. M. (1989). *Procédé de biolavage*. Patent application, September 1989.

Besson, G., Martin, G. (1986). *Procédé d'épuration et de désodorisation de gaz et installation pour la mise en œuvre de ce procédé*. Patent number 2 559 122, no. 8 518 327, licensed to Société MURGUE-SEIGLE, 9 boulevard Monge, 69330 MEYZIEU, France.

Brauer, H. (1986). Biological purification of waste gases. *Int. Chem. Eng.* **26**, 3, 387–395.

Cooper, D. E., Rands, M. B., Woo, C. P. (1975). Sulfide reduction in fellmongery effluent by red sulfur bacteria. *J.W.P.C.F.* **47**, 8, 2088–2100.

Dalouche, A. (1989). Désodorisation d'atmosphères chargées de composés soufrés. Thèse d'Etat, Rennes.

Dalouche, A., Lemasle, M., Le Cloirec, P., Martin, G., Besson, G. (1989). Utilisation de biofiltres pour l'épuration de gaz chargés en composés azotés et soufrés. *Proceedings of the 8th World Clean Air Congress*. The Hague, 11–15 September, 1989. Vol. 4, pp. 379–384.

Depeyre, K., Isambert, Creti, C., Yang, Xl. (1987). Un procédé de désulfuration des gaz chargés en H$_2$S. *Proceeding, Journées d'études ISF*, France, Paris, 2 April, 1987.

De Savorin, D. J. (1986). Method for the biological purification of contaminated gases, European patent no. 85, 20 20.29.6.

Disks, R. M. M., Ottengraf, S. P. P. (1989). Process technological view on the elimination hydrocarbons from waste gases. *Proceedings of the 8th World Clean Air Congress*. The Hague, 11–15 September 1989. Vol. 4, pp. 405–410.

Fisher, K., Bardtke, D. (1984). Biodesodorization: the biofilter and its design. *Proceedings of the Société belge de filtration*, 25–27 April, 1984.

Furusawa, N., Togashi, I., Harai, R., Shoda, M., Kubota, H. (1984). Removal of hydrogen sulfide by a biofilter with fibrous Peat. *J. Ferment. Technol.* **62**, 6, 589–594.

Gust, V. M., Sporenberg, F., Schippert, E. (1979). Grundlagen der biologischen Abluftreinigung. *IV Staub Reinhalt Luft*. **39**, 9, 308–314.

Heinicke, R. (1987). Biological method for eliminating grease and odors from sewage systems U.S. Appl. 4666606, 19 May 1987.

Ishizaki, A., Sidhu, H. S., Lim, C. I., Lim W. H. (1984). Desodorizing of sludge solids by actinomycetes. *Agri. Biol. Chem.* **51**, 4, 1155–1157.

Jol, A., Dragt, A. H. J. (1989). Biological elimination of ammoniac in ventilation air from livestock production. Man and his ecosystem. *Proceedings VIII World Clean Air Congress and Exhibition*. The Hague, September 1989. Vol. 2, pp. 275–280.

Kanezashi, H., Okad, K. (1980). Biological desodorisation. Application to shell molding plants. *Akushi non Kenkyu* **9**, 44, 28–38.

Kirchner, K., Schlacheter, U., Rehn, H. J. (1989). Biological purification of exhaust air using fixed bacterial monocultures. *Proceedings VIIIe World Clean Air Congress and Exhibition*, The Hague, September 1989. Vol. 4, pp. 367–372.

Klaus, G., Kutzner, H. J. (1980). Microbiological degradation of odorous substances from wastes air streams. *Landwirtsch. Forsch. Sonderh.* **37**, 541–550.

Koch, W., Liebe, H. G., Striefler, B. (1982). Betriebserfahrungen mi bio-filtern zur Preduzierung geruschsivitensiver emissionen. *Staub. Reinhalt Luft* **42**(12), 488–493.

Koch, W., Liebe, H. G., Striefler, B. (1983). Biofilter auch bei Kottrockmings anlagen bewahrt W/D Wasser. *Luft und Betrieb* **7–8**, 30–33.

Köhler, H. (1987). Reduction of odorous emission from foundries and paint-shops with the aid of biological exhaust air treatment. *80th Annual Meeting Air Pollution Control Association*, New York, 22–26 June, 1987, pp. 495–499.

Lyn, S. K. (1986). Microbiological desulfuration of gases. European patent no. 86, 1 131 092.

Martin, G., Le Cloirec, O., Le Masle, M., Cabon, J. (1989). Réduction de produit odorants sur tourbes. *Proceedings of the 8th World Clean Air Congress*, The Hague, 11–15 September 1989. Vol. 4, pp. 373–378.

McKim Matthew, P. (1985). Biological contact gas scrubber for waste gas purification, European patent no. 85, 30 3792 7.

Ohta, Y., Sato, H. (1985). An artificial medium for deodorant microorganism, *Agric. Biol. Chem.* **49**, 4, 1195–1196.

Ohta, Y., Fukhoka, M., Fujii, K., Sajo, H. (1984). Rapid microbial deodorization of agricultural and animal wastes. *Europ. Congr. Biotechnol.* **3**, 29–134.

Ohta, Y., Ikeda, M., Henmi, Y. (1979). Studies on desodorization of malodorous substances by microorganisms. Rapid desodorization of chicken feces by microorganisms. *Hakko Kogaku Kaishi* **57**(5), 372–379.

Ohta, Y., Toshiyuki, M. (1987). Microbial deodorization of floating scum of wastewater from fish-processing facility. *Hakko Kogaku Kaishi* **6**, 517–524.

Pomeroy, R. D. (1982). Biological treatment of odorous air. *J.W.P.C.F.* **54**(12), 1541–1545.

Tabasaran, O., Affoyon, L., Rettenberger, G. (1979). Enisatz von biofiltern zur Deponiegasdesodorierung. *Muell Abfall* **11**(5), 132–135.

Van Geelen, M. A., Van der Hoek, K. W. (1977). Odor control with biological air washers. *Agric. Environ.* **3**, 2–3, 217–212.

Wada, A., Shuda, M., Kubota, H., Kobayashi, T., Katayama, Fujimara, Y., Kuraishi, H. (1986). Characteristics of H_2S oxidizing bacteria inhabiting a great biofilter. *J. Ferment. Technol.* **64**(2), 161–167.

13

Origin and Elimination of Tastes and Odors in Water Treatment Systems

J. Mallevialle and I.H. Suffet

13.1 Introduction

The resolution of taste and odor (T&O) problems that occur in the field of potable water production has evolved in recent years toward more and more sophisticated techniques. In the past, plant managers tried to modify treatment operations empirically. Today, in some of the newest plants, processes or process combinations have been implemented as preventative measures. These are intended to significantly decrease the probability of organoleptic problems in the water supply.

In an ideal system, a water supplier must be able to identify the compounds responsible for T&O in real time and implement an appropriate treatment before the compound(s) reach the consumer's faucets. In reality, T&O elimination remains a combination of know-how and scientific approaches and is indeed a very complex water quality problem to understand and control. This complexity is due to several factors: possible appearance of many types of T&O problems (e.g., hydrocarbon, medicinal, . . .) and the potential presence of numerous trace organic compounds (at nanogram to several micrograms per litre), which can lead to synergetic and antagonistic effects, which in turn result in transient incidents. In all probability, the Expert Systems technique represents one of the methods of the future to fully exploit this combination of know-how and scientific approaches (Anselme et al., 1989).

T&O can be generated at three different levels: the resource, the water treatment plant (notably by the formation of oxidation by-products) and finally, the distribution network. Solutions obviously depend on what chemical(s) are causing the problem and at what concentration levels the chemical(s) are present.

13.2 Global Approach Methodology

T&O problems can, in certain cases, be relatively simple to solve if they are approached sytematically. The first thing to do is to try and determine the origin of the problem as rapidly as possible by taking taste-and-odor sensory panel measurements on the resource, after each treatment step and at different points on the distribution network. Then, several solutions can be considered.

1. When the resource quality is the cause, it is advisable to determine if the problem is due to one or many compounds which are present in concentrations above their taste or smell detection threshold. If this is the case, it would be possible to treat the problem at the resource level (e.g., take action on the natural environment for the control of microbial metabolites), search and shutdown pollution sources of anthropogenic origin, or adopt a specific treatment at the potable water treatment plant.

2. In numerous cases, however, tastes and odors are produced by a complex mixture of organic compounds that are present in concentrations less than their sensory detection threshold. It is extremely difficult to develop curative action at the resource level and at the treatment plant level. One can only consider empirical solutions such as reagent dose increase at the treatment level. If the T&O problems reoccur frequently, it would be useful to generate a data base relating flavor profile analysis (FPA) (list of descriptors such as chlorine, hydrocarbons, etc.) to gas chromatographic profiles (lists of compounds identified by mass spectrometry) determined under very controlled laboratory conditions (Mallevialle and Suffet, 1987, Bartels et al., 1986). This is the only approach that makes it possible to optimize the operation of a treatment line with the least empirical data. It also enables the operator to consider the temporary use of an alternative resource or a dilution with a water of different origin as described by McGuire et al. (1981, 1983) and Means and McGuire (1986).

3. When T&O are generated during potable water production, as a general rule it is easy to pinpoint the treatment phase(s) where the problem is occuring, but it is much more difficult to identify the responsible compounds with certainty and precision. Once again, the solution rests with a combination of know-how and the scientific approach as will be shown by examples given below.

4. In the case when the distribution network is responsible, the phenomena can be described by many factors.

- *Bacteria regrowth problems*. The solution is inevitably the optimization of the treatment line in order to lower the concentration of assimilable organic carbon as much as possible and increase the level of residual chlorine or implement rechlorination on the network.
- *By-product formation due to slow kinetic reactions between residual disinfectant and organic compounds in the water*. The determination of which compounds are responsible is usually difficult because one is often faced with synergetic effects. The solutions are treatment line optimization for more efficient dissolved organic carbon elimination, low dose chlorination at various points on the network, or the use of another disinfectant.
- *Interaction with pipes or reservoir coating if the identification methodology has been relatively well conducted for different types of material*. Curing the problem is expensive because it may be necessary to change the incriminating material. It is thus preferable to take preventative measures by testing the materials before their implementation and by scrupulously respecting installation protocol (Anselme et al., 1987; Bruchet et al., 1988; Krasner and Means, 1986).
- *Pollutant diffusion*. This occurs generally when solvents or hydrocarbons occur in the soil where polyethelene or PVC pipes are installed. If the problem is easily discovered the remedy is likely to be costly (pipe replacement, soil purification, etc.) (Marshall et al., 1982; Vonk, 1985).

13.3 Treatment at the Resource Level

In this section, only actions undertaken to resolve the problem of microbial metabolites generated by various organisms living in raw water reservoirs will be developed (Mallevialle and Suffet, 1987). A bibliographic review has recently been published by the Research Foundation of the American Water Works Association (Casitas Municipal Water District, 1987). The following checklist includes the principal actions to take in order to ensure an efficient control of this type of resource.

1. *Catchment basin protection*. It is advisable to set local laws in order to allow the purchase of neighboring property, to restrict industrial or agricultural activity and to control the discharge of nutrients such as nitrogen and phosphorus.

2. *Monitoring system*. It is recommended that a sampling program be established to analyze plankton and benthic fauna as well as different physical-chemical parameters. It is particularly necessary to have access to reliable analytical techniques for sensory analysis and identification of characteristic metabolites.

3. *Position of the water intake.* It is important to be able to position the water intake at different heights on both sides of the thermocline in order to pump high-quality water. However, this technique is not valid during reservoir overturn periods.

4. *Reservoir water level.* It is possible to lower the water level to expose algae attached to sediments and rocks in shallow areas and to let them dry out. This technique, nevertheless, has two inconveniences, a rapid lowering of the water level can affect recreational activities on the basin (swimming, boating, etc.) and it leads to a reduction of storage capacity, which can prove to be important.

5. *Aeration.* The hypolimnic zone can be destratified and oxygenated by aeration. This technique, which has been shown to be very effective in certain cases, cannot be considered to be a universal "miracle" remedy.

6. *Copper sulfate addition.* This can allow planktonic algae control by copper sulfate application in liquid or small crystal form and blue algae of the *Oscillatoria curviceps* type by using large sized pieces. This treatment can be done with boats or helicopters, but may lead to conflicts with people in charge of recreational activities.

7. *Biological activity control.* This has to do with removing problems caused by algae, to allow natural microorganisms to degrade microbial metabolites and to monitor the mineral composition of the water, such as alkalinity, in order to avoid algal growth.

One or the other of these techniques are used in numerous countries. It should be noted that certain water distribution services have organized themselves so as to make use of all of them (McGuire et al., 1983; Means and McGuire, 1986; Means et al.,1984).

13.4 Production Plant Treatment Processes

Practically all the different treatment processes used to produce potable water have an effect on T&O: in a positive way, by eliminating microbial compounds or their precursors, or in a negative way, by leading to microbial by-product formation. It is thus necessary to consider the whole treatment line in order to understand T&O treatment.

13.4.1 Coagulation/Settling/Filtration

The main objective of the processes is to remove suspended particles and colloidal substances, which, for the most part, are not linked to removal of the organoleptic chemicals in the water. However, the following points should be noted:

1. Clarification processes with settling or flotation remove planktonic algae that could lyse and leak T&O compounds such as geosmin and MIB (Ashitani et al., 1988). These algae cause a certain number of practical

problems that are not always easily resolved, such as floc flotation in settling tanks. Covering the process is thus recommended.

2. Sludge collection and disposal systems must be designed in such a way so as to avoid fermentations, which can generate septic type T&O.

3. Coagulant doses must be optimized for maximum removal of the organic matrix from water, which, under the subsequent action of an oxidant or disinfectant, leads to the formation of aldehyde type by-products or chlorinated phenolic type odors (Anselme et al., 1988).

4. Powdered activated carbon (PAC) addition at the clarification treatment step allows the removal of a certain number of microbial compounds such as hydrocarbons and certain solvents. The PAC dose necessary to obtain efficient pollutant removal can be calculated by Freundlich's coefficients, which can be found in the literature (Sontheimer et al., 1988).

5. When an oxidant like chlorine is used to maintain cleanliness of clarification equipment, it can generate intense taste problems (see Section 13.4.3), some of which are excellent examples of the effects of synergism. Many authors (Suffet et al.,1986; Anselme et al., 1985) have observed that an increase in contact time with chlorine between prechlorination and sand filtration led to a significant increase in the frequency of earthy/ musty T&O. This phenomenon can be explained either by synergism between the residual free chlorine and certain products with an earthy/ musty odor in the water or by chlorinated by-product formation with an earthy/musty odor such as the formation of chloroanisols.

13.4.2 Aeration

In the case where the compounds responsible for odors or aromas have a relatively high vapor pressure, any treatment involving aeration such as a waterfall or bubbling will lead to a total or partial removal. For such a treatment to be efficient, the compounds must have a Henry's constant greater than 10^{-3} m^3 atm/mol. Compounds such as H_2S, chlorinated solvents, or hydrocarbons can be removed. With Henry's constants around 10^{-5} m^3 atm/mol, geosmin and methylisoborneol are hardly removed (Lalezary et al., 1984). This air carrying effect can be found in processes for iron removal or in biological nitrification.

13.4.3 Oxidation

Oxidants or disinfectants (ozone, halogens, chloramines, chlorine dioxide) react with all or part of the organic compounds occurring in water. The degradation level—in other words, the nature of the by-products—depends on several factors such as:

• Oxidant strength
• The type of reaction (additive/substitution) leading to the oxidation of chemical bond in the compound

- The structure of the compound
- Environmental factors such as pH, temperature, the presence of compounds likely to interfere with the oxidant, etc., are important.

The efficiency of each oxidant for T&O removal will thus depend on the balance between positive and negative effects.

13.4.3.1 Ozonation

A review of different mechanisms of ozone action on aqueous organic pollutants by Doré (1985), based on work completed by Hoigné and Bader (1976; Hoigné, 1977a, 1977b, 1977c, 1978, 1984) shows that ozone can react directly with aqueous organic matter to form carboxylic acids and/or carbonyls by acting as

1. A dipole on $C = C$ double bonds
2. An electrophilic agent on aromatics by ring hydroxylation
3. A nucleophilic agent on $C = N$ double bonds

Ozone is one of the most efficient agents for T&O removal, but intermediate reaction products are formed during water treatment (Mallevialle, 1982). The type and quantity of these disinfectant by-products depends upon ozone dosage, reaction time, radical inhibiting agents/scavengers, and pH. Aliphatic and aromatic aldehyde formation ($>$C-6), is frequently reported in the literature (Anselme et al., 1988; Schalekamp, 1983) to develop fruity, fragrant, and orangelike odors (Suffet et al., 1986).

Figure 13.1 shows an example of the production of a fruity odor through the ozonation line at the Morsang-sur-Seine plant in the Parisian region. These fruity odors are caused by the formation of about 0.1 μg/L of aliphatic aldehydes (2.5 mg/L ozone dosage, 10 min contact time, 1.5–2 mg/L dissolved organic carbon). These fruity odors are generally easily removed by granulated activated carbon (GAC) filtration after ozonation treatment (Anselme et al., 1985).

In all illustrated cases where the compounds responsible for T&O are identified, it is possible to evaluate their ozone removal potential. For this purpose, it is sufficient to refer to the different publications that list the kinetic reaction constants (e.g., Doré, 1985; Hiogné, 1984; Bartels et al., 1986; Lalezary et al., 1986a). However, when there is time it would be preferable to redetermine the kinetics constants on the specific water, as is shown in a series of publications dealing with geosmin and methylisoborneol removal. In 1986, Lalezary et al. showed that these two compounds were not attacked by ozone in distilled water. In 1987 and 1988, respectively, Glaze et al. and Terashima showed that the same compounds were largely removed when they were ozonated in surface water. The explanation currently given comes from the fact that the natural organic matrix can play the role of "promoter"

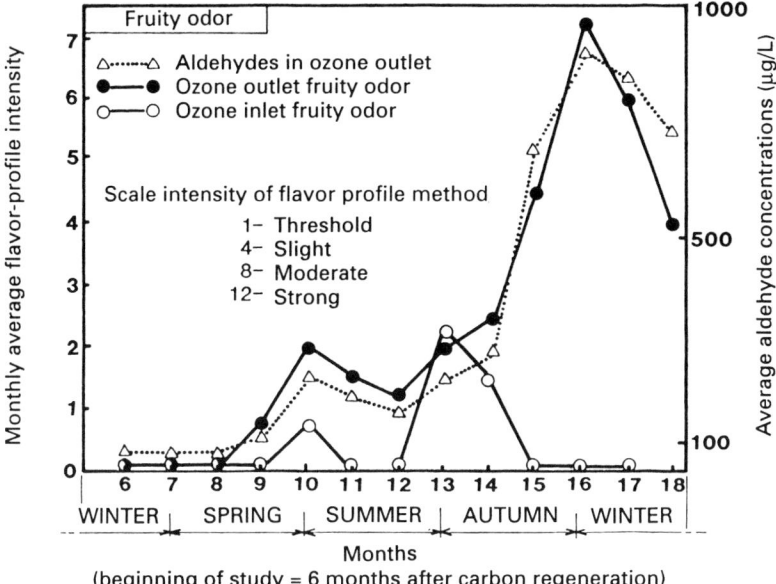

Figure 13.1.   Evolution of fruity odor and aldehyde concentration.

in the generation of hydroxyls, radicals, and OH° that are capable of degrading these two compounds. This hypothesis is confirmed by results obtained by Glaze (1987) and Duguet et al., (1989), who have demonstrated that the H_2O_2/ozone combination (0.4 to 1 ratio) was very efficient for the removal of geosmin and methyl isoborneol.

In the years to come, the ozone/H_2O_2 combination will most likely be used as one of the major tools in the fight against chronic tastes and odors or even accidental pollution. It is best to consider ozonated water as a reagent that can be put into use quickly in crisis situations (Duguet et al.,1989).

13.4.3.2 Chlorination (Chlorine and Chloramines)

The free halogens (e.g., chlorine) and chloramines that are used as water disinfectants can leave undesirable tastes and odors in the water. Bryan et al. (1973) evaluated the effect of chlorine, bromine, and iodine on the taste of water at different pH values (5, 7, and 9). Free chlorine has a taste threshold that is variable according to pH (0.075 mg/L at pH = 5, 0.156 mg/L at pH = 7 and 0.450 mg/L at pH = 9), whereas bromine and iodine have taste thresholds that are more consistent. The threshold scale according to pH for bromine goes from 0.168 to 0.226 mg/L and for iodine from 0.147 to 0.204 mg/L.

Krasner and Barrett (1984) determined the sensory threshold values for various chlorine based disinfectants (Table 13.1). Hypochlorous acid and hypochlorite ions both have similar chlorinous taste and odor. Monochloramine solution give a chlorinelike odor and taste, while the flavor profile panel tended to describe the dichloramine odor and taste as swimming-pool-like, chlorinelike, or bleachy, particularly at high concentrations. Because of the presence of ammonia in raw waters or the addition of ammonia through the use of chloramines as a disinfectant, it is interesting to compare the evolution of chloramines, the free-chlorine residual and the flavor profile intensities for a given NH_3/Cl ratio, as shown in Figure 13.2 (Krasner and Barrett, 1984). Based on those curves, one would not expect monochloramine to cause a taste problem by itself if the treated water has a residual of less than 1.5 mg/L.

T&O problems that develop in water utilities that use chlorinous disinfection are frequently an indirect consequence of chlorination. Chlorine can react with naturally occuring organics in two basically different ways. First, chlorine can oxidize organics by accepting electrons from the organic substrate. Second, chlorine can substitute into the organic matrix or, by addition, lead to the formation of chlorinated organic products. Only a small percent of the free chlorine that is consumed as the chlorine demand of natural waters actually substitutes onto organic compounds; the majority is simply reduced to chloride (Johnson and Jensen, 1986). Substitution reactions on carbon atoms in benzene rings is generally slow but the mechanism can be accelerated if the carbon atom is activated. This activation occurs if neighboring carbon atoms are bonded to electron-donating groups, such as oxygen, for example, in phenols.

The odor threshold values of these chlorinated by-products are generally significantly lower than the original products as shown in Table 13.2 (Mallevialle and Suffet, 1987). Their influence on the organoleptic quality of waters would thus be considerable as shown in the table.

Substitution on nitrogen atoms also occur (e.g., formation of organic chloramines). Many nitrogen-containing organic compounds react in a breakpoint-type reaction in which nitrogen to carbon bonds are oxidized off

Table 13.1. Sensory Thresholds of Different Chlorine Compounds

	Thresholds (in mg/L as Cl_2)	
Compounds	Odor	Taste
Hypochlorous acid	0.28	0.24
Hypochlorite ion	0.36	0.30
Monochloramine	0.65	0.48
Dichloramine	0.15	0.13

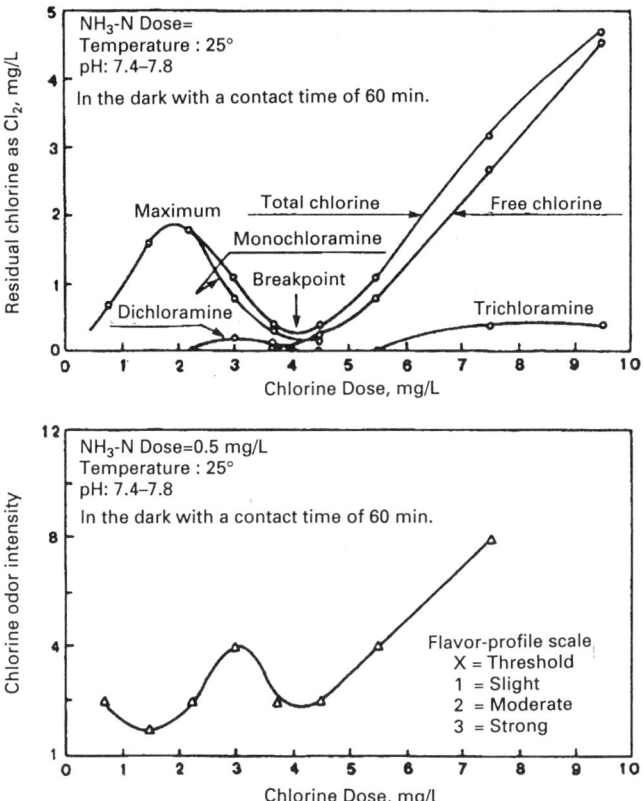

Figure 13.2. Chlorine residual as a function of chlorine dose **(top)** and effect of chlorine dosage on odor intensity **(bottom)**.

of the organic carbon. For example, some amino acids form stable mono-chloramines as intermediates when chlorinated.

A great deal of attention has recently been focused on the problem of trihalomethane (THM) formation in chlorination of natural waters (Rook, 1980). The THM concentrations measured in chlorinated waters are generally between 0.001 and 0.2 mg/L, while the odor threshold value is 0.1 mg/L for chloroform and 0.3 mg/L for bromoform (Van Gemert and Nettenbreijer, 1977).

However, THMs are only the tip of the total organic halogen (TOX) iceberg (Johnson and Jensen, 1986). Some of the compounds that constitute TOX have been identified, such as di- and trichloroacetone (Suffet et al., 1976) and di- and trichloroacetic acid (Christman et al., 1983). More research should be undertaken to correlate T&O problems with the major chlorinated products, most notably those produced by reactions of chlorine with natural

Table 13.2. Comparison Between the Detection of Odor Threshold Values of Chlorinated and Nonchlorinated Phenols (Source: Van Gemert and Nettenbreijer, 1977)

Compound	Threshold Value (mg/L)
Phenol	1–0–5.9
4-Chlorophenol	0.0005–1.2
2,4-Dichlorophenol	0.002–0.21
Anisole	0.05
2,3,6-Trichloranisole	3×10^{-10}
2,4,6-Trichloranisole	3×10^{-8}

substances in waters. For example, the detection odor threshold of dichloroacetic acid is 0.2 mg/m³ air (Van Gemert and Nettenbreijer, 1977).

In many cases of underground waters containing traces of bromide and iodide at concentration levels around 0.1 mg/L, disinfection with chlorine leads to oxidizing the bromide and iodide ions into bromine and iodine, which react with the organic matrix to form brominated and iodinated THMs that are responsible for intense pharmaceutical tastes and odors (Bruchet et al., 1989). The presence of iodinated THMs at concentrations between 0.001 and 0.010 mg/L seems to be the predominant factor in the degradation of organoleptic properties of waters. The detection odor threshold of iodoform is, for example, 20 ng/L.

To summarize, treatment by chlorination (chlorine or chloramines) seems to create more problems than it resolves. However, it should be noted that certain "fishy" or "muddy" odors due to anaerobic conditions can be eliminated by the presence of free chlorine (Wajon et al., 1985; Krasner et al., 1986).

13.4.3.3 Oxidation/Disinfection by Chlorine Dioxide

The main advantage of chlorine dioxide is that it produces considerably fewer chlorinated by-products, both volatile and nonvolatile, than chlorine does. The small amounts of chlorinated organic compounds formed during chlorine dioxide disinfection can probably be attributed to secondary reactions taking place between organic products and free chlorine released during ClO_2 decomposition. As for the mechanics of this process, most studies indicate that chlorine dioxide reacts primarily as a one-electron acceptor. Consequently, the reactions of ClO_2 are much more specific than those of chlorine and do not lead to the development of as many chlorinous T&O.

In a few cases (Mallevialle and Suffet, 1987) ClO_2 disinfection of potable water has been observed to produce strong fishy odors. This phenomenon is particularly noticable, for example, in the shower.

In a comparative study of the efficiency of different oxidants (chlorine, chlorine dioxide, ozone, and permanganate) to remove geosmin, 2,3,6-trichloranisole (TCA), isopropyl-3-methoxypyrazine (IMP), 2-isobutyl-3-meth-oxypyrazine (IBMP), and methylisoborneol, Lalezary et al. (1986a) showed that chlorine dioxide was the most efficient. Nevertheless, it should be noted that these experiments were done with distilled water solutions.

In summary, chlorine dioxide constitutes an excellent alternative to chlorination treatment in all cases where chlorine generates T&O (Walker et al., 1986). However, questions do remain about the health effects produced by chlorite and chlorate that are produced during the use of chlorine dioxide.

13.4.3.4 Oxidation Using Potassium Permanganate

Many water purveyors have described the efficiency of this oxidation method. In the study mentioned above, Lalezary et al. (1986a) have shown that with a contact time of two hours, only the IMP, IBMP and TCA are somewhat removed. The removal was apparently due to an adsorption on the manganese dioxide which is formed by permanganate reduction at neutral pH. Most studies with permanganate are not clear-cut as it is used in combination with other operations.

13.4.4 Activated Carbon Adsorption

Activated carbon, either in powdered form (PAC) or granular form (GAC), has been successfully used to treat tastes and odors in many water treatment facilities (Lalezary et al.,1986b; Gammie and Giesbrecht, 1986; Yagi et al., 1983; Fiessinger and Richard, 1975a, 1975b; Graese et al., 1987). Many studies indicate that PAC is not as effective as GAC, but the cost is lower. In fact, as shown by Feissinger and Richard (1975a), when the required dose of PAC is greater than 20 mg/L, it is generally preferable to rely on GAC filtration with contact times ranging from 5 to 30 min.

Several examples can be quoted from the literature. In pilot studies involving PAC, Lalezary et al. (1985) observed geosmin and methyl isoborneol removal from 66 ng/L to a few ng/L by additions of PAC doses ranging from 5 to 23 mg/L. This result is confirmed by Yagi et al. (1983) who found, however, that more than 100 mg/L were necessary to remove 100 ng/L of geosmin. Montiel (1983) indicates that PAC is very efficient removal of T&O, but doses greater than 200 mg/L are necessary to lower the taste threshold in some instances.

In fact, here again, if one wants to scientifically assess the efficiency of activated carbon, it is necessary to identify the T&O compounds and check isotherm data. Only on-site experiments can account for parameters such as competition effects, carbon saturation, biological degradation, slow adsorption, etc. (Sontheimer et al., 1988). At present, activated carbon constitutes the best tool to tackle T&O problems either as a crisis reactant (PAC) or as a preventive measure (GAC).

13.4.5 Biological Treatments

In some cases, water treatment lines involve processes in which biological activity occurs (slow sand filtration, GAC, denitrification,. . .) that may have a significant influence on tastes and odors. The biological degradation of organic products leads to the formation of by-products that are not yet well known. Some of these by-products, such as phenols, aldehydes, or carboxylic acids, produced by the biodegradation of aromatic compounds such as the ever present alkylbenzenes, could play an important role in T&O generation within the treatment line.

An example of chloroanisole formation during a slow sand filtration operation of clarified Seine river water is given by Montiel and Ouvrard (1986). According to several authors (e.g., Bemelmans and Ten Never de Brauw, 1974; Rigaud et al., 1984) some organisms could make methyl-substituted chlorophenols into chloroanisoles which are responsible for musty odors with odor thresholds less than 0.1 ng/L.

Namkung and Rittman (1987) recently published a survey of biological processes for the removal of T&O especially due to geosmin, MIB, phenol, and naphthalenes. Bank filtration (Sontheimer, 1980) and dune filtration (Hrubeck and de Kruijf, 1983) have been shown to be efficient in the removal of some T&O organic micropollutants.

If one takes into account the fact that ozonation increases the biological activity that develops within a carbon filter, it is found that ozone/GAC combination is very efficient in the removal of earthy and musty T&O (Anselme et al., 1985; Sasaki et al., 1987; Terashima, 1988; Vik et al., 1988). However, Yagi et al (1983 and 1988) and Ashitani et al (1988) also showed that geosmin and MIB could be degraded by microorganisms fixed onto activated carbon beds.

13.4.6 A Case Study

At a great number of sites, T&O are actually generated by a complex mixture of trace organics of natural or anthropogenic origin occurring at concentrations lower than their sensory detection limit. Moreover, the method of flavor profile analysis showed that on a given water, an aroma or an odor could be decomposed into several descriptors of different intensities (Mallevialle and Suffet, 1987). How do we understand and solve such complex problems?

A case study was completed on one of the treatment lines (3,300 m³/h) of the potable water production plant located at Mosang-sur-Seine that includes the following treatments: prechlorination (0.5–1 mg/L), coagulation/settling (40–80 mg/L, AlCl₃—Superpulsator Degrémont), sand filtration, ozonation (1–5 mg/L, 10 min minimum contact time), GAC filtration (F 400), and final chlorine disinfection. Samples were taken weekly after each treatment step for a period of 1 year. Two types of analysis were performed: a sensory technique (flavor profile analysis) and an organic compounds identification technique (closed-loop stripping extraction and GC-MS) (Anselme, 1987; Mallevialle and Suffet, 1987).

Without presenting details, it can be said that about twenty T&O descriptors were detected. Among the most frequent were musty, muddy, plastic, chlorine, fishy, and fruity. During the study, about two hundred organic compounds were identified (aliphatic and aromatic hydrocarbons, various solvents, aldehydes, esters, alcohols, etc.) but always at such low concentrations that it was impossible to determine the responsible molecule(s). Thus, the authors have treated the totality of the data statistically in order to develop correlations such as: higher than C-6 aldehydes and fruity, alkylbenzenes and earthy or musty, pyrazines and fishy, iodized haloforms and pharmaceutic, alkylphenols and plastic, etc. This correlation approach has made it possible to begin to understand and optimize the influence of different treatment processes.

An example of the results is shown in Figure 13.4. Insofar as intensities from 1 to 2 are only detected by trained specialists, one can conclude that the ozone/GAC combination (Figure 13.3) is particularly efficient in removing earthy and musty aromas.

Figure 13.4 summarizes the results obtained on the totality of the descriptors. In conclusion, the ozone/GAC combination is shown to be the most efficient treatment method for resolving T&O problems due to complex mixtures of trace substances.

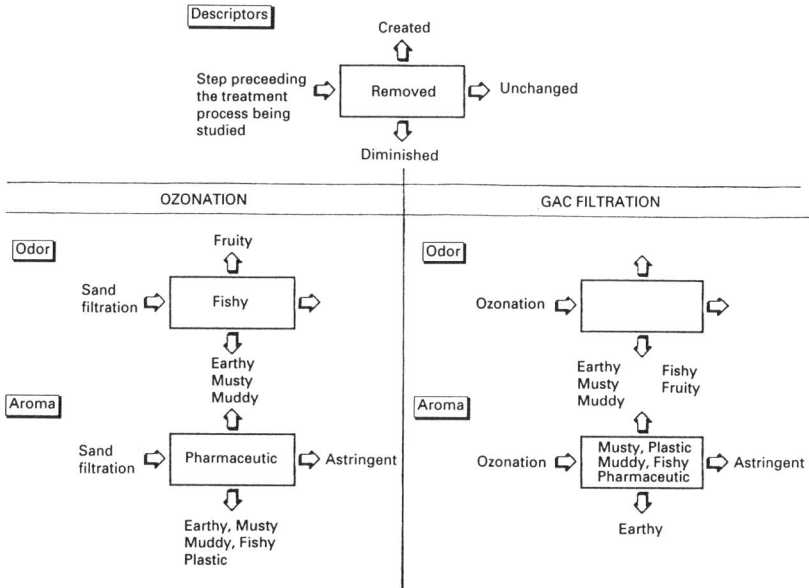

Figure 13.3. Modification of odor and aroma intensity caused by ozonation and GAC filtration (Morsang/Seine).

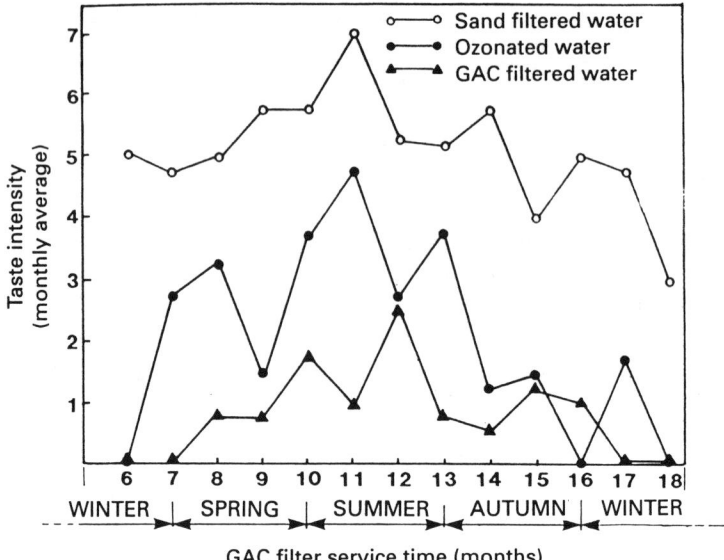

Figure 13.4. Efficiency of the combination of ozone/GAC filtration treatments for the removal of earthy and musty aromas.

13.5 Conclusion

Beyond any doubt, one of the most difficult problems faced by water distributors is to guarantee the organoleptic properties of the product provided to the consumer. Indeed, how is it possible to control a phenomenon that is not well understood? Today, the solution is still a combination of know-how and developing scientific approaches.

It is rare to have a simple T&O problem to resolve; thus, it is best to have a systematic approach to the problem. The first step necessarily involves the command of a certain number of scientific tools, notably analytical (sensory techniques, trace organics identification). This will make it possible to have as thorough a knowledge of the water quality as possible in the water resource as well as in the treatment line and in the distribution network. These results then constitute a data base that will help the interpretation of the observed changes when T&O are detected. Once the T&O causes are known adequate solutions can be proposed quickly.

As far as the water resource is concerned, the solutions are easy to list but much more difficult to implement, as the solutions often depend on other factors outside of the water distributor's responsibility. On the other hand, as far as the water treatment plant is concerned, the examples described in this chapter show that each of the processes that can be used have positive effects and negative effects that will have to be taken into account for the whole treatment line. The oxidation/adsorption coupling (ozone only or in combination with hydrogen peroxide—activated carbon) is, as of today, the most efficient process to lower the probability of the occurence of T&O problems. As far as the distribution network is concerned, the understanding and the solution to the problems will be greatly facilitated by the help of consumer sensory panels, who constitute an excellent detection network.

Bibliography

Anselme, C. (1987). Étude et caractérisation des problèmes de mauvais goûts de l'eau potable. Thèse de doctorat preésentée à l'Université de Paris VII le 16 October 1987.

Anselme, C., Bruchet, A., N'Guyen, K., Mallevialle, J. (1987). Caractérisation des produits de relargage de canalisations en polyéthylène défectueuses. *L'Eau, l'Industrie, les Nuisances* **108**, 75, March.

Anselme, C., Davagnier, M., Mallevialle, J., Bordet, J. P. (1989). Système expert d'aide au diagnostic pour la résolution de problemes de goûts et odeurs indésirables dans l'eau potable. *Conférence spécialisée IWSA-AIDE "Organic Micropollutants"*, September 1989, Barcelona, Spain.

Anselme, C., N'Guyen, K., Mallevialle, J. (1985). Influence des traitements de désinfection et d'oxydation sur les qualités organoleptiques de l'eau: cas de l'usine de Morsang/Seine, proc. *38èmes journées int. du Cebedeau*, Bruxelles, Belgium.

Anselme, C., Suffet, I. H., Mallevialle, J. (1988). Effects of ozonation on tastes and odors, *J. AWWA* **10**, vol. 80, October 1988.

Ashitani, K., Hishida, Y., Fujiwara, K. (1988). Behavior of musty odorous compounds during the process of water treatment. *Wat. Sci. Tech.* **20**, 8/9, 261–267.

Bartels, J. H. M., Brady, B. M., Suffet, I. H. (1986). Taste and odor in drinking water supplies. *Phase I and II. AWWA Research Foundation*, Denver, Colorado.

Bemelmans, M. H., Ten Never de Brauw, M. C. (1974). Chloroanisoles as Off-Flavor Components in Eggs and Broilers. *J. Agric. Food Chem.* **22**, 1137.

Bruchet, A., Leroy, P., Mallevialle, J. (1988). Controlling drinking water coating materials: implementation of a test Protocol and examples of application. *Water Quality Technology Conference (AWWA)*, St. Louis, Missouri, November.

Bruchet, A., N'Guyen, K., Anselme, C., Mallevialle, J. (1989). Identification and Behaviour of Iodinate Haloform Medicinal Odor. Proc. Sunday Seminar on Taste and Odor, *AWWA* Conf., Los Angeles, CA, June 1989.

Bryan, P. E. et al. (1973). Taste Thresholds of Halogens in Water. *J. AWWA* **65**, 5, 363.

Casitas Municipal Water District (1987). *Current methodology for the control of algae in surface reservoirs.* AWWA Research Foundation, Denver, CO.

Christman, R. F., et al. (1983). Identity and yields of major halogenated products of aquatic fulvic acid chlorination. *Env. Sci. Tech.* **17**, 10, 625.

Doré, M. (1985). The different mechanisms of the action of ozone on aqueous organic pollutants. In *Proceedings of the International Conference on the Role of Ozone in Water and Wastewater Treatment*, edited by R. Perry and A.E. McIntyre. SP Press, London, England.

Duguet, J. P., Bruchet, A., Mallevialle, J. (1989). New advances in oxidation processes: the use of the ozone/hydrogen peroxide combination for the micropollutants removal in drinking water. *Proc. Conférence spécialisée IWSA-AIDE "Organic Micropollutants,"* Barcelona, September 1989.

Fiessinger, F., Richard, Y. (1975a). La technologie du traitement des eaux potables par le charbon actif granulé: I Le choix du charbon. *T.S.M. L'eau* **7**, 271.

Fiessinger, F., Richard, Y. (1975b). Même titre III. La régénération du charbon. *T.S.M.* **10**, 415.

Gammie, L., Giesbrecht, G. (1986). Full-scale operation of granular activated carbon contactors at Regina/Moose. *JAW. Proc. AWWA, Ann. Conf.*, Denver, CO.

Glaze, W. H. (1987). Drinking water treatment with ozone. *Env. Sci. Tech* **21**, 3, 224–230.

Glaze, W. H., Schep, R., Ruth, R., Madjzoob, S., Chauncey, W. (1987). Removal of taste and odor compounds by oxidation processes: summary report, phase I. Submitted to Metropolitan Water District of Southern California, July 15.

Graese, S. L., Snoeyink, V. L., Lee, R. G. (1987). Granular activated carbon filter-adsorber systems. *J. AWWA* **79**, 12, 64–74.

Hoigne, J. (1984). In *Ozone and its Practical Application*, edited by R. G. Rice and A. Netzer. Ann Arbor Science Publishers, Ann Arbor, MI.

Hoigne, J., Bader, H. (1976). The Role of Hydroxyl Radical Reactions in Ozonation Processes in Aqueous Solutions. *Wtr. Res.* **10**, 377.

Hoigne, J. (1977a). Ozonation of Water: Selectivity and Rate of Oxidation of Solutes. *3rd Congress of I.O.A.* Paris.

Hoigne, J. (1977b). Rate Constants for Reactions of Ozone With Organic Pollutants. *I.O.A. Symp. Adv. Ozone Technol.*, Toronto, Canada.

Hoigne, J. (1977c). Beeinflussung der Oxidationswirkung von Ozon und oh-Radikalen Durch Carbonat. *Vom Wasser* **48**, 283.

Hoigne, J. (1978). Ozonation of Water; Kinetics of the Oxidation of Ammonia by Ozone and Hydroxyl Radicals. *Prog. Wtr. Technol.* **12**, 79.

Hrubeck, J., De Kruijf, H. A. M. (1983). Treatment methods for the removal of off-flavours from heavily polluted river water in the netherlands: a review. *Wat. Sci. Tech.* **15**, 6/7, 301.

Johnson, J. D., Jensen, J. N. (1986). T.H.M. and TOX formation: routes, rates and precursors. *J. AWWA* **78**, 4, 156.

Krasner, S. W., Barrett, S. E. (1984). Aroma and flavor characteristics of free chlorine and chloramines. *Proc. AWWA WQTC*, Denver, CO.

Krasner, S. W., Barrett, S. E., Dale, M. S., Hwang, C. J. (1986). Free chlorine versus monochloramine in controlling off-tastes and odors in drinking water. *Proc. AWWA Ann. Conf.*, Denver, CO.

Lalezary, S., Pirbazar, M., McGuire, M. J., Krasner, S. W. (1984). Air stripping of taste and odor compounds from water. *J. AWWA* **76**, 3, 83.

Lalezary, S., et al. (1985). Pilot plant studies for the removal of geosmin and 2-MIB by powdered activated carbon. *Proc. AWWA, Ann. Conf.*, Washington, D.C.

Mallevialle, J., Suffet, I. H., eds (1987). *Identification and Treatment of Tastes and Odors in Drinking Water.* AWWA Research Foundation, Denver, CO.

Marshall, J., Hope, P. S., Ward, H. (1982). *Sorption and diffusion of solvents in highly oriented polyethylene.* Polymer Repts, vol. 23

McGuire, M. J., Krasner, S. W., Hwang, C. J., Izaguirre, G. (1981). Closed-loop stripping analysis as a tool for solving taste and odor problems. *J. AWWA* **73**, 530–537.

McGuire, M. J., Krasner, S. W., Hwang, C. J., Izaguirre, G. (1983). An early warning system for detecting earthy-musty odors in reservoirs. *Wat. Sci. Tech.* **15**, 6/7, 267–277.

McGuire, M. J., Jones, R. M., Means, E. G., Izaguirre, G., Preston, A. G. (1984). Controlling attached blue green algae with copper sulfate. *J. AWWA* **76**, 5, 60–65.

Means, E. G., Preston, A. E., McGuire, M. J. (1984). Scuba diving as a cost-effective tool for managing water quality problems. *J. AWWA* **76**, 10, 86–92.

Means, E. G., McGuire, M. J. (1986). An early warning system for taste and odor control. *J. AWWA* **78**, 3, 77.

Montiel, A. (1983). Municipal drinking water treatment procedures for taste and odour abatement: a review. *Wat. Sci. Tech.* **15**, 6/7, 279.

Montiel, A., Ouvrard, J. (1986). Origine et identification de certains goûts de moisi dont l'intensité augmente dans le réseau de distribution. *Conf. 66e Congres AGHTM,* Barcelone.

Namkung, E., Rittmann, B. E. (1987). Removal of taste and odor causing compounds by biofilms grown on humic substances. *J. AWWA* **79**, 7, 197–112.

Rigaud, J., et al. (1984). Incidence des composés volatils issus du liège sur le "goût de bouchon" des vins. *Science des Aliments* **4**, 81.

Rook, J. J. (1980). Possible pathways for the formation of chlorinated degradation products during chlorination of humic acids and resorcinol. In *Water Chlorination: Environmental Impact and Health Effects,* edited by R. L. Jolley, W. A. Brungs, and R. B. Cummings. Ann Arbor Science Publishers, Ann Arbor, MI.

Sasaki, T., Kobayashi, K., Ueda, S. (1987). Application of odor removal process to an existing water treatment systesm. Presented at the Second Int. Symposium on Off-Flavours in the Aquatic Enviornment, Kagoshima, Japan, October 12–16.

Schalekamp, M. (1983). All about ozone. its advantages and disadvantages in treating water. *Aqua* **3**, 89.

Sontheimer, H., Crittenden, J., Summers, S. (1988). Activated carbon for water treatment. Appendix A, DVGW, Forschungstelle, Engler Bunte Institute, Karlsruhe, Germany.

Sontheimer, H. (1980). Experiences with riverbank filtration along the Rhine river. *J. AWWA* **72**, 7, 386.

Suffet, I. H., Brener, L., Silver, B. (1976). The identification of 1,1,1-trichloroacetone in drinking water, a known precursor of the haloform reaction. *Env. Sci. Tech.* **10**, 1273.

Suffet, I. H., Anselme, C., Mallevialle, J. (1986). Removal of tastes and odors by ozone. Sem. Ozonation and water Treatment. *Proc. AWWA, Ann. Conf.,* Denver, CO.

Terashima, K. (1988). Reduction of musty odor substances in drinking water.: A pilot plant study. Presented at the Second International Symposium on Off-flavours in the Aquatic Environment, Kagoshima, Japan. *Wat. Sci. Tech.* **20**, 8/9, 275–281.

Van Gemert, L. J., Nettenbreuer, A. H. (1977). *Compilation of Odour Threshold Values in Air and Water.* TNO, Zeist, Netherlands.

Vik, E. A., Storhaug, R., Naes, H., Utkilen, H. C. (1988). Pilot scale studies on geosmin and MIB removal. *Wat. Sci. Tech.* **20**, 8/9, 229–236.

Vonk, M. W. (1985). The diffusion of water and solvents into high density polyethylene. *IWSA Conference,* Monastir, Tunisia.

Wajon, J. E., Alexander, R., Kagi, R. I., Kavanaugh, B. (1985). Dimethyltrisulphide and objectionable odours in potable water. *Chemosphere* **14**, 1, 85–89.

Walker, G. S., Lee, F. P., Aieta, E. M. (1986). Chlorine dioxide for taste and odor control. *J. AWWA* **78**, 3, 84.

Yagi, M., Kajino, M., Matsuo, V., Ashitani, K., Kita, T., Nakamura, T. (1983). Odor problems in Lake Biwa. *Wat. Sci. Tech.* **15**, 6/7, 311.

Yagi, M., Nakashima, S., Muramotos, S. (1988). Biological degradation of musty odor compounds, 2-MIB and geosmin, in a bio-activated carbon filter. *Wat. Sci. Tech.* **20**, 8/9, 225–260.

14

Odor Elimination in Wastewater Treatment Plants and Sewage Networks

H. Paillard and G. Martin

14.1 Introduction

In wastewater treatment, emission sources may be numerous and the composition and intensity of gas may vary.

The points at risk for olfactive nuisance at the network level as well as the wastewater treatment plant level have been described in Chapter 8. It should be recalled that organic matter is composed of proteins, fats and sugars. These substances are easily biodegradable in aerobic and anaerobic processes. In a network, an anaerobic metabolism may develop leading to the emission of reduced nitrogen and sulfurous and hydrocarbon compounds. In wastewater treatment plants, emissions can be produced by degassing and, above all, by anaerobic transformation of by-products (sludges, residues). Wastewater collection networks often represent a major source of bad odors that are given off at pumping stations and mainly at the inlet of the wastewater treatment plant (holding well and pretreatments). Within the wastewater treatment plant, all the treatment steps, especially those concerning sludges, are equal possible sources of olfactive nuisances. Solving an odor problem begins with analytical research of malodorous sources following a method capable of identifying and quantifying the main compounds or families of malodorous products present and evaluating their odorous levels.

If necessary, preventative solutions, aimed at limiting the causes of malodorous emissions, are applied. Preventative measures against olfactive nuisances concern the fight against anaerobic conditions. These methods may be applied upstream on the network. This would be, for example, a residence time limitation in network pipes under pressure or better pipe and pumping station maintainence in order to remove deposits and grease that favor anaerobic fermentation. At the wastewater treatment plant, prevention of malodorous emissions may consist of the modification of operating parameters (aeration rate, residence times in thickeners, and in sludge tanks).

14.2 Limiting Malodorous Emissions at the Network Level

14.2.1 Prevention of Malodorous Compound Formation

In a network, reduced sulfur compounds (a majority of H_2S) are the main compounds at the origin of bad odors. Their formation mechanism has been described in Chapter 8.

The diagram in Figure 14.1 summarizes the different H_2S formation steps and emissions in a network.

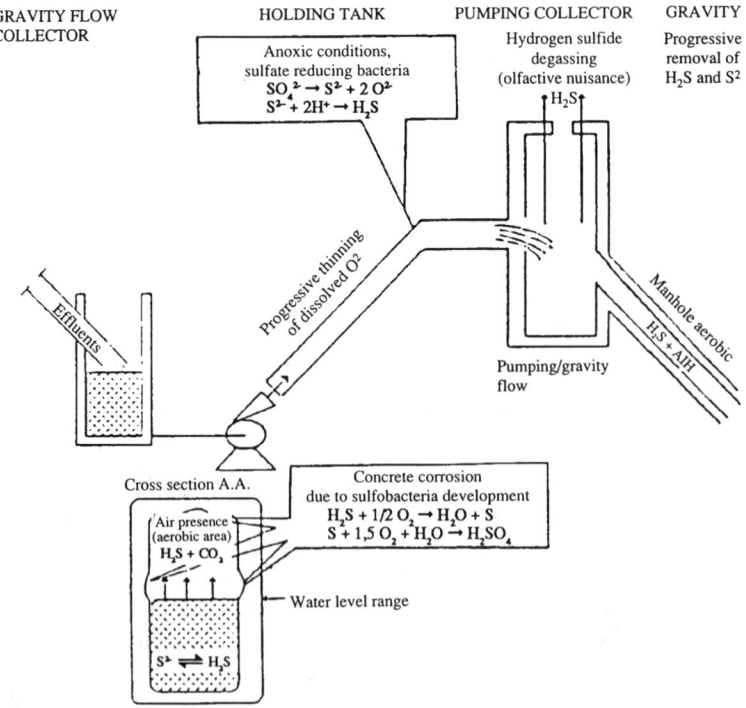

Figure 14.1. Different steps of odor production in wastewater treatment networks (Malberti et al., 1989).

14.2.1.1 Advantages of Sulfur Formation Prediction Models

The risk of bad odor formation after effluent anaerobic residence time must be provided for during the design stage of a sewage network. For example, in the case of pressure flow pipes, pumping distances should be limited and the pipe diameters should not be oversized in order to limit residence time. There are mathematical models using emipric equations that are theoretically capable of sulfur production. They are generally based on the biofilm (Holder, 1989) theory and have been established in the laboratory. As a first approach, they make it possible to evaluate the risk of sulfur formation as a function of the effluent's nature, hydraulic conditions, and temperature. Equations with results close to sulfur concentrations in water are the Pomeroy and Parkhurst (1972) prediction model in the case of networks operated by gravity flow, and the Boon and Lister (1975) model modified by Bertin (1988) in the case of pressure flow pipes.

1. *Pomeroy's model (gravity flow pipes)*

$$\frac{dS}{dt} = M' \, (DBO_5) \, 1.07^{\,(T-20)} \, r^{-1} - N \, (pu)^{3/8} \, (S) dm^{-1}$$

with dS/dt = variation in sulfide mg/L/h concentration, S = sulfide concentration mg/L, T = effluent temperature in °C, p = pipe slope in m/m, u = flowrate in m/s, r = hydraulic radius in m, dm = average hydraulic depth in m, Ts = residence time in h. M' and N are experimental coefficients: $M' = 0.32 \times 10^{-3}$ and $N = 0.64$ for calculating maximum sulfide values; $M' = 0.32 \times 10^{-3}$ and $N = 0.96$ for calculating average sulfide values.

2. *Boon and Lister's model (pressure flow pipes)*

$$\frac{dS}{dt} = 0.928 \times 10^{-3} \, DCO \, 1.07^{\,(T-20)} \, (1 + 0.37 \, D) r^{-1}$$

with $S = dS/dt \, Ts$ or with $S = dS/dt \, Lg \, Ts^{1.2}$ (Bertin, 1988).

These models can still be improved. They are useful in many ways: besides predicting olfactive nuisances on specific points in existing or still to be constructed networks, they may also make it possible to predict corrosion risks due to H_2S or to adjust antisulfide treatments applied to the effluent (oxidant or iron salts) as a function of residence time or temperature, for example.

14.2.1.2 Precautions to be Taken with Hydraulics

In the case of gravity flow pipes, a self-cleaning rate higher than 0.5 m/s is desirable (Colin and Munk Koefed, 1987) in order to avoid deposit accumulation favorable to sulfide production. In France, few problems linked to sulfide formation in gravity flow pipes have been reported.

In very hot weather, sulfide emissions have been observed in gravity flow pipes transporting mixed unsieved and nondegreased urban and industrial effluents (food industry). However, in Europe the vast majority of network odor problems are situated at the level of pumping stations receiving an effluent from pressure pipes with too long a residence time. The risks of H_2S formation are high for residence times longer than 3 h for urban wastewaters. Even on an already existing network, it is possible to reduce water residence time by reducing the pipe diameter with an internal PVC or polyethelene coating or by increasing the number of pumping stations in order to reduce the length of sections under pressure.

In the case of a network to be constructed, numerous odor problems can be avoided at the design stage by taking into account, for instance, the two preceding items and by planning double piping on a site with highly variable seasonal populations, thus with a highly variable pollution load.

14.2.1.3 Example of the Hyères Network

In the Hyères network, odors and corrosion due to H_2S formation have been a problem for over 20 years. Many treatment studies and tests carried out since 1977 have made it possible to locate network sections at risk. They have led to the definitive application of several remedies adapted case by case (Walls et al., 1986), notably the following hydraulic solutions:

1. Construction of a double pipe in ∅150 in place of a delivery pipe in large diameter ∅250 in order to reduce residence time during low flowrate periods
2. Construction of an intermediate pumping station on a 2 km delivery section

14.2.2 Curative Treatments

Preventative methods are sometimes insufficient and effluents may go into anaerobic fermentation at specific points on the network (deposit areas, delivery sections, high organic load, long residence time, high temperatures). As shown in Figure 14.1, odor and corrosion problems will be observed at pressure-gravity flow pipe junctions. Possible solutions are either treating odorous air collected at this junction, or treating the effluent with a reagent injected upstream or just before leaving the pressure flow pipe.

Treating the polluted air is not a solution put into practice very often because it only resolves the problem locally. Implemented techniques are either activated carbon adsorption, chemical scrubbing in a packed tower, or, exceptionally, biological treatment.

Generally, effluent treatment is used in most cases. Several curative processes capable of removing already formed odorous compounds exist, either by

1. Oxidation of sulfides into elemental S or SO_4^{2-} according to the oxidation potential and pH (see Figure 14.2): use of H_2O_2, Cl_2, O_2, or $KMnO_4$
2. Sulfide precipitation into FeS with iron salts

Some processes make it possible to avoid malodorous compound formation by raising the redox potential and oxygen content; for example, preventative treatments using oxygen or hydrogen peroxide. (See Figure 14.2.) Characteristics of different treatments available, established from in situ application results (Burgh and Gaume, 1982; Cadena and Peters, 1988; Le Goallec, 1989; Musquere et al., 1983; Paillard et al., 1988; Pomeroy, 1970; Rudolph, 1981; Walls et al., 1986) are described in Table 14.1.

Iron salts have the disadvantages of flocculating in the network and forming sludges, being difficult to prepare (ferrous sulfate in powder), coloring the effluent and sludges black (FeS), and, for Fe II, increasing the effluent's reducing character. However, these treatments are less expensive. Chlorine and permanganate reagents are not used very often in Europe because they are expensive and cause undesirable secondary effects on downstream biological treatments, especially chlorine. H_2O_2, thanks to its positive secondary effects (oxygen, redox) is used more and more often as a curative treatment in injection at the downstream extremity of the pressure flow pipe. Aeration poses implementation problems in pipes under pressure. As for nitrates, a rather precise dosage is required in order to avoid excesses.

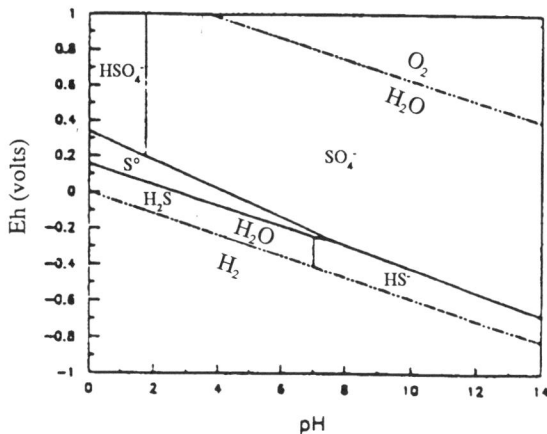

Figure 14.2. Different forms of sulfur as a function of pH and redox potential.

Table 14.1. Characteristics of Different Network Treatments. Calculations for liquid O_2 and aeration based on average network residence time conditions of 3 h (maximum 12 h).

Type of Treatment	Physical and Chemical Characteristics	Theoretical Dosage Calculated on H_2S	Actual Dosage	Implementation
Curative:				
Clairtan	$FeClSO_4$ + Aq liquid d = 1.5	7.2 g/g S^{2-}	11 g/g S^{2-}	Dosing pump
Ferric chloride	$FeCl_3$ $6H_2O$ liquid d = 1.46 41%	6.8 g/g S^{2-}	10 g/g S^{2-}	Dosing pump
Ferrous sulfate	$FeSO_4$ $7H_2O$ loose powder bagged powder	8.7 g/g S^{2-}	10 g/g S^{2-}	Preparation tank + dosing pump percolater filter
Hydrogen peroxide	Liquid H_2O_2 d = 1.135	2.85 g/g S^{2-}	$\approx$5.7 g/g S^{2-}	Dosing pump
Chlorine	NaOCl at 48° chlorometric	8.8 gCl_2/g S^{2-}	5 to 15 gCl_2/g S^{2-}	Dosing pump
Potassium permanganate	Powder	13.2 g/g S^{2-}	13.5 g/g S^{2-}	Preparation tank + dosing pump
Curative and preventative:				
Liquid oxygen	Liquid O_2 under pressure	2 g/g S^{2-}	15 g O_2/m³ h^{-1} residence time (45g/m³)	Relief valve diffuser
Hydrogen peroxide	Liquid H_2O_2	2.85 g/g S^{2-}	8.5 g/g S^-	Dosing pump
Preventative:				
Aeration	Compressed air	2 gO_2/g S^{2-}	180 to 200 L/m³	Compresser diffuser
Nitrate salts	$NaNO_3$ powder	. . .	10 g/g S^{2-}	Preparation tank + dosing pump
	$Ca(NO_3)$ powder	. . .	12 g/g S^{2-}	

14.2.2.1 Several Application Examples

To solve odor and concrete corrosion problems on the Sanary and Bandol networks, several H_2O_2 treatment stations have been implemented. In the underwater pipe that crosses the bay of Bandol (2.2 km long, 500 mm diam), concentrations can reach 15 mg/L in S^{2-} at night (6–12 h residence time) and up to 10 mg/L in S^{2-} during the day (3 h average residence time). An 80–120 mg/L of 35% H_2O_2, controlled by the effluent pressure pumps, makes it possible to maintain the sulfide content at less than 1 mg/L in the downstream effluent in the pressure flow pipe.

In the case of the Hyères network, several remedies (Walls et al., 1986) have been applied to resolve odor problems. Out of 28 pumping stations, 9 main stations are treated. On each of two main branches, one station is treated with H_2O_2 (60–120 mg/L of 50% H_2O_2) and the other five with ferrous sulfate. The H_2O_2 and ferrous sulfate solutions are prepared each week in a central workshop common to all treatment points. The average ferrous sulfate rate is about 110 g/m³. These treatments can reduce sulfide concentrations from 5 to 20 mg S^{2-}/L to less than 1 mg S^{2-}/L. The combination of H_2O_2 and $FeSO_4$ treatments makes it possible to lower operating costs, reduce the black coloration due to FeS, and to partially benefit from H_2O_2 oxidation ability on odorous compounds other than H_2S (mercaptans and amines, for example).

14.2.3 Network Treatment Optimization

14.2.3.1 Use of Prediction Models

In practice, it is difficult to precisely adjust the reagent dose, since sulfide concentrations can vary widely during the day as shown in Figure 14.3. Most of the time, reagents are overdosed in order to obtain satisfactory treatment efficiency. Operators sometimes use Pomeroy's or Boon and Lister's prediction models to evaluate treatment doses. These calculations remain qualitative and generally overestimate the quantity of sulfides formed.

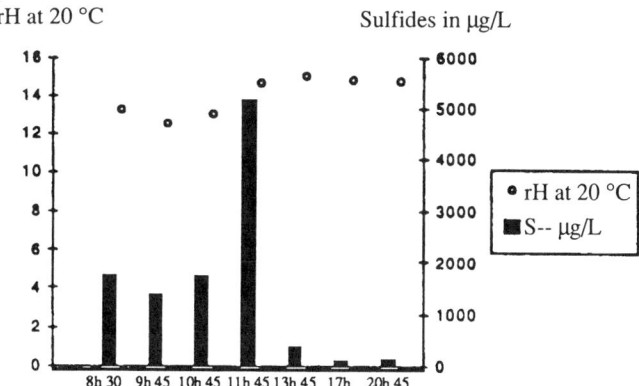

Figure 14.3. Evolution of sulfur concentrations formed as a function of intake time at the last pumping station at the intake of the Sanary-Bandol wastewater treatment plant.

Currently, numerous research efforts with the objective of noticeably improving the validity of these prediction models are being made. If several recent studies confirm that temperature, organic load, and residence time are major parameters intervening in modeling (Bertin, 1988; Stringfellow et al., 1988), there are still other parameters to be confirmed, such as oxygen concentration leaving the pressure flow pipe, suspended matter and deposits, the effluent's potential redox, sulfate concentration, etc.

14.2.3.2 Automatic Treatment Regulation Systems

Parallel to modeling, another approach consists of developing automatic reagent dosage regulation systems using captors capable of measuring either H_2S in the air, sulfides in wastewaters, or redox potential (Paillard and Bourdon, 1988; Paillard and Blondeau, 1988; Le Goallec, 1989).

Reliable analyzers that measure H_2S in the air exist, which makes it possible to continuously monitor treatment efficiency and eventually to adjust it manually. It seems difficult to automatically control a treatment with continuous measurement because the correlation between the sulfides in the water and the H_2S in the air does not appear to be sufficiently precise.

Automatic reagent dosage regulation from continuous measurements of sulfides in the water appears to be much more reliable. Nevertheless, very few analyzers are designed to work in an agressive medium such as raw wastewater. Such a system, developed by OTV, has been functioning since 1988 at the inlet of the Angers wastewater treatment plant, where sulfide concentrations in the effluent can reach 5 mg S^{2-}/L in summer. This makes it possible to adjust the antisulfide reagent dosage (Clairtan) injected downstream from the plant as a function of sulfide concentration in the water. An analysis is carried out every 5 min and the measurement cell is cleaned at least every week. Measurement precision is currently about $\pm$ 0.2 mg S^{2-}/L.

Furthermore, a study conducted in England (Crook and Mc Ewen, 1987) has shown that automatic H_2O_2 treatment monitoring was possible with continuous measurement of downstream potential redox. Monitoring with redox measurement tests are currently being carried out on the Brest (CEO/CGE) and La Baule (CISE) networks.

14.3 Limiting Olfactory Nuisances at the Wastewater Treatment Plant Level

14.3.1 Odorous Emissions

The configuration of wastewater treatment plants varies widely. However, it is possible to represent the main treatment stations and principal emission sources in Figure 14.4.

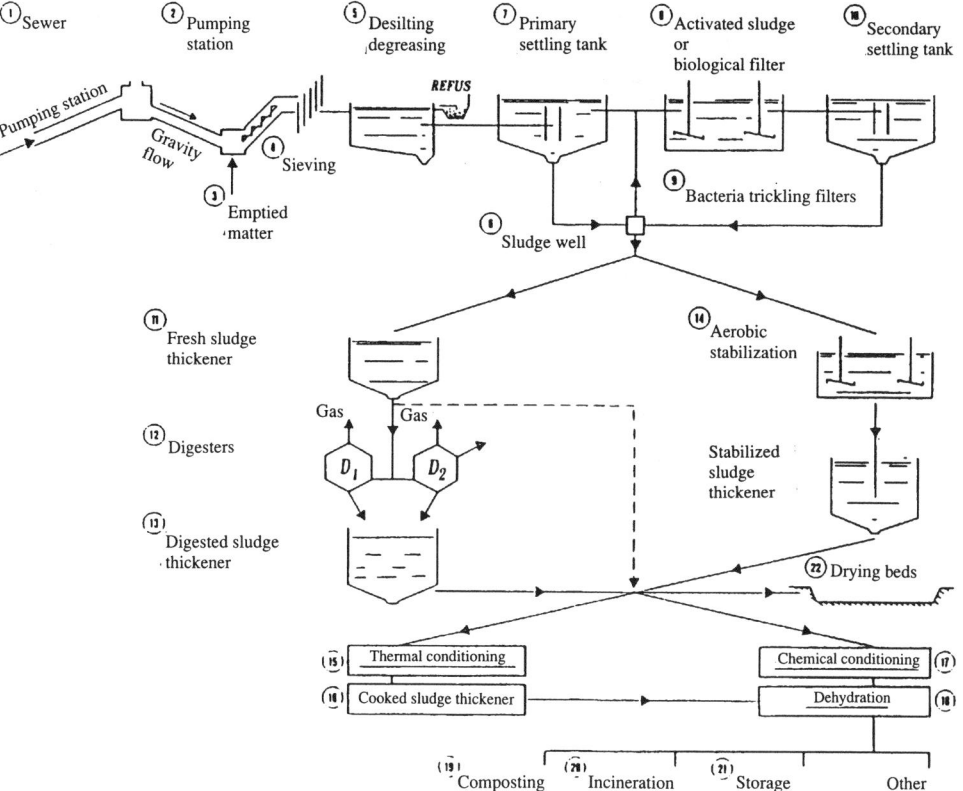

Figure 14.4. Main possible stations on an urban wastewater treatment line.

14.3.1.1 Water Line

Points 1 and 2 in Figure 14.4 correspond to network transportation and the pumping station examined in Section 14.1. The water line in the rest of the figure can be explained as follows:

3. *Emptying matter transfer:* high sulfurous product content; watertight; danger
4 and 5. *Pretreatment:* exposition of water and sewage to the atmosphere produces odors. Odors of wastes, silt, H_2S, RSH, NH_3, aldehydes, and solvents; example of H_2S concentration: 0–30 mg/m^3.
6. *Sludge well:* corresponds to sludge exposition to air and odor emissions; sulfurous products; concentrations vary from 1 to 1,000 mg/m^3 of total sulfurs with ventilation
7. *Primary settling tank:* Slight odor depending on the water septicity; H_2S: 0.2–1.3 mg/m^3 of air over the water surface towards 50 cm.

8. *The aeration basin:* not considered to be malodorous (aerated). Aerosol effects would be more of a problem in turbine. However, some air diffusion systems, shallow or producing large bubbles (INKA sieves, for example) favor sulfurous product degassing. Total sulfur concentrations can reach close to 1 mg/m³. Taking the large surfaces of aeration basins into account, pollutant flux can be high and become the origin of olfactive nuisances

9. *Bacterial trickling filters:* the process relies on anaerobic conditions on the support and bacterial trickling filters favor odors; H_2S; covering.

10. *Secondary settling tank:* normally aerobic with little or no odor.

14.3.1.2 Sludge Line

At the sludge line in Figure 14.4 we have the following:

11. *Thickener.* close to the sludge well; covering; aeration; consider deodorization; 0.5–30 mg/m³ in total sulfur if covered and ventilated; up to 200 mg/m³ in H_2S if covered and unventilated

12. *Digester:* Normally closed; gases are burned in a boiler or flare; gases are rich in H_2S and NH_3; sulfur removal pretreatment necessary; example:

    150–15,000 mg/m³ of H_2S
    0–100 mg/m³ of RSH
    1–150 mg/m³ of NH_3
    3–10 mg/m³ of amines

13. *Digested sludges thickener:* during thickening, digested sludges are less malodorous than fresh sludges

14. *Aerobic stabilization:* badly conducted, can generate odors; the duration of treatment being long (e.g., 15 days) and noncontinuous (with refermentation risk), there are risks of odor emissions in badly operated plants; enzymatic aerobic stabilization or thermophilic aerobic stabilization allowing a shorter residence time (12–48 hours), generate less odor than classic aerobic stabilization; however, these two new processes are currently only in the pilot stage in France

15. *Thermal conditioning:* gases highly charged with sulfurs, aldehydes, ketones, and organic acids.

16. *Cooked sludge thickener:* Considerable nuisances possible; H_2S: 1–30 mg/m³, mercaptans and disulfides; NH_3: 1–500 mg/m³, aldehydes and ketones

17. *Chemical conditioning:* the emission is different depending on the process. For example, if lime addition is used, there are nitrogenous compound emissions

18. *Dehydration:* the room may be disagreeable for personnel; odorous emissions will depend on the nature of sludges (stabilized or not), type of conditioning and the dehydration system used; press filter example:

    1–50 mg/m³ NH_3
    1–10 mg/m³ H_2S

    A flotation apparatus can also be a nuisance source (H_2S degassing, mercaptans, disulfides)

19. *Composting:* nitrogenous odors

20. *Incineration:* if $T > 700°C$ and the incoming sludge is at a sufficiently low temperature, odors are slight

21. *Storage–drying beds:* the type of sludge plays a large part and odors appear if fermentation-digestion are restarted; odor depends on water content.

14.3.2 Prevention of Degassing and Formation of Malodorous Compounds

14.3.2.1 Hydraulic Remedies

In some cases, simple precautions make it possible to reduce volatile compound emissions responsible for bad odors. This is a matter of limiting, among others, the following:

• Waterfalls and violent turbulence, notably in the effluent intake tank, in treatment works feed canals and at the level of settling tank weir head that should be as low as possible.
• Purification by-products; residence time in storage works.

Sludges and pretreatment waste residence times limited to 24 h in sludge thickeners and storage are desirable but difficult to respect in practice.

14.3.2.2 Chemical Remedies Applied to the Effluent and/or Sludges

In many cases, anaerobic fermentations are difficult to avoid. Thus, reduced sulfur compound formation is observed (H_2S in majority, mercaptans, and disulfides) in sludge and waste silos. The compounds are sometimes already present in the effluent when it arrives at the plant. The operator can limit these malodorous compound emissions by applying an "antisulfide" treatment directly to the effluent at the plant intake and on the fluid return at the plant head (overflow from the sludge thickener and storage) or in the thickener. In the last case, iron salts are generally used to precipitate sulfides. Higher effluent or sludge pH by soda addition or, more often, lime can limit sulfurous product release.

In the case of wastewaters, the pH must not exceed 8.5 in order to remain compatible with a biological treatment and 9 in order to respect waste norms. In a thickener, raising the pH to 9 by lime addition limits bad odors by solubilizing an important part of reduced sulfur compounds and reducing biological fermentations. Lime is also widely used to stabilize sludge; it makes it possible to reach a pH close to 11, strongly reducing sulfurous compound emissions and inhibiting bacterial activity. Nevertheless, such a pH favors emissions of other odorous compounds such as ammoniac and amines.

14.3.2.3 Collection of Odorous Streams and Ventilation of Premises

To avoid the propagation of bad odors emitted at different treatment steps, odorous sources should be kept in tight enclosures. The actual tendancy is oriented toward the complete covering of all treatment works due to concern over site integration with the surrounding landscape. However, covering pumping stations, pretreatments, and sludge treatment lines are insufficient in many cases. After confinement, odors must be evacuated by forced ventilation in order to maintain volatile toxic compound concentrations (H_2S, RSH, and CH_4, for example) below the value limits relative to worker's health (Bouscaren, 1984). Air extraction has the effect of maintaining the buildings in depression and bad odor leaks toward the neighbors are thus avoided. Ventilation must allow sufficient dilution of the air in the buildings in order to guarantee satisfactory work conditions (odor, safety) and limit the corrosive action (Paillard and Bourdon, 1988) of reduced sulfur compounds (H_2S in particular). In the case of sources with high odor levels, a second enclosure inside the treatment building causing the problem is desirable.

The ventilation flowrates practiced in treatment rooms depend on the odorous compound concentrations present and the type of treatment, as shown by Table 14.2. Depending on the sensitivity of the area receiving the olfactive nuisances, either a direct release of collected air with simple dilution into the atmosphere (such as in the case of the Marseille plant) or a physical-chemical or biological deodorization treatment can be considered.

14.3.3 Malodorous Compound Concentrations at Intake of Deodorization Systems

An analysis campaign carried out by Anjou Recherche in 1988–1989 on more than 10 urban wastewater treatment plants equipped with deodorization systems has made it possible to identify and quantify the major compounds in collected air. The results are presented in Table 8.5.

The compounds present in the highest concentrations are sulfurous compounds, mainly H_2S. Nitrogenous products are present in significant con-

Table 14.2. Polluted Air Extraction Rate for Each Treatment Room in More Than 10 French Wastewater Treament Plants.

Treatment Stage or Area	Average Ventilation Rate (10 plants) in V/V.H	Maximum Ventilation Rate in V/V.H	Ventilation Rate at Antibes (172,000 I.E.) in V/V.H	Ventilation Rate at Monaco (90,000 I.E.) in V/V.H
Pumping station	4	6	5	5
Pretreatments	3.2	8	8	5
Primary settling tanks	3.3	6	2	5
Biological treatment	1	2.4	2.4	4
Thickener	6	10.8	10	3
Sludge press	6	9.5	5	5
Sludge centrifuge	1	1	—	—
Sludge storage	4.4	6	5	3.5
Total air flowrate extract (Nm³/h)			97,500	58,000

centrations when sludges are treated with lime (dehydration and stabilization).

14.3.4 General Points on Deodorization Techniques

Preventative methods in the fight against bad odors may lead to insufficient results. In these cases, curative techniques that consist of trapping and/or destroying the odorous compounds after their collection are implemented.

14.3.4.1 Masking

Masking consists of acting on the sense of smell by transforming the disagreeable odor into a neutral odor (neutralizing agent) or into an agreeable odor (masking agent). Masking agents, generally based on resinous essences (terpenes) or other aromatic compounds (Vanilline, Eugenol,. . .), are sometimes used punctually in wastewater treatment plants. They are not recommended because of their cost and uncertain efficiency when the intensity of the odorous source and meteorological conditions are variable (W.P.C.F., 1979) and risks undergone by personnel when toxic products (H_2S) are present in high concentrations.

14.3.4.2 Thermal Oxidation With or Without Catalyzers

Thermal oxidation is carried out between 700 and 1,000 °C for residence times of 0.1–1 second in ovens (Pantel, 1987). In the presence of a catalyst (platinum, chrome oxide, or copper), temperatures are about 300–450 °C. These catalysts are progressively poisoned by sulfurous compounds. Thermal oxidation is little used in wastewater treatment plants as it is a very efficient but very expensive deodorization process in terms of energy. In France, the only known application of this process concerns the acceptable recycled air flowrate to the incineration plant burner. For example, at the Lyon-St. Fons wastewater treatment plant, the polluted air from the sludge treatment room with a flowrate of about 1,500 m³/h is burnt in the sludge incineration oven, the sludges being previously dehydrated on a filter press.

14.3.4.3 Activated Carbon Adsorption

This is a very widely used deodorization technique in the industry. It is thoroughly described in Chapter 11. In wastewater applications, it is only economically feasible for small gas flowrates ranging from 2,000 to 3,000 Nm³/h.

In the presence of H_2S and NH_3, special activated carbons, impregnated with an aldehyde or a carboxylic acid, respectively (Le Cloirec and Martin, 1987) should be used. The activated carbon efficiency tends to decrease with increasing air humidity (>75% relative humidity). In wastewater treatment plants, very humid air should be reheated by about 10 °C without exceeding 50 °C.

14.3.4.4 Biological Purification

The principle of biological deodorization is described in Chapter 12. This process can be realized either with biofilters or bioscrubbers. As far as industrial biofilters used in wastewater treatment plants are concerned, their main characteristics are as follows (Ottengraf et al., 1984; Ottengraf, 1987; Martin and Besson, 1988; Besson et al., 1989):

1. Superficial filtration load of 100–200 m³/m²/h
2. Diversified filtering bacterial support: compost, peat, bark, carbon
3. 20–40 second residence time with maximum material height of 1 m
4. Addition of C, N, and P nutritional complement and humidity (B.E.N. process, Besson and Martin, 1985, licensed to Société MURGUE-SEIGLE) by sprinkling solution.

Gas scrubbing in activated sludge recycled in a packed tower has already been applied industrially in the United States (Pomeroy, 1982) in existing biofilters. The efficiency of this process is still in question and its cost is higher than that of peat beds. Semi-industrial tests currently underway in England show the interest of this process, but little data is available at this time on performances of packed towers on activated sludge.

From an economic point of view, biodeodorization on peat beds may be considered the most economical treatment for wastewater treatment plants with air flowrates lower than 15,000 Nm³/h. For 6,000 Nm³/h air flowrates, for example, its working cost is five times lower than that of physical-chemical deodorization in scrubbers and its investment cost is slightly lower (see Table 14.8).

Current research makes it possible to hope for improved biofilter performances in the near future by multiplying filtration velocities by two to five times of what they are now.

14.3.4.5 Gas-Liquid Adsorption with Chemical Scrubbing

This is by far the most frequently used technique in wastewater treatment plants. This process is described in detail in Chapter 10. In order to be compatible with the high air flowrates to be treated in a wastewater treatment plant (3,000–200,000 Nm³/h), the scrubbers used are packed towers working in countercurrent.

Following the acid or alkaline nature of the compound to be removed, a neutralizing agent (alkaline or acid) is added to the scrubbing solution in order to accelerate the transfer of the compound into the sprinkling water. In this way, the size of the scrubbers can be reduced and their efficiency increased. In the same way, oxidant addition (ozone, chlorine, or hydrogen

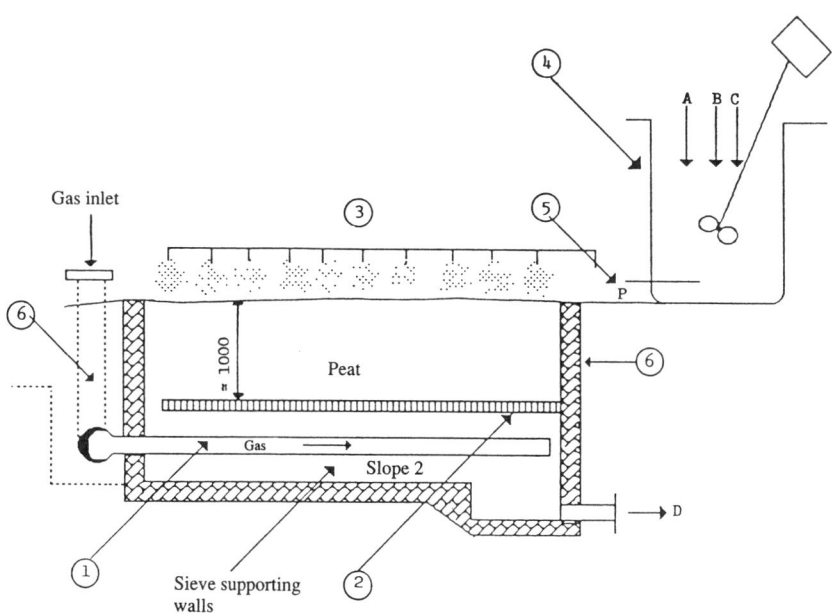

Figure 14.5. Diagram of biodeodorization principles on peat beds (Martin, 1984). **1:** Air arrival inlet; **2:** slab support; **3:** water spraying ramp + nutrients; **4:** nutrient tank; **5:** pump; **6 (A + B + C):** Carbon + Nitrogen + Phosphorus; surface: 68 m²; air flowrate: 10,000 Nm³/h; material height: 80 cm; residence time: 20 s.

peroxide) makes it possible to accelerate the transfer of highly volatile compounds into the scrubbing water, which, continuously regenerated, can thus be continuously recycled in the tower.

Gas-liquid transfer efficiency in a packed tower depends on the nature and concentration of the gaseous pollutant in question, but also on the following parameters (Carleton, 1979; Jarosz, 1984):

1. The degree of water nebulization over the packing
2. The packing contact surface (generally between 100 and 200 m^2/m^3)
3. The number of transfer units, which partially depends on the packing height (1–2 m)
4. Air residence time (0.5–4 seconds per tower) and its velocity of flow
5. The relationship between the flowrates of sprinkling water and air to be treated (about 1–5 L of water/Nm^3 of air)
6. The scrubbing solution pH (continuously adjusted with acid or alkaline) and its redox potential (generally continuously adjusted with an oxidant)

The number of scrubbers and type of scrubbing solution used depend on the nature and concentration of the odorous compounds present. In general, two to three scrubbing towers in a series are used in a wastewater treatment plant. As a function of the compound families present, the sprinkling solutions may be

1. Sulfurous compounds: alkaline solution (NaOH) or, even better, alkaline oxidizing solution
2. Nitrogenous compounds: acid solution (H_2SO_4) or, even better, acid oxidizing solution
3. Aldehydes, acids, and ketones: alkaline oxidizing solution or sometimes a slightly alkaline and reducing solution (sodium bisulfite) for certain compounds.

14.3.5 Description of Industrial Units

14.3.5.1 Biodeodorization

Only peat bed biodeodorization has been the object of significant industrial use at the present time. For the moment, this treatment is applicable to the deodorization of polluted air flowrates of more than 15,000 Nm^3/h. In wastewater treatment plants in France, there are many industrial units currently treating from 1,000 to 10,000 Nm^3/h of air; notably, those at Carry Sausset, Saint Cyr/Mer, Toulon Est, and La Londe les Maures.

For example, the biodeodorization unit at Toulon Est consists of a 68 m^2 peat bed situated on the roof of the plant, capable of treating 10,000 Nm^3/h of polluted air. Nutrient addition is carried out by sequenced sprinkling. Bed humidity is maintained by an automatic sprinkling control through a continuous measurement of the peat humidity level. The pollutants to be removed are essentially reduced sulfur compounds. After a filter maturing period of approximately 2 months, necessary for maximum biomass growth, H_2S and total sulfurous compound removal yields reach over 98%.

The peat bed at Carry Sausset, with a 60 m^2 surface, can treat 10,000 Nm^3/h of malodorous air coming from the totality of covered treatment works of the plant. The compounds to be removed are sulfurous and nitrogenous products. Nutrients are added several times a day with the sprinkling water. Removal yields obtained on the main families of malodorous compounds are given in Table 14.3.

14.3.5.2 Deodorization with Chemical Scrubbing

14.3.5.2.1 Deodorization by Scrubbing with Chlorinated Water

14.3.5.2.1.1 Description of Different Possible Configurations. As a function of the purification level to be reached, that is to say, the sensitivity of the receiving area and the nature of the odorous compounds to be removed, one to four towers can be implemented. In most cases, the major odorous compound to be removed is H_2S. In this case, one alkaline pH scrubbing tower or, even better, a scrubbing tower with a slightly alkaline (pH = 9) oxidizing solution is sufficient. Oxidant use makes it possible to sensibly reduce chem-

Figure 14.6. View of the Toulon Est peat bed situated on the roof of the wastewater treatment plant. Surface: 68 m^2; air flowrate: 10,000 Nm^3/h; material height: 80 cm; residence time: 20 s.

Table 14.3. Working Results on 24 h of Biodeodorization Treatment at Carry Sausset in August 1988

Compounds	Upstream Peat Bed	Downstream Peat Bed	Yield %
Total sulfides mg H_2S/m^3	4.10	0.02	98.7
Hydrogen sulfide mg/m³	3.30	<0.01	>99
Methyl mercaptan mg/m³	1.07	<0.01	>99
Dimethyl sulfide mg/m³	<0.03	<0.03	—
Ammoniac mg N/m³	1.20	0.02	98.3
Nitrogen mg N/m³	0.20	0.01	95.0
Olfactometry: Olfactive threshold in standard odor units per m³ of air	4064	154	96.4

ical product costs as shown by Figure 14.7 (Morton and Card, 1987) and increase deodorization efficiency on compounds with a very high Henry's constant, such as mercaptans and amines, for example.

Chlorine oxidation can be carried out by gaseous chlorine or bleach. Gaseous chlorine is injected in the bottom of the tower with a hydroinjector. When the gaseous chlorine hydrolyses, the medium becomes acidic, leading to the consumption of soda required to maintain the pH compatible with a

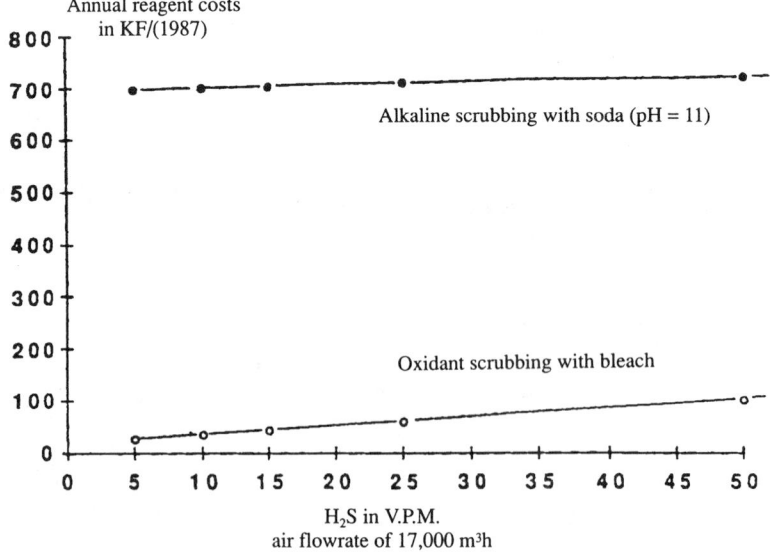

Figure 14.7. Yearly reagent costs for H_2S removal for a deodorization by scrubbing with packed towers system (soda + bleach) at Cape May (Morton and Card, 1987).

maximum sulfurous product (pH > 9) absorption. Bleach is injected with a dosing pump from a stock of bleach at 48 ° Chlorometric or may be produced in situ by NaCl brine electrolysis. When bleach hydrolyses, it gives off OH⁻ ions, which helps maintain alkaline pH in the scrubbing bath used to remove sulfurous products. Thus, soda consumption is limited in order to maintain alkaline pH in a sulfurous compound removal tower.

Extended storage of bleach (more than 60 days) is not recommended, since NaOCl concentration drops with temperature, time, and storage conditions. The result is that NaOCl production in situ is currently enjoying some success in spite of much higher investment costs. The functioning principles of an electrochlorinator are given in Figure 14.8.

Between 5 and 6 kW are necessary to produce 1 kg of NaOCl. In some installations, metal electrodes seem to corrode rapidly. Graphite electrodes, which have a lower yield, have a much higher resistance. The scrubbing bath is continuously circulated between the electrodes and its salt concentration is between 20 and 100 g/L.

The lower the pH is, the stronger the chlorine reactivity is with respect to different odorous compounds. However, in the case of sulfurous compounds, for example, it is necessary to work with an alkaline pH to absorb the pollutants. Chlorine consumption strictly depends on scrubbing water pH and the compound to be oxidized. In the case of nitrogenous compounds, chlorine cannot be used as it can provoke the formation of chloramines sometimes much more odorous than the initial compounds. These compounds are removed by acid scrubbing alone.

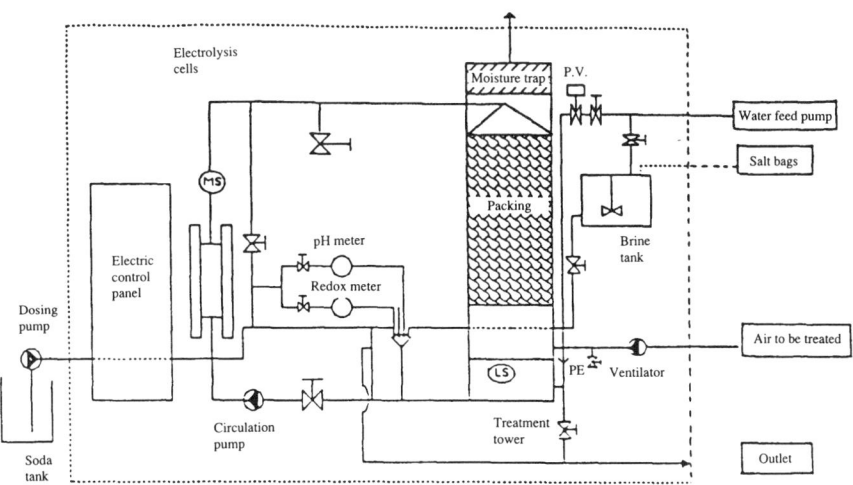

Figure 14.8.   Diagram of the deodorization by electrochloration system at the Chateauroux wastewater treatment plant (Tabaize, 1988).

Oxidation reaction stoichiometries to be remembered under industrial implementation conditions are given in Table 14.4. With chlorine, H_2S is well removed starting from pH = 9. To remove mercaptans, which are always in the minority with respect to H_2S, pH = 11 must be reached. At this pH, chlorine consumed by H_2S removal is four times higher than at pH = 9 and the risks of packing scaling by carbonate deposits are much more important. In practice, when H_2S and mercaptans are simultaneously present, two towers in a series are used to remove them: the first one at pH = 9 in oxidizing conditions and the second at pH = 11, also with an oxidant.

14.3.5.2.1.2 Application Examples of Deodorization by Scrubbing with Chlorinated Water.

Several application examples are detailed in Table 14.5. Generally, the number of towers in a series are three at the most. However, in the case of the Monaco plant, deodorization treatment is carried out with four towers in a series because of the site's extreme sensitivity from the point of view of odorous emissions. Indeed, the wastewater treatment plant is situated in a building located in the middle of the urban area.

The treatment in place is described in Figure 14.9. The air to be deodorized comes from completely covered water and sludge treatment works and has a flowrate of 50,000 Nm^3/h. Water treatments consist of a pretreatment, physical-chemical plate settling, biological treatment with "Biocarbone" filters, and a complete sludge treatment line.

The treatment characteristics are as follows:

Tower 1: Acid scrubbing to remove nitrogenous compounds at pH = 3–4

Tower 2: Oxidant scrubbing with electrolytic bleach at pH = 9 to oxidize sulfurous products (H_2S, disulfides)

Tower 3: Slightly oxidizing alkaline scrubbing at pH = 11 to remove mercaptans, residual H_2S (less than 5%) not removed in tower no. 2, and traces of volatile fatty acids

Tower 4: Reducing scrubbing with sodium bisulfite in neutral or slightly alkaline medium in order to remove the remains of fatty acids, aldehydes, ketones, and residual chlorine coming from tower no. 3.

Table 14.4. Oxidation Reaction Stoichiometry for Several Malodorous Compounds.

Compound M	Cl cons./ M degraded Mole Per Mole	Oxidation pH	References
Hydrogen sulfide	4	10	(WPCF, 1979)
Hydrogen sulfide	1	7	(WPCF, 1979)
Methyl mercaptan	4	11	(WPCF, 1979)
Aliphatic aldehyde	<0.1	7	(MERLET, 1986)
Primary alcohol R-CH$_2$OH	<0.1	7	(MERLET, 1986)
Aliphatic ketone R-CO-R	3	7	(MERLET, 1986)

Table 14.5. Several Examples of Chemical Scrubbing Deodorization Installations (Anjou Recerche Document). Removal Yields in % Are Indicated in Parentheses.

Wastewater Treatment Station	Angers	Antibes	Toulon Est	Sanary	Saumur*
Size	180,000	172,000	75,000	60,000	50,000
Treated air flowrate Nm³/h	7,000	97,500	19,600	15,700	8,000
Number of towers in series	2	3	2	2	2
Tower surface in m²	1.1	6	9	7.3	11
Packing height in m	1	1	1.5	1.5	1
Type of packing	Novalox	Multicellular	Etapack	Etapack	Novalox
Oxidant	Electrolytic chlorine	Electrolytic chlorine	Ozone	Bleach	Electrolytic chlorine
Treatment capacity:					
Average (in g/h)	—	2,000	190	1,300	500
Maximum (in g/h)	500	7,500	280	—	—
1st tower pH	9.7	3	4	6	2.7
2nd tower	9.4	8.5 to 9	9	9	8.5
3rd tower	—	10.5 to 11	—	—	—
Outlet concentration (24 h average)					
Total sulfides mg:					
H_2S/m³	0.06 (98%)	—	0.003 (>99%)	0.04 (>99%)	0.02 (<99%)
H_2S mg/m³	0.04 (99%)	—	<0.01 (>99%)	—	—
Ammoniac mg: N/m³	0.04 (68%)	—	0.14 (42%)	0.03 (93%)	0.08 (83%)
Organic nitrogen: total mg N/m³	0.05 (50%)	—	0.05	0.06 (65%)	—
Olfactometry (punctual)					
Olfactive threshold in S.O.U./m³		107	—	145	207

*According to Morichon and Serpaud (1989).

435

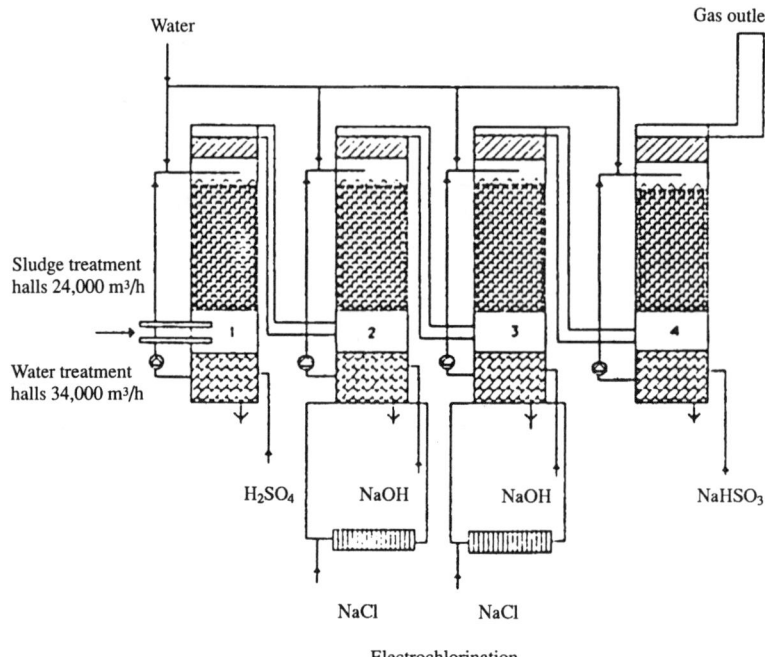

Figure 14.9.   58,000 Nm³/h odor treatment unit at the Monaco wastewater treatment plant (90,000 EqHts)

Electrolytic NaOCl production is subject to redox potential measurements in tower nos. 2 and 3. Continuous pH regulation is maintained in all four towers as well as redox regulation in tower no. 4.

Softened water makeup is provided to tower nos. 2 and 3 in order to limit scaling. During the summer, average reagent consumptions are estimated as follows:

Salt NaCl	50 kg/d
Soda 100%	48 kg/d
Sulfuric acid 100%	28 kg/d
Sodium bisulfite 100%	8 kg/d
Softened water	10 m³/d

14.3.5.2.2 Deodorization by Scrubbing with Ozonated Water

The use of ozone in deodorization is not recent. Indeed, gaseous phase oxidation has been widely used in the United States and in England since the 1950s. This process is no longer applied in France as it requires high residence times ranging from 30 to 40 seconds.

On the other hand, wet ozonation allows contact time reduction to less than 4 seconds in the deodorization units. Absorption then oxidation of odorous compounds are obtained by the scrubbing of the gases with ozonated water in packed towers similar to those used in the case of deodorization by scrubbing with chlorinated water.

Generally, one or two scrubbing towers in a series are implemented in wastewater treatment plants. According to the nature of the compounds to be removed, the sprinkling solutions are

1. Sulfurous compounds: alkaline solution (NaOH) or, better yet, ozonated alkaline solution
2. Nitrogenous compounds: acidic solution (H_2SO_4) or, better yet, ozonated acidic solution
3. Aldehydes, acids, and ketones: ozonated alkaline solution

The main advantages of ozone used as an oxidant in wet process deodorization are

1. To allow very high oxidation rates, hence very short contact times
2. To react under a wide range of pH and to oxidize a great number of compounds
3. To lead to the formation of soluble and nonmalodorous reaction by-products, which are also nontoxic for an activated sludge

Oxidation mechanisms of the main malodorous compounds occuring in a wastewater treatment plant (sulfurous and nitrogenous compounds) are explained in Chapter 10 (Hoigné et al., 1985; Laplanche et al., 1987).

Oxidation reaction stoichiometries to be remembered under industrial implementation conditions are given in Table 14.6. Wet process ozone deodorization will be efficient only if ozone is almost completely transferred

Table 14.6. Oxidation Reaction Stoichiometry of Several Malodorous Compounds by Ozone

Compound M	O_3 cons./M degraded Mole Per Mole	pH	References
Hydrogen sulfide*	1	>9	(Le Sauze et al, 1989)
Methyl mercaptan*	0.9	>9	(Le Sauze et al, 1989)
Ethylamine	2	<6.5	(Laplanche et al, 1987)
Trimethylamine	2	<6.5	(Laplanche et al, 1987)
Aliphatic aldehydes R-CHO	1	?	(Anderson, 1983)
Primary alcohols R-CH₂OH	1	?	(Anderson, 1983)
Aliphatic ketones R-CO-R'	2	?	(Anderson, 1983)

*For a 15 g/m³ O_3 concentration in the air, the stoichiometry increases when this concentration increases.

into the water used for scrubbing the polluted gas. Two techniques for the injection of ozonated air yield a satisfactory transfer: the injection into the scrubbing water tank using a hydroinjector or the injection at the bottom of the scrubbing water tank using porous plates as described in Figure 14.10.

Using ozone as an oxidizing agent, a single tower operated at pH = 9–9.5 is enough to remove H_2S and mercaptan amounts occuring in wastewater treatment plants. When amines and ammoniac occur, mainly when lime treatment is implemented in the water line or in the sludge line, a first ozonation tower operated at a pH below 6 must be placed upstream from the sulfurous compound removal tower.

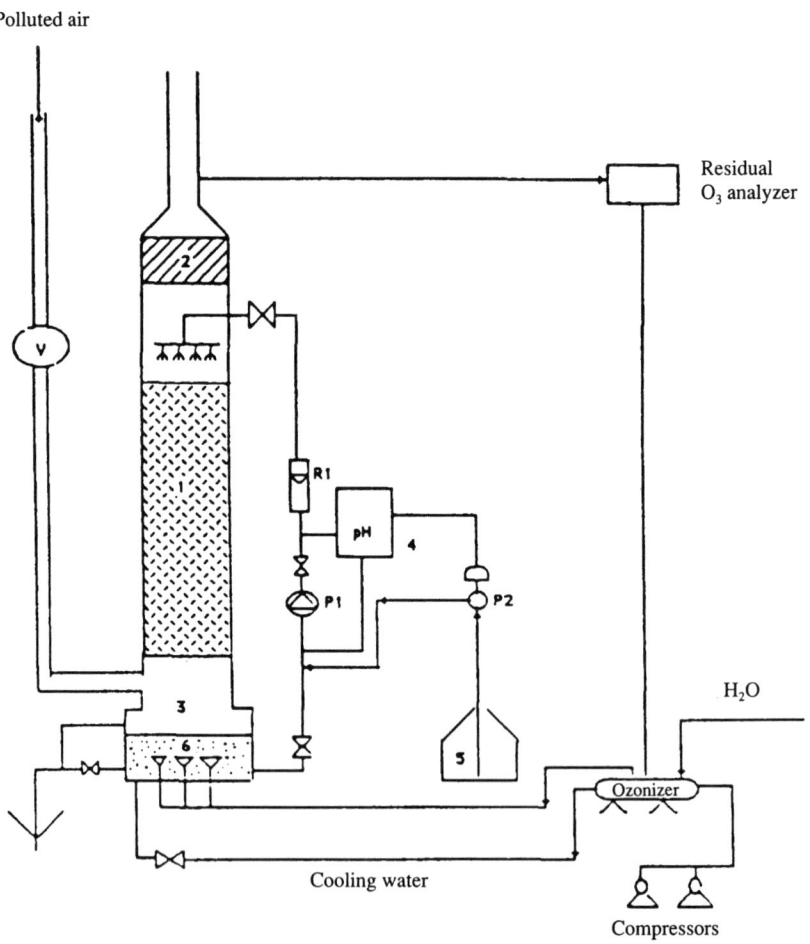

Figure 14.10. Diagram of a deodorization tower using the wet ozonation process. **1:** Packing; **2:** Moisture trap; **3:** Tank bottom; **4:** pH probe; **5:** Soda tank; **6:** Ozone diffusers; **P1.** Recirculation pump; **P2.** Soda pump; **R1.** Scrubbing water flow meter; **V.** Ventilator.

In order to maintain the treatment efficiency with time, a continuous pH control is required. The ozone dosage can be controlled by the ozone content of the treated air as well. This amount should not exceed 1 mg O_3/Nm^3 of air if secondary nuisances due to residual ozone and extra reagent wastes are to be avoided.

Ozonation rates amount to an average of 10–12 mg O_3/Nm^3 and a maximum 24–26 mg O_3/Nm^3. Table 14.5 shows some examples of industrial units in wastewater treatment plants. One should note that the wet ozonation process is also widely used in industry to deodorize slaughterhouses, fish processing plants, and meat quartering plants.

14.3.6 Economic Comparison of Different Deodorization Processes

As far as small deodorization units are concerned, a comparison has been established by Martin and Besson (SAPS, 1988) for the typical case of very polluted gaseous effluent issued from a meat quartering plant (see Table 14.7). The costs are therefore not directly applicable to deodorization units for wastewater treatment plant gaseous effluents, which are much less polluted. This is why they are expressed in adimentional units with respect to a 100 basis.

From an economic point of view, it should be noted that bideodorization is the most interesting technique. For air flowrates exceeding 15,000 Nm^3/h, bideodorization investment costs are, as of today, very high and only scrubbing towers are economically feasible for these large gaseous flowrates. The oxidizing agent must then be chosen as a function of different

Table 14.7. Economic Comparison of Different Deodorization Processes (SAPS, 1988) for a 6,000 Nm^3/h Installation Functioning 5,000 h/y in a Meat Quartering Plant (Comparative Costs base 100).

Processes	Thermal Oxidation	Catalytic Oxidation	Two-Level Scrubbing	Biological Filter Gerfo-Martin
Investment cost	100	76	42	39
Running cost 5,000 h/y working time	54	53	30	6
Total annual cost (depreciation on 10 years)	64	68	35	13
Efficiency on non condensible (98%)				
Ammoniac	98%	92%	>99%	>95%
Amines	99.9%	98%	>99%	>99%
Sulfurous products	99%	96–99%	95–99%	>99%
Aldehydes	80–99%	50–99%	60–90%	not sampled

criteria like reagent storage, investment costs, running costs, and oxidation efficiency.

As shown in Table 14.8, the kind of oxidizing agent used does not influence the running costs. Reagent costs are low with respect to ventilation costs. Therefore, it is important to minimize headlosses in the scrubbing towers by the use of the proper packing material and by reducing the number of scrubbing towers as much as possible. Morichon and Serpaud (1989) estimated the total running costs of a deodorization unit by scrubbing (electrolytic chlorine, two towers in a series) to 1.33 FF/1,000 m³ of treated air including maintainence and personnel costs.

14.3.7 Deodorization Treatment Optimization

14.3.7.1 Automatic Control of Reagent Addition

The odorous flux of a wastewater treatment plant varies during the day following the variation of the effluent pollution load; they also vary with the seasons. Specific maneuvers during the operation of a plant may provoke fast and important odorous fluxes. Since absorption towers have very short residence times (4–5 seconds maximum), it is indispensable to keep the ab-

Table 14.8. Economic Comparison of Different Oxidants Used in Scrubbing Towers. Case of a 100,000 Nm³/h Deodorization System for the Removal of Nitrogenous Compounds (NH_3 and Amines) and Sulfurous Compounds (H_2S + Mercaptans). Cost of Reagents and Energy, 1988.

Oxidant	Bleach	Electrolytic Chlorine	Ozone
Number of necessary towers	3	3	2
Storage oxidants	yes in the form of	salt in bags or bulk	none
Other oxidants	NaOCl 48° H_2SO_4 and NaOH	H_2SO_4 and NaOH	H_2SO_4 and NaOH
Working costs in reagents and energy (in centimes/1,000 Nm³ treated)			
Reagents for pH correction (H_2SO_4 and NaOH) + salt	4.0	9.7	4.0
Oxidants	3.0	12.0	
Make up water	5.7	5.8	5.7
Energy for general ventilation and water sprinkling	76	77.2	70*
Total in centimes per 1,000 Nm³	94.2	95.7	91.7

*Possibly excessive estimate because the headloss is lower (one tower less).

sorption solution's pH and oxidizing potential constant with time. Automatic pH control is now current practice in industrial plants. As far as the oxidizing agent is concerned, in the case of chlorine, for instance, a continuous control of the redox potential is generally implemented.

However, control with a continuous monitoring of residual chlorine seems to be more reliable. In the case of ozone, the oxidizing agent addition is controlled by the continuous measurement of residual ozone in the air at the vent of the scrubbing tower (Anderson, 1983). Its value should not exceed 1 mg O_3/Nm^3 of air if secondary nuisances due to residual ozone and extra reagent wastes are to be avoided.

Biofilters such as peat beds, for instance, are less sensitive to pollution load surges since the residence time in the filtering bulk is longer (20–30 seconds) and the buffering capacity is more important. Nevertheless, the process reliability can be increased by continuous monitoring of humidity. Water and associated nutrient addition by sprinkling can then be controlled by the measurement of humidity as in the case of the Toulon Est peat bed.

14.3.7.2 Deodorization System Maintenance

In the case of deodorization with packed towers, deconcentration blow-downs of the scrubbing solution, which progressively concentrate organic pollutants and salts, are required to maintain the process efficiency with time. The makeup water may be softened in order to avoid carbonate scaling on the packing material. Water consumption (make up water + evaporation + ?) amounts to 1–2% of the recycled water flowrate.

14.3.7.3 Deodorization Treatment Safety and Performance

In in-town wastewater treatment plants, deodorization units should reach removal yields in excess of 98% for sulfurous and nitrogenous compounds. Rather than referring to performances in terms of yield, it is preferable to establish maximum allowable pollutant concentrations in the treated gas emitted into the atmosphere. Physical-chemical and olfactometric measurements performed in parallel on 15 deodorization units made it possible to establish threshold concentrations of major compounds that correspond to the absence of olfactive nuisances for the surrounding neighborhood (Table 14.9). At the outlet of the exhaust stack, residual odorous compounds are diluted into the atmosphere due to atmospheric dispersion and their concentration decreases even more to reach perception threshold concentrations already given in Table 8.1.

In practice, maximum atmospheric dispersion is sought for safety purposes in order to minimize the nuisance in the case of deodorization unit breakdown and therefore obtaining treated air as neutral as possible for the first exposed neighbors. Exhaust gas velocities of 10–15 m/s and a minimum exhaust stack height of 3–4 m are recommended as a function of the site.

Table 14.9. Desirable Threshold Concentrations for the Principal Odorous Compounds at the Wastewater Treatment Plant Outlet in Order to Guarantee the Absence of Nuisances for the Neighborhood

Compound in mg/Nm³ of Air	Desirable Concentration at Deodorization Outlet in mg/Nm³ of Air
Hydrogen sulfide H_2S	≤0.1
Total sulfides (in H_2S)	≤0.15
Mercaptans (in CH_3SH)	
Dimethylsulfide (CH_3-S-CH_3)	≤0.07 on the total
Dimethyldisulfide (CH_3-S-S-CH_3)	
Ammoniac (NH_3)	≤5
Amines (in CH_3NH_2)	≤0.1

Bibliography

Anderson, R. V. (1983). Sewage treatment processes. The solution to the odour problem. *Water Serv.* **87,** 1046, 163–166.

Bertin, L. (1988). Hydrogène sulfuré et réseaux d'assainissement. Rapport d'étude Agence de Bassin Adour Garonne.

Besson, G., Martin, G. (1985). Procédé d'épuration et de désodorisation de gaz et installation pour la mise en œuvre de ce procédé. French patent number 8518327, 11 Dec. 1985, licensed to Société MURGUE-SEIGLE.

Besson, G., Lebault, J. M., Martin, G. (1989). La biodégradation avec équilibre nutritionnel de gaz chargés en polluants biodégradables à fortes teneurs. In *Proceedings 8th World Clean Air Congress,* Vol. 3. The Hague, Elsevier, pp. 385–392.

Boon, A. G., Lister, A. R. (1975). Formation of sulphide in rising main sewers and its prevention by injection of oxygen. *Progress in Water Technology* **7,** 2, 289–300.

Bouscaren, R. (1984). Les produits odorants. Leurs origines. *TSM l'Eau* **6,** 313–320.

Burgh, A. J., Gaume, A. N. (1982). Attacking odors with H_2O_2. *Water Eng. Management* 26–28.

Cadena, F., Peters, R. W. (1988). Evaluation of chemical oxidizers for hydrogen sulfide control. *J.W.P.C.F.* **60,** 7, 1259–1263.

Carleton, A. J. (1979). Absorption of odours. Summary report in no. LR 301, Warren spring Laboratory.

Colin, F., Munk, Koefed, N. (1987). Formation de l'H_2S dans les réseaux d'assainissement. Conséquences et remèdes. Agence de Bassin Rhône-Méditerranée-Corse.

Crook, B. V., McEwen, B. A. (1987). Operational benefit of using reduction-oxidation potentiel in septicity control. *Water Pollut. Control.* **86**(1), 20, 33.

Hoigné, J., Bader, H., Haac, W. R., Staehelin, J. (1985). Rate constants of reactions of ozone with organic and inorganic compounds in water. Part III. *Water Res.* **19**(8), 993–1004.

Holder, G. A., Vaughan, G., Drew, W. (1985). Kinetics studies of the microbiological conversion of sulphate to hydrogen sulphide and their relevance to sulphides generation within sewers. *Wat. Sci. Tech.* **17,**(2-3), 183–196.

Jarosz, J. (1984). Désodorisation par lavage chimique. *TSM l'Eau* **6**, 325–329.

Laplanche, A., Wei, Y., Martin, G., Langlais (1987). A process of washing and ozonation to deodorize an atmosphere contaminated by amines. In *Proceedings of the 8th Ozone World Congress*, Zurich.

LeCloirec, P., Martin, G. (1987). Procédés de désodorisation d'effluents gazeux, Patent no. 83/10934.

Le Goallec, O. (1989), (personal communication).

Le Sauze, N., Laplanche, A., Martin, G., Paillard, H. (1989). A process of washing and ozonation to deodorize an atmosphere contaminated by sulfides. In *Proceedings of the IOA Congress, Wasser*, Berlin, 89, 4.1–4.2.

Malberti, E., Volle, A., Renard, P. (1989). Étude comparative des différentes méthodes de lutte contre la formation de l'hydrogène sulfuré dans le collecteur syndical des effluents du SIVOM des communes du canton de Villefranche sur Mer. *TSM l'Eau* **2**, 81–91.

Martin, G., Le Cloirec, P., Laplanche, A., Lemasle, M., Gillet, M. (1986). Étude de situations de pollution odorante dans divers effluents industriels. *7 ème congrès pour l'air pur*, Sydney 25–29 août.

Martin, G. (1984). La biodésodorisation, cas d'usine de traitements de sous-produits d'origine animale. *TSM L'Eau* **6**, 338–341.

Martin, G., Besson, G. (1988). Biodésodorisation des gaz. *Colloque AFITE*, November 1988.

Merlet, N. (1986). Contribution à l'étude de formation des trihalométhanes et des composés organohalogénés non volatils lors de la chloration des molécules modèles. Thèse de Docteur ès Sciences. Université de Poitiers, no. 426.

Morichon, R., Serpaud, B. (1989). Traitement des odeurs de station d'épuration par procédé d'électrochloration. Exemple de la ville de Saumur. *T.S.M. L'eau* **1**.

Morton, C., Card, T. (1987). Design of packed towers for odor control. In *Proceedings of the 60th AWPCF Conference*, Philadelphia, PA, Oct.

Musquere, P., Reinbold, M., Jarassier, A. (1983). H_2S dans les réseaux d'assainissement. *La technique de l'Eau et de l'Assainissement*, no. 437.

Ottengraf, S. P. P. (1987). Biological systems for waste gas elimination. *TIBTECH* **5**, 133–137.

Ottengraf, S. P. P., Van den Oever, A. H. C., Kempenaars, F. J. C. M. (1984). Waste gas purification in a biological filter bed. In *Innovations in Biotechnology*, edited by E. H. Horwink and R. R. Van der Meer. Elsevier Science Publishers, B.V.

Paillard, H., Blondeau, F. (1988). Les nuisances olfactives en assainissement: causes et remèdes. *TSM l'Eau* **2**, 80–88.

Paillard, H., Bourdon, J. F. (1988). Olfactory nuisances in Wastewater: Causes and cures, *W.P.C.F. 61st Annual Conference*, Dallas, Texas, Oct. 2.

Pantel, A. (1987). Réduction et neutralisation des odeurs: l'incinération. *Journée IIGGE du 8 janvier 1987*, Lyon.

Pomery, R. D. (1970). Sanitary sewer design for hydrogen sulfide control. *Public Works* 93–130.

Pomeroy, R. D. (1982). Biological treatment of odorous air. *J.W.P.C.F.* **54**, 12, 1541–1545.

Pomeroy, R. D., Parkurst, J.D. (1972). Self purification in sewers. In *Proceedings of the 6th WPCF Conference*, Jerusalem, June. Pergamon Press, New York.

Rudolph, K. V. (1981). L'aération des conduites sous pression en vue d'éviter les corrosions et les odeurs d'eaux résiduaires. *Korrespondenz Abwasser* **28**(11), 789–794.

Stringfellow, W. T., Connel, N. R., Felin, C. F., Coleman, W. P. (1988). Variables influencing sulfide concentration in a gravity flow collection system. *J.W.P.C.F.* **60**(12), 2111–2114.

Tabaize, M. (1988). Un nouveau procédé de désodorisation pour stations d'épuration et de relèvement d'eaux usées. *L'Eau, L'Industrie, Les Nuisances* **118**, 44–46.

Walls, G., Volle, A., Iwema, A., Hoyaux, B. (1986). Étude de la formation de sulfures dans un réseau d'assainissement du littoral. Application au cas de la ville d'Hyères. *TSM L'Eau* May 231–244.

Walls, G., Hoyaux, B., Iwema, A. (1987). Comparaison des méthodes de lutte contre l'attaque d'un réseau d'assainissement par les sulfures; cas de la ville d'Hyères., *67e Congrès AGHTM,*Nice.

W.P.C.F. (1979). Odor control for wastewater facilities. Manual of practice no. 22. *W.P.C.F.* Washington, DC.

15

Regulations Concerning Odors

A. Milhau, M. Hamelin, and V. Tatry

15.1 Preamble

Nuisances caused by odors are a social problem that dates from the beginning of time. In France, it was the decree of October 14, 1810 that put a regulation framework into place that made the reduction of these nuisances possible. It was updated by the law of December 19, 1917 and finally replaced by the law of July 19, 1976, which is relative to registered sites for the protection of the environment. This interest in eliminating nuisances and bad odors has increased considerably in the past few years, at the same time as demands for a better quality of life.

Technical progress now allows the reduction of emissions that seemed inevitable in the past. Thus, legal pressure can be applied in an efficient way.

However, certain difficulties remain:

1. It is enough to consider the reduced number of adjectives in the everyday language to describe an odor in comparison to those used to describe a visual sensation. This weakness results in great difficulty when expressing a concept identifying an odorous effluvium and communicating it. In order to compensate for this deficit, measurement methods use, for example, comparisons of odor, intensity, or nuisances.
2. Odor perception fluctuates not only between individuals (differences of olfactory sensitivity, education, and psychological factors particular to each subject) but also for the same person with time:

- Habituation occurs in the case of a continuous background odor; the brain neutralizes the nervous sensation corresponding to the odor and no longer responds except to variations;
- Adaptation is associated with lessened olfactory captor sensitivity concerning the type of odor the person is constantly subjected to.

3. Locating odor emission sites is most often difficult, even more so in the case of diffused sources.
4. A low dose of odorous products can be enough to create an important nuisance in the environment; the concentration scale ranges from ppb (parts per billion) to a hundred ppm (parts per million).
5. The profitability of deodorization processes is absent in nearly all cases; this would suggest that the process choice and plant design should be undertaken with that much more care.

Nevertheless, valid measurement and odor characterization methods have recently become available:

1. A physical-chemical method, which consists of looking for chemical elements that are likely to create an olfactory sensation in the atmosphere
2. A method using an olfactometer and a panel of experts in order to determine the presence of odor and its intensity both at emission and in the environment
3. An inquiry method around a specified site to determined the nuisance experienced by the local population

Their implementation still remains relatively complex and difficult.

In spite of these problems inherent to odor pollution, texts concerning olfactory nuisances more or less directly, either in the environment or in the workplace, have been drafted. Some of them are very general and others much more precise. The goal of this chapter is to present a few of them.

15.2 Emission Regulations

It is possible to distinguish three types of restriction among the different texts of regulations on odors. They provide for

1. Prohibition to harm local populations, which is rather general
2. A limitation of odorous emissions
3. A conversion of the existing site (covering of odorous materials, sufficient chimney height,. . .).

15.2.1 In France

15.2.1.1 The Legislative Framework

Regulations on odors are essentially covered by law 76-663 of July 19, 1976, which was modified relative to registered sites for the protection of the environment and on its application decree of September 21, 1977.

It concerns sites operated or kept by all persons, physical or moral, public or private, that are likely to create a health hazard or inconvenience for the community. These sites are defined in the nomenclature of registered sites established by decree in the Council of State.

According to the gravity of the hazard or inconvenience presented, these sites are subject to either of the following:

1. Authorization prior to implementation, which comes at the end of a procedure including presentation by the project officer, an impact study, and a study concerning the hazards involved, which is subjected to public inquiry. After consulting the municipal councillors of the concerned communities and the departmental health council, the prefect rules on the authorization request.
2. Registration in which the prefect specifies the general prescriptions that the site will be subject to. The prefect can eventually modify regulations that have been established at the national level in order to adapt them to local particularities.

(These procedures are described in detail in the appendix at the end of this chapter.)

The authorization procedure corresponds to sites involving the highest number of risks or inconveniences. Within the framework of this authorization procedure and for certain industrial activities, ministerial decrees and technical investigations have been published by the Ministry of the Environment. These technical investigations specify the elements that must be included in prefectorial authorization decrees and in some cases suggest limited values that take the state of the technique into account. The clauses that tend to reduce olfactory nuisances often have a general character. However, some of them are more detailed, most notably those that are featured in the technical investigation of June 27, 1977 concerned with meat quartering.

Besides procedures related to registered sites, there are two important texts that are concerned with atmospheric pollution and odors:

1. Law no. 61-842 of August 2, 1961: this law regulates smoke or odorous gas emissions into the atmosphere. Modified on July 7, 1980, it created the Agence Pour la Qualité de l'Air (Agency for Air Quality), which was replaced by the law of December 19, 1990 and the Agence de l'Environnement et de la Maîtrise de l'Energie [ADEME (Agency for Environment and Energy Management)]. This new agency is a fusion of three

separate agencies: the Agence Pour la Qualité de l'Air (Agency for Air Quality), the Agence Pour la Récuperation et l'Elimination des Déchets (Agency for Waste Recovery and Disposal), and l'Agence Française Pour la Maîtrise de l'Energie (French Agency for Energy Management). Its responsibilities are to facilitate and implement surveillance, protection, and information initiatives concerning atmospheric pollution matters.

2. Departmental sanitary regulations: are not applicable to registered sites. This is a series of regulations mainly concerned with reducing causes of insalubrity. It is usually suppressed by law no. 86–17 of January 6, 1986 (article 67) and replaced by a series of decrees made by the Council of State after being advised by the Conseil supérieur d'hygiène de France (Superior Public Health Council of France). These decrees set the "general rules and regulations of hygiene and all other measures taken to preserve health." State representatives in the department and mayors can complete the decrees cited above by orders whose purposes are "to take the necessary measures to ensure the protection of public health in the department and the community." However, all decrees that should replace departmental sanitary regulations have not been promulgated and, as a consequence, the departmental sanitary regulations continue to be applied. The problem of odors may be taken into account within the regulations by such measures as those concerning

- The salubrity of habitations, agglomerations, and the totality of the environment in which human beings live
- Atmospheric pollution of domestic origin.

15.2.1.2 Olfactory Nuisances and the Law Concerning Registered Sites

The law relative to registered sites for the protection of the environment and its decree of application allows the study and reduction of odors, as for other nuisances, within the respective framework of impact studies and the prefectorial application decree.

15.2.1.2.1 Impact Study

The following information is necessary in order to study the impact of odorous emissions on the environment:

1. Estimation of gaseous wastes (flowrate, concentration of different chemical substances, temperature,. . .)
2. Outlet height
3. Plant layout
4. Site topology and local meteorology

With these elements, it is then possible to estimate the on-site dilution rate at the perception threshold. If the estimated olfactory nuisances remain too high for the neighborhood, the initial conditions must be modified to reduce the odors and steps must be taken to comply with the new recommended site parameters. This type of study is difficult to carry out because it relies on extrapolation of real data from previous realizations.

Existing uncertainties about gaseous emissions and meteorological data also lead to a lack of precision concerning odors at ground level. Several years ago, odor impact studies did not have the required precision. Presently, several detailed on-site measurement campaigns make it possible to model the principal phenomena. Therefore, the impact study concerning olfactory nuisances is notably lower.

15.2.1.2.2 Prefectorial Decree

Considering all the elements and expert witnesses, the prefect may demand the following, by decree:

1. Result obligations:

   - Minimum purification yield on the different odorous products present
   - Dilution factor at the perception threshold
   - Odor flowrate
   - Emission limits

2. Implementation of preventative measures:

   - Chimney/smokestack height
   - Covering of storage areas
   - Purification systems

3. The implementation of monitoring, allowing the verification of restriction compliance or nuisance evaluation for the neighborhood. This accomplishment concerning odors remains difficult and monitoring is rarely featured in the restrictions.

The industrialist retains free choice of the methods that will be implemented to comply with the decree.

Some individual prefectorial decrees impose the respect of value limits. For example, the dilution factor at the perception threshold can be limited. It should be noted that a prefectorial decree imposes at least a daily measurement of the dilution factor at the perception threshold on the industrialist within the framework of self-monitoring.

A tendancy to take olfactory nuisances into account in a more efficient way is asserting itself. This new orientation was confirmed with the publication of the technical investigation related to meat quartering workshops and the norm X 43101 describing the measurement method of the odor of a gaseous effluent by determination of the dilution factor at the perception threshold. Technical investigation imposes minimum purification yields for the different families of odorous compounds emitted and requires an emission threshold, which is expressed in relation to dilution at the perception threshold.

15.2.1.3 Ministerial Decrees, Memoranda, and Technical Inquiries

By taking general application regulation texts that mention olfactory nuisances in chronological order, it would appear that the problem is treated with increasing precision. Thus, in earlier inquiries, odors are only mentioned and the clauses described are mostly conversion steps. For example,

1. Avoid creating a source of nuisances for the neighborhood with storage areas, ponds, etc.
2. Construct chimneys high enough to allow proper dispersion of emissions
3. Comply with a minimum distance with respect to a third party
4. Respect combustion parameters indicated for incinerators in order to destroy the organic molecules present

Implicated texts are as follows:

1. The memorandum of June 6, 1972 regarding urban waste incineration plants (restrictions established by this text have been reinforced by the decree of June 9, 1986 regarding household waste incineration)
2. The memorandum of August 17, 1973 regarding sugar processing plants, sugar processing and distilling plants, and sugar beet processing plants
3. The memorandum of January 3, 1975 regarding nuisance reduction for potato starch processing plants

With the technical inquiry of August 12, 1976 regarding pig farms in addition to the preceeding restrictions, new more demanding clauses are imposed such as the following:

1. If there is land disposal, the wastewaters will be

   • Aerated during their storage
   • Disposed of by burial
   • Surface spread and buried immediately by plowing
   • Deodorized before surface spreading

2. An adapted ventilation system with respect to neighboring habitations must be adopted.

Also, the technical inquiry of June 27, 1977 relative to meat quartering includes a section concerning "odor reduction" that is very complete and well adapted. Besides already existing clauses in previously quoted technical inquiries, other detailed restrictions are established.

1. Limits to be respected: these limits are as applicable to unheated gaseous emissions (coming from storage) as to heated gaseous emissions (coming from cookers). This concerns several parameters:

- The dilution rate to the perception threshold of the effluents: at the time of emission, it must be lower than 200; in other words, if the gaseous flux was diluted 200 times, 50% of the members of a panel would not detect the odor;
- The odor flowrate: it has been fixed at a maximum level of 1 million m³/h under all circumstances;
- The minimum purification yields: for meat quartering plants, the chemical products responsible for odors have been separated into three main groups:

Products	Reduced sulfur	Ammoniac/amines	Aldehydes/ketones
Required yields	>98%	>98%	>95%

2. Inspection realization: after meeting with the industrialist, the registered site inspector decides on the verification frequency, which, besides self-monitoring, must be carried out at least once a year by an approved laboratory. In addition, the site must be equipped in such a way that the measurements can be carried out. There are three verification categories:

- Verifications carried out by an approved laboratory: measurements of parameters limiting odors take a long time to perform and require precise procedures. Therefore, measuring compounds such as hydrogen sulfide, aldehydes, ketones, and amines necessitates efficient analytical equipment.
- Verifications realized within the framework of self-monitoring: for the same parameters there are measurement procedures that are less precise, but simpler and more rapid. Thus, for the same example as above, the INERIS (Institut National de l'Environnement Industriel et des Risques) studied the possibility of using colorimetric tubes, which would allow the industrialists to perform the measurements themselves;
- Unexpected verifications, realized by a laboratory designated by the administration, requested by the registered site office.

3. Deodorization systems: many systems meet the treatment requirements and are listed in the technical inquiry on meat quartering with several related explications.
4. Production procedure modifications: the inquiry strongly suggests separate treatment for heated and unheated odorous gases. They must be efficiently collected.

- Heated odorous gases: the emission outlets specified in article 11 of the technical inquiry (dryers, various valves, etc.) must be covered by hoods or covers in order to avoid too much dilution with ambient air. The circuit material must not be corroded by gaseous flux contact. Also, gaseous effluents generated by the production process are re-

quired to be as unvarying as possible in order to improve the efficiency
of the odor reduction system.

- Unheated odorous gases: in order to reduce odors, raw material storage time should be reduced and the different areas concerned should
  be properly covered with hoods or covers.

More recent texts including a section concerning odors related to meat
quartering have been published. However, even if they include some of the
indicated clauses, they are not as complete:

1. Memorandum and technical inquiry of December 17, 1981 concerning
   pollutions and nuisances due to beef cattle farming
2. Memorandum and decree of October 1, 1982 concerning chicken farming
   sites
3. Memorandum and decree of February 1, 1983 concerning slaughterhouses

Besides texts concerning the food industry, it is interesting to note the
recent existence of two texts related to the reduction of organic compound
emissions:

1. Memorandum and technical inquiry of April 5, 1988 concerning graphic
   copying workshops
2. Memorandum and technical inquiry of August 25, 1988 concerning pre-
   laquering workshops

In these texts, emission limits for nonmethanic hydrocarbons (expressed as
methane-equivalent) are established. Even if these measures are not directly
aimed at odors, the fixed restrictions tend to strongly reduce on-site olfactory nuisances due to the emission reduction of molecules that are frequently
odorous.

15.2.1.4 Fiscal and Financial Advantages

15.2.1.4.1 Fiscal Advantages

Enterprises that construct buildings that are designed to satisfy the obligations of the law of August 2, 1961 relative to the fight against atmospheric
pollution and odors can exceptionally depreciate these installations over a
period of twelve months and are completed between January 1, 1990 and
December 31, 1991 (general tax code article 39, section F, article 21 of the
fiscal law for 1993).

The following are considered immovable installations:

1. The buildings themselves
2. Constructions embedded in such a way that they cannot be moved without deteriorating their location
3. Material resting on special foundations that are part of the building

Installations concerned by the aforementioned depreciation may have half of their rental values, used to establish local taxes, taken into account under application of article 1518 A of the general tax code, modified by the fiscal law of 1990. This clause may be applied starting from January 1, 1991. Since January 1992, it is possible to depreciate the overall rental value. Previously, rental values were taken into account at two thirds of their total amount.

15.2.1.4.2 Financial Advantages

ADEME, a national public establishment with an industrial and commercial character, has the task of creating, facilitating, and carrying out the demonstration or development initiatives of air pollution prevention techniques. Technically and financially, ADEME also manages grants given to industrialists according to the parafiscal law on atmospheric pollution.

Thus, ADEME can offer industrialists who wish to install an odorous emission reduction process financial grants up to 50% of the investment cost of a project in the form of subventions or refundable loans. However, the project must have an innovative or exemplary character.

15.2.1.5 Sanctions and Penal and Administrative Responsibilities

15.2.1.5.1 Registered Sites

The law of July 19, 1976 and its application clause of September 21, 1977 have established a complete set of penalties:

1. Fines up to 1 million francs
2. Prison penalty for the operator up to 1 month
3. Operation shutdown
4. Formal demand to complete necessary works within a defined time period

Most notably, these measures penalize lack of authorization request, lack of registration, or noncompliance with prefectorial decrees.

The operator remains legally responsible for damages or nuisances to the inhabitants or the environment. For this reason, complaints based on unusual neighborhood problems can be lodged before the civil court by the inhabitants themselves or environmental defense associations.

The courts have already had occasion to condemn operators for nuisances provoked by odors:

1. Odors emanating from a workshop (civil cassation court, June 6, 1972)
2. Smoke and odorous product emissions coming from a coffee roasting plant (Paris Appeals Court, February 19, 1974)
3. Odors coming from a compost manufacturing plant (Orléans Appeals Court, October 13, 1987)

15.2.1.5.2 Unregistered Sites

In the case of nuisances or damages undergone by the inhabitants or the environment, the mayor may make a written report. The mayor may even impose a shutdown of the site in case of emergency.

Article 26 of the law relative to registered sites allows prefects, after being advised (except in urgent cases) by the mayor and departmental health council, to instruct the operator to take the necessary measures so that the duly noted hazards or inconveniences cease. If the operator fails to take the necessary steps to comply within the given time limit, the prefect may apply the administrative sanctions provided for registered sites.

In the case of olfactory nuisance, a private individual may

1. Lodge a complaint with the prosecutor of the Republic, who will then decide what action to take
2. Summon the operator responsible for the nuisance before the magistrates' court with the help of a bailiff
3. Lodge a complaint before the administrative court against the mayor of the community where the olfactory nuisances are generated for

   • Not having enforced a municipal decree that would have allowed the reduction of said nuisance,
   • Not having written such a decree to begin with

4. Request that the administrative court apply article 26 of the law on registered sites.

15.2.2 In Other Countries

An exhaustive description is difficult since few countries have written regulation texts especially concerned with odors. The following list gives norms, general outlines, or existing rulings proposing emission limits for various industrial branches and only for three groups of odorous compounds (reduced sulfur compounds, aminated compounds, aldehydes and ketones). It should be noted that the simultaneous presence of several organic compounds may be at the origin of odors.

One can have access to exhaustive lists of various molecules mentioned in national regulations by interrogating data banks:

1. EMILIE (maximum allowed emission values)
2. AIRQUAL (maximum allowed air quality values)

These two data banks are managed by the IFE (Institut Français de l'Energie/The French Energy Institute) and may be consulted by Minitel in the future.

15.2.2.1 Canada

In Canada there are general outlines which were signed on September 22, 1979, and applicable to paper pulp plants (kraft process) after the previously stated date. The proposed maximum values concern reduced sulfur compounds, a term used to designate the totality of reduced sulfur compounds, such as hydrogen sulfide and methyl mercaptan as well as sulfur and dimethyldisulfide.

The maximums are expressed in g of H_2S/ton of produced pulp. According to the site characteristics and the emission locations under consideration, maximums range from 15 to 225 g H_2S/ton.

15.2.2.2 South Korea

According to norms established in 1977, ammoniac and formaldehyde concentrations are respectively limited to 250 and 50 ppm for all industrial processes.

15.2.2.3 The United States

The United States has chosen to regulate emissions according to the type of industrial process. Notably, the norm established April 20, 1986 defined emissions of reduced sulfur compounds for the paper pulp industry using the kraft process. The maximum values are expressed in ppm for dry gases and range from 5 to 8 according to site configuration.

15.2.2.4 Japan

The norm of September 1976 fixes the maximum emission rates of "q" gaseous pollutants according to the following formula:

$$q = 0.108 \times H_e^2 \times Cm \ (m^3/h)$$

where H_e is the effective outlet height in meters and C_m is the pollutant concentration in ppm.

It should be noted that the outlet must be higher than 5 m:

For ammoniac:	C_m ranges between 1 and 5 ppm
For hydrogen sulfide:	C_m ranges between 0.02 and 0.2 ppm
For trimethylamine:	C_m ranges between 0.005 and 0.007 ppm

15.2.2.5 Germany

Germany has opted for emission regulations per type of pollutant. Technical inquiries are regrouped in the "TA Luft" (Technische Anleitung zur Reinhaltung der Luft) in which an authorization procedure for new installations is described. Most notably, this requirement includes an environmental impact study.

Paragraph 3-1-9 is specific concerning odorous substances and recommends the following precautions:

1. Partially or totally wall up the site
2. Keep areas closed during depression
3. Store raw materials, products, and wastes in appropriate areas
4. Disperse odorous gases correctly by means of adequate chimney height

15.2.2.6 The United Kingdom

The technical inquiry (Best Practicable Means No. 10) of October 1981 limits amine and trimethylamine wastes at emission for the amine producing chemical industry:

Maximum value for amines: 5 ppm
Maximum value for trimethylamine: 1 ppm

Since the United Kingdom is divided into four nations, each one of them has its own regulations. Thus, most clauses defining nuisances in England and Wales are included in part III of the Public Health Act of 1936 while others are included in parts II and XI as well as in other laws.

The law of 1936 allows

1. Either local authorities or individuals to take action against an industry emitting one or several characteristic nuisances—odors considered as nuisances can thus be controlled;
2. The designation of a plant as a malodorous industry

   • When it is on a list of notoriously malodorous branches of industry (list included in section 107 of the law of 1936);
   • When a local authority has requested its inclusion and the secretary of State's office has confirmed said inclusion.

The status of malodorous industry obliges the industrialist to make efforts to reduce the olfactory nuisance. Nonobservation of specified regulations leads to being fined.

15.2.2.7 Switzerland

The order of December 16, 1985 applicable beginning March 1,1986 establishes limits for amines and organic compounds.

1. Amines: In the case of foundry core manufacturing in metallurgy, the maximum amine concentration at emission is 5 mg/m^3 for a dry gas at 0 °C and at 1,013 mbar;
2. Organic compounds: For all industrial processes except those regulated by other laws, the maximum emission value in the case of an hourly mass

flowrate higher than 2 kg/h is 0.10 g/m^3 of organic compounds at 0 °C at 1,013 mbar and for a dry gas.

The simultaneous presence of several organic compounds must be taken into account as well. The norm of December 16, 1985 includes a list of the various products concerned and mentions molecule associations that can be the origin of olfactory nuisances.

15.2.3 Regulation Prospects

The evolution of regulations will depend notably on progress made in olfactometric analysis. New measurement methods will lead to new ways of expressing value limits in regulations.

Thus, a new odor measurement method based on the nuisance created has been adjusted and tested on a site heavily affected by olfactory nuisances. These experiments were carried out in France on the Etang de Berre site near Marseille, in Normandy near Le Havre, in northern France near Dunkerque, and in Holland.

The measurement procedure used, which was the same for both sites, is as follows: a jury made up of a large number of persons (more than 100) is chosen from among the inhabitants of the region concerned. The designated residents must be uniformly distributed on the site. They are asked to go outside their homes on a specified day and hour and sniff. If there is no odor perception, the questionnaire stops at that point; otherwise, the subject must indicate if the odor is "not irritating," "slightly irritating," "irritating," "very irritating," or "extremely irritating." The subject then sends the response to a designated laboratory by mail. A statistical study is carried out and the nuisance indexes according to geographical area are determined.

These studies have been shown to be conclusive. The Dutch government is already developing a project in which olfactory nuisance reduction goals will be considered to be reached from the moment a data index has not been surpassed on a site during a determined percentage of time.

However, this type of inquiry is difficult to manage. Studying the questions is a long and complex process and it is also necessary to motivate the jury during a considerable length of time.

15.3 Odors in Work Areas

Before apprehending odor problems in work areas, it should be noted that since 1913, article R 232-12 of the Work Code limits atmospheric pollution in work areas by stating that "air in workshops must be renewed in such a way that it remains in a state of purity necessary to the workers' health." However, this clause is rather general and the problem is to determine what concentration of a chemical product can provoke intoxication in a human organism. This leads to the concept of exposition limit values.

15.3.1 Exposition Limits

Three main groups of exposition limits can be distinguished:

1. Average exposition values (AEV): These values represent the time average of the concentrations to which workers are actually exposed during their work activity 8 hours a day and 5 days a week;
2. Short-term limit values (SLV): these are concentrations to which a worker can only be exposed for periods shorter than 15 minutes;
3. Upper limit values: these are values that cannot be exceeded, even instantaneously.

Many countries have published exposition limit lists, the majority of them relying on works carried out by a reduced number of countries: the United States, the USSR, Germany, and Sweden. However, the list established by the ACGIH (American Conference of Governmental Industrial Hygienists) in the United States is the one that has influenced other countries the most.

Differences in exposition limits for the same chemical compound can exist between two countries; for example, between Russia and America. This is due to different basic concepts. For Russians, effects on the organism due to inhalation must not appear, whereas, for Americans, limited or reversible effects may be tolerated. In addition, the defined concentration must correspond to a "technical feasibility."

In France, two concentrations are used: AEV and SLV. These values are published by the INRS and are only indicative. However, in case of "definitive excessive values creating a dangerous situation," the work inspector can order the strict application of the above values according to article L 231-S of the Work Code. In addition, for certain substances, regulation values are set within the framework of decrees, clauses, and articles of the Work Code; the cases of asbestos, benzene, and vinyl chloride can be cited, for example.

15.3.2 Odors and Safety

As stated at the beginning of this section, the worker's health must be preserved. The World Health Organization gives the following definition of human health: "A state of physical, mental and social well-being and not only an absence of illness or infirmity." According to this definition, an appropriate ventilation system must allow the reduction of ambiant odor to an acceptable level.

However, it may be prejudicial to exclude the use of the sense of smell as a means of detection, since multiple accidents have been avoided in this way. This is also why a characteristic odor has been added to some nonodorous toxic substances. The whole problem is to know whether or not the use of the sense of smell is reliable. A priori, workers may be afflicted with anosmia or hyposmia. Besides this problem, habituation and adaptation phenomena described at the beginning of this chapter may decrease the interest of using olfaction for danger prevention and may even increase risks.

After various studies carried out by the INRS, no connection between the olfactory thresholds and the AEV, SLV, or even the lower explosivity limits (LEL) have been found.

Study ND 1590-124-86, carried out at the Institut National de la Recherche sur la Sécurité (National Institute of Research on Safety) by M. Rousselin and M. Falcy, entitled "The nose, chemical products and safety," compared criteria of danger and toxicity (explosive limit, toxic dose, lethal dose, AEV) to the olfactory detection threshold. Safety ratios have been established from these values. The authors concluded that, among 217 listed substances, only six allow the use of olfaction as a means of detection with a certain amount of safety:

1. Ethyl acetate
2. Chloroethane
3. Methanol
4. Octane
5. Tetrahydrofurane
6. Xylene

However, it is possible to use other chemical product odors than the six listed above for industrial safety purposes in specific plants and under specific conditions.

Thus, in most cases and according to the studies, odors used as early warning systems do not seem to be very reliable. Considering the nuisances that they create, their removal seems rather desirable. Nevertheless, the main objective in workplaces would be to comply with the AEV and SLV that the INRS publishes every 3 months.

15.4 Conclusion

As has been noted throughout this book, odors constitute a set of complex phenomena and the removal of the nuisances that they generate is generally not an easy problem to solve. Nevertheless, technical solutions exist for their reduction or elimination and several countries have implemented regulatory means to decrease these nuisances. In France, article 17 of decree 77 dated September 12, 1977 specifies that, "prescriptions take into account the efficiency and economy of available techniques on one hand and on the other hand the quality, the calling and the use of the surrounding environment." This quotation perfectly illustrates the spirit of French regulations in the field of environmental protection and, more specifically, in the case of olfactory nuisances: the application of the best available technology that is technically sustainable.

Thus, within the framework of the fight against olfactory nuisances, regulations according to existing laws that decrease technical inquiries include the required tools to adapt themselves to the technical evolution and to enforce the application of adequate systems.

15.5 Appendix: Authorization and Registration Procedure— Directions for Use

15.5.1 The Authorization Procedure

When industrialists plan to create plants subject to authorization, they must submit a request dossier to the prefect containing the following information:

1. The petitioner's identity (name or corporate name, legal form, address, petitioner's position)
2. The site where the plant will be built
3. The nature and number of planned activities as well as the nomenclature heading(s) under which the site must be listed
4. The manufacturing processes implemented by the petitioner, the materials used and manufactured products

The following documents must be included with the authorization request:

1. A 1/25,000 or a 1/50,000 scale map indicating the projected plant location
2. A minimum 1/2,500 scale map of the plant's surroundings as far as a distance that, without being less than 100 m, must be at least equal to the tenth of the posting radius noted in the nomenclature at the heading of the corresponding plant; on this map all the buildings, including their usage, railways, public roads, wells, canals, and waterways should be noted
3. A complete minimum 1/200 scale map detailing the plant and specifying the usage of the buildings and the surrounding grounds and the sewer layout to a minimum distance of 35 meters from the plant
4. An impact study as provided for by the law of July 10, 1976 (article 2) relative to the protection of the environment and for which the contents are specified for registered sites by the decree of September 21, 1977. The objective of this study is

   • To define the effects that the environment will undergo. In order to do this, the study must approach and anticipate the totality of possible nuisances by determining the nature, importance, and effects on the site
   • To specify the measures planned to suppress and limit the previously established nuisances

5. A study determining the hazards that the plant may present and measures to reduce or diminish them

6. A notice relative to the projected plant's conformity with hygiene and
   safety rules for personnel and legislative prescriptions and regulations

If it is necessary for the petitioner to obtain a construction permit, proof of
this request must be included in the dossier.

When the prefect judges that the dossier is complete:

1. This information is communicated to the administrative court's presiding
   judge,
2. The prefect proposes opening and closing inquiry dates, and
3. The petitioner is simultaneously informed

Within 15 days, the administrative court's presiding judge chooses either

1. An inquiry commisioner or
2. The members of an inquiry commision, from which the judge designates
   the president.

As soon as the prefect is informed of the administrative court's presiding
judge's choice(s), the prefect announces the opening of the public inquiry by
decree, which also specifies

1. The object and the date of the inquiry: the duration is 1 month, except in
   the case of a prorogation request, which may not exceed 15 days,
2. The location, days, and times so that the public may familiarize itself
   with the dossier and express its opinions on a register provided for this
   purpose,
3. The name or names of the inquiry commisioner(s) and the times and
   places it will be possible to meet them,
4. The perimeter in which the notice will be posted. This notice must be
   posted at least 15 days before the opening of the inquiry. Also, the inquiry
   announcement must be published in newspapers distributed in the interested department(s) within the same time limit as previously stated.

All publicity costs are the petitioner's responsibility. The municipal councils
of communities included in the previously cited perimeter are called to give
their opinion(s) at the opening of the inquiry and up to 15 days after its
closing.

At the time of the inquiry's opening, the prefect sends copies of the authorization demand to the concerned departmental services: equipment, ag-
riculture, sanitary, and social action and civil safety. The inquiry dossier and
opinions are sent to the registered site inspector who establishes and transmits a report and propositions to the departmental hygiene council. The petitioner may be heard by the council.

Depending on the council's opinion, the prefect writes a decree that may
be criticized by the petitioner. Three months after receiving the dossier, unless there is a supplementary deadline, the prefect must have established a

decree authorizing the exploitation of the site, or its extension in which the following is prescribed:

1. Limitations that have to do with

   • Gaseous or liquid effluents
   • Solid wastes
   • Nuisances (noise and odors)

2. The means, nature, and frequency of measures or analysis necessary for monitoring.

The decree is then sent to the concerned city hall and extracts of the decree are posted, and, finally, a public notice is published in two newpapers.

Any notable modification to the plant must be brought to the prefect's attention before its realization.

15.5.2 The Registration Procedure

A registration relative to a plant must be addressed to the prefect of the department where it is to be built before being put into operation. A dossier including the following information must be presented:

1. The petitioner's identity (name or corporate name, legal form, address, petitioner's position)
2. The site where the plant will be built
3. The nature and number of planned activities as well as the nomenclature heading(s) under which the site must be listed
4. The petitioner must include a 100 m radius survey map and a complete minimum 1/200 scale map accompanied by a key and, if need be, descriptions of the plant and specifying the usage of the buildings and the surrounding grounds to a minimum distance of 35 m from the plant, as well as neighboring property, wells, canals and waterways, and sewers.

The method and conditions of water use, treatment, and drainage of wastewaters and all manner of emanations as well as elimination of the plant's wastes and residues must be specified.

When the prefect judges that the dossier is complete:

1. Receipt of the registration is acknowledged,
2. The general prescriptions applicable to the plant is communicated to the petitioner,
3. A copy of the registration and the text of general restrictions is sent to the city hall in the community where the plant will be in operation.

A copy of the receipt is posted at the city hall for a minimum duration of 1 month, with mention of the possibility of a third party to consult the general prescription text on the spot. A copy of prefectorial decrees is addressed to the city halls of concerned departments and extracts of them are published in two local newspapers.

If petitioners want to have some of the restrictions applicable to the plants modified, they address requests to the prefect who rules on the request by decree. Once again, the different decrees are established according to the registered sites' inspection report and after the departmental hygiene council's opinion is given.

The petitioner

1. May be heard by the departmental hygiene council,
2. Must be informed of council meetings,
3. Receives a copy of the registered site inspection's propositions.

15.5.3 Clauses Applicable to Registered Sites Submitted to Authorization or Registration

Any and all modifications made by the industrialist to the site, its operation, or to its immediate surroundings leading to a notable change in the initial dossier's elements must be made known to the prefect before their realization; the prefect may demand that a new authorization or registration be made. Any transfer of a registered site to another location necessitates a new authorization or registration.

The authorization or registration ceases to be valid when the site has not been put into service within 3 years or when operation has been discontinued for longer than 2 consecutive years, except in case of absolute necessity.

Bibliography

Rousselin, X., Falcy, M. (1986). Le nez, les produits chimiques et la sécurité. *Cahier de Notes Documentaires de l'INRS*, ND 1590-124-86, 331–344.

International Union of Air Pollution Prevention Association (1988). Clean air around the world. The law and practice of air pollution control in 14 countries in 5 continents. 146 p.

Journal Officiel de la République Française (1987). Installations classées pour la protection de l'environnement. *Brochure 1001 I textes généraux—nomenclature—1001 II arrêtés types—1001 III arrêtés, circulaires et instructions*, 5ème édition.

INDEX

Abortion, spontaneous, in mice, 17
Absorption, of gas in water, 380, 384
Absorption sampling, 274–276
Absorption tower
    packed, 294–297
        plugging with dust, 322
    technology of, 294–306
    for treatment of mixtures, series of, 321
Acetone, removal of, in a fixed bacteria system, 387
Acetophenone, ortho-amino, similarity of odor to keto-9-decene-trans-oic acid, 156
Acetylcholinesterase, in olfactory glomeruli, 17
Acid-base scrubbing, 283, 292–293, 319–320
Acid-base washing, of adsorbents, 353–354
Acid dissociation constant K_a, 167
Acidic character π, modulation by steric hindrance and unsaturation, 171–172

Actinomycetes
    for deodorizing sludge, 393
    odors associated with, 247, 251–252
Action potentials, 112–113
Activation process
    chemical, carbon adsorption materials, 335
    physical, carbon adsorption materials, 334
Activation, response by, 10
    qualitative discrimination as, 11
Activity, chemical concept, 93–96, 152–153
    and solubility, 100
Activity coefficient γ, 95, 102, 152–153
ADAPT (computer program), 168–169
Adaptation
    in crossmodality experiments, 51
    and discrimination threshold, 31–32, 36–37, 43
    and evoked cortical potentials, 113
    and perceived intensity, 53–54
    and stimulus effectiveness, 123
Addition models, 191
Adsorbents, naturally occurring, 346